The Geometry of Hamilton and Lagrange Spaces

Fundamental Theories of Physics

An International Book Series on The Fundamental Theories of Physics: Their Clarification, Development and Application

Volume 118

The Geometry of Hamilton and Lagrange Spaces

by

Radu Miron

Al. I. Cuza University,
Iaşi, Romania

Dragoş Hrimiuc

University of Alberta,
Edmonton, Canada

Hideo Shimada

Hokkaido Tokai University,
Sapporo, Japan

and

Sorin V. Sabau

Tokyo Metropolitan University,
Tokyo, Japan

KLUWER ACADEMIC PUBLISHERS
DORDRECHT / BOSTON / LONDON

A C.I.P. Catalogue record for this book is available from the Library of Congress.

ISBN 0-7923-6926-2

Published by Kluwer Academic Publishers,
P.O. Box 17, 3300 AA Dordrecht, The Netherlands.

Sold and distributed in North, Central and South America
by Kluwer Academic Publishers,
101 Philip Drive, Norwell, MA 02061, U.S.A.

In all other countries, sold and distributed
by Kluwer Academic Publishers,
P.O. Box 322, 3300 AH Dordrecht, The Netherlands.

Printed on acid-free paper

Printed in the Netherlands.

Contents

PREFACE

The title of this book is no surprise for people working in the field of Analytical Mechanics. However, the geometric concepts of Lagrange space and Hamilton space are completely new.

The geometry of Lagrange spaces, introduced and studied in [76],[96], was extensively examined in the last two decades by geometers and physicists from Canada, Germany, Hungary, Italy, Japan, Romania, Russia and U.S.A. Many international conferences were devoted to debate this subject, proceedings and monographs were published [10], [18], [112], [113],... A large area of applicability of this geometry is suggested by the connections to Biology, Mechanics, and Physics and also by its general setting as a generalization of Finsler and Riemannian geometries.

The concept of Hamilton space, introduced in [105], [101] was intensively studied in [63], [66], [97],... and it has been successful, as a geometric theory of the Hamiltonian function the fundamental entity in Mechanics and Physics. The classical Legendre's duality makes possible a natural connection between Lagrange and Hamilton spaces. It reveals new concepts and geometrical objects of Hamilton spaces that are dual to those which are similar in Lagrange spaces. Following this duality Cartan spaces introduced and studied in [98], [99],..., are, roughly speaking, the Legendre duals of certain Finsler spaces [98], [66], [67]. The above arguments make this monograph a continuation of [106], [113], emphasizing the Hamilton geometry.

*

*   *

The first chapter is an overview of the geometriy of the tangent bundle. Due to its special geometrical structure, TM, furnishes basic tools that play an important role in our study: the *Liouville vector field* C, the *almost tangent structure* J, the concept of *semispray*. In the text, new geometrical structures and notions will be introduced. By far, the concept of *nonlinear connection* is central in our investigations.

Chapter 2 is a brief review of some background material on Finsler spaces, included not only because we need them later to explain some extensions of the subject, but also using them as duals of Cartan spaces.

Some generalizations of Finsler geometry have been proposed in the last three decades by relaxing requirements in the definition of Finsler metric. In the Lagrange geometry, discussed in Chapter 3, the metric tensor is obtained by taking the Hessian with respect to the tangential coordinates of a smooth function L defined on the tangent bundle. This function is called a *regular Lagrangian* provided the Hessian is nondegenerate, and no other conditions are envisaged.

Many aspects of the theory of Finsler manifolds apply equally well to Lagrange spaces. However, a lot of problems may be totally different, especially those concerning the geometry of the base space M. For instance, because of lack of the homogeneity condition, the length of a curve on M, if defined as usual for Finsler manifolds, will depend on the parametrization of the curve, which may not be satisfactory.

In spite of this a Lagrange space has been certified as an excellent model for some important problems in Relativity, Gauge Theory, and Electromagnetism. The geometry of Lagrange spaces gives a model for both the gravitational and electromagnetic field in a very natural blending of the geometrical structures of the space with the characteristic properties of these physical fields.

A Lagrange space is a pair $L^n = (M, L(x,y))$ where $L = L(x,y)$ is a regular Lagrangian.

For every smooth parametrized curve $c : [0,1] \to M$ the action integral may be considered:

$$I(c) = \int_0^1 L\left(x(t), \frac{dx(t)}{dt}\right) dt.$$

A geodesic of the Lagrange Space (M, L) is an extremal curve of the action integral. This is, in fact, a solution of the Euler–Lagrange system of equations

$$\frac{d}{dt}\left(\frac{\partial L}{\partial \dot{x}^i}\right) - \frac{\partial L}{\partial x^i} = 0, \quad \dot{x}^i = \frac{dx^i}{dt},$$

where $(x^i(t))$ is a local coordinate expression of c.

This system is equivalent to

$$\frac{d^2x^i}{dt^2} + 2G^i\left(x, \frac{dx}{dt}\right) = 0,$$

where

$$G^i(x,y) = \frac{1}{4} g^{ij}\left(\frac{\partial^2 L}{\partial y^j \partial x^k} y^k - \frac{\partial L}{\partial x^j}\right)$$

and

$$g_{ij} = \frac{1}{2}\frac{\partial^2 L}{\partial y^i \partial y^j}.$$

Here G^i are the components of a semispray that generates a notable nonlinear connection, called *canonical*, whose coefficients are given by

$$N^i{}_j = \frac{\partial G^i}{\partial y^j}.$$

This nonlinear connection plays a fundamental role in the study of the geometry of TM. It generates a splitting of the double tangent bundle

$$TTM = HM \oplus VM$$

which makes possible the investigation of the geometry of TM in an elegant way, by using tools of Finsler Spaces. We mention that when L is the square of a function on TM, positively 1–homogeneous in the tangential coordinates (L is generated by a Finsler metric), this nonlinear connection is just the classical Cartan nonlinear connection of a Finsler space.

An other canonical linear connection, called *distinguished*, may be considered. This connection preserves the above decomposition of the double tangent bundle and moreover, it is metrical with respect to the metric tensor g_{ij}. When L is generated by a Finsler metric, this linear connection is just the *famous Cartan's metrical linear connection* of a Finsler space.

Starting with these geometrical objects, the entire geometry of TM can be obtained by studying the curvature and torsion tensors, structure equations, geodesics, etc. Also, a regular Lagrangian makes TM, in a natural way, a hermitian pseudo-riemannian symplectic manifold with an almost symplectic structure.

Many results on the tangent bundle do not depend on a particular fundamental function L, but on a metric tensor field. For instance, if $\gamma_{ij}(x)$ is a Riemannian metric on M and σ is a function depending explicitly on x^i as well as directional variables y^i then, for example,

$$g_{ij}(x,y) = e^{2\sigma(x,y)}\gamma_{ij}(x)$$

cannot be derived from a Lagrangian, provided $\dfrac{\partial\sigma}{\partial y^i} \neq 0$. Such situations are often encountered in the relativistic optics. These considerations motivate our investigation made on the geometry of a pair $(M, g_{ij}(x,y))$, where $g_{ij}(x,y)$ is a nondegenerate, symmetric, constant signature d–tensor field on TM (i.e. $g_{ij}(x,y)$ transform as a tensor field on M). These spaces, called *generalized Lagrange spaces* [96], [113], are in some situations more flexible than that of Finsler or Lagrange space because of the variety of possible selection for $g_{ij}(x,y)$. The geometric model of a generalized Lagrange space is an almost Hermitian space which, generally, is not reducible to an almost Kählerian space. These spaces, are briefly discussed in section 3.10.

Chapter 4 is devoted to the geometry of the cotangent bundle T^*M, which follows the same outline as TM. However, the geometry of T^*M is from one point of view different from that of the tangent bundle. We do not have here a natural tangent structure and a semispray cannot be introduced as usual for the tangent bundle. Two geometrical ingredients are of great importance on T^*M: the *canonical 1-form* $\omega = p_i dx^i$ and its exterior derivative $\theta = dp_i \wedge dx^i$ (the *canonical symplectic* *strucutre* of T^*M). They are systematically used to define new useful tools for our next investigations.

Chapter 5 introduces the concept of *Hamilton space* [101], [105]. A *regular Hamiltonian* on T^*M, is a smooth function $H : T^*M \longrightarrow \mathbb{R}$, such that the Hessian matrix with entries

$$g^{ij}(x,p) = \frac{1}{2}\frac{\partial^2 H(x,p)}{\partial p_i \partial p_j}$$

is everywhere nondegenerate on T^*M (or a domain of T^*M).

A Hamilton space is a pair $H^n = (M, H(x,p))$, where $H(x,p)$ is a regular Hamiltonian. As for Lagrange spaces, a canonical nonlinear connection can be derived from a regular Hamiltonian but in a totally different way, using the Legendre transformation. It defines a splitting of the tangent space of the cotangent bundle

$$TT^*M = HT^*M \oplus VT^*M,$$

which is crucial for the description of the geometry of T^*M.

The case when H is the square of a function on T^*M, positively 1-homogeneous with respect to the momentum p_i, provides an important class of Hamilton spaces, called *Cartan spaces* [98], [99]. The geometry of these spaces is developed in Chapter 6.

Chapter 7 deals with the relationship between Lagrange and Hamilton spaces. Using the classical Legendre transformation different geometrical objects on TM are nicely related to similar ones on T^*M. The geometry of a Hamilton space can be obtained from that of certain Lagrange space and vice versa. As a particular case, we can associate to a given Finsler space its dual, which is a Cartan space. Here, a surprising result is obtained: the L-dual of a Kropina space (a Finsler space) is a Randers space (a Cartan space). In some conditions the L-dual of a Randers space is a Kropina space. This result allows us to obtain interesting properties of Kropina spaces by taking the dual of those already obtained in Randers spaces. These spaces are used in several applications in Physics.

In Chapter 8 we study how the geometry of cotangent bundle changes under symplectic transformations. As a special case we consider the homogeneous contact transformations known in the classical literature. Here we investigate the so–called "homogeneous contact geometry" in a more general setting and using a modern approach. It is clear that the geometry of T^*M is essentially simplified if it is related to a given nonlinear connection. If $f \in Diff(T^*M)$, the push forward of a nonlinear connection by f is no longer a nonlinear connection and the geometry of T^*M is completely changed by f. The main difficulty arises from the fact that the vertical distribution is not generally preserved by f. However, under appropriate conditions a new distribution, called oblique results. We introduce the notion of *connection pair* (more general than a nonlinear connection), which is the keystone of the entire construction.

The last two decades many mathematical models from Lagrangian Mechanics, Theoretical Physics and Variational Calculus systematically used multivariate Lagrangians of higher order acceleration, $L\left(x, \frac{dx}{dt}(t), ..., \frac{1}{k!}\frac{d^kx}{dt^k}(x)\right)$, [106].

The variational principle applied to the action integral

$$I(c) = \int_0^1 L\left(x(t), \frac{dx}{dt}(t), ..., \frac{1}{k!}\frac{d^kx}{dt^k}(t)\right) dt$$

leads to Euler–Lagrange system of equations

$$E_i(L) := \frac{\partial L}{\partial x^i} - \frac{d}{dt}\frac{\partial L}{\partial y^{(1)i}} + \cdots + (-1)^k \frac{1}{k!}\frac{d^k}{dt^k}\frac{\partial L}{\partial y^{(k)i}} = 0,$$
$$y^{(2)i} = \frac{dx^i}{dt}, \ldots, y^{(k)i} = \frac{1}{k!}\frac{d^kx^i}{dt^k},$$

which is fundamental for higher order Lagrangian Mechanics. The energy function of order k is conservative along the integral curves of the above system.

From here one can see the motivation of the Lagrange geometry for higher order Lagrangians to the *bundle of acclerations of order* k, (or the osculator bundle of order k) denoted by T^kM, and also the L-dual of this theory.

These subjects are developed in the next five chapters.

A *higher order Lagrange* space is a pair $L^{(k)n} = (M, L(x, y^{(1)}, ..., y^{(n)}))$, where M is a smooth differentiable manifold and $L : T^kM \to \mathbb{R}$ is a regular Lagrangian or order k, [106]. The geometry of these spaces may be developed as a natural extension of that of a Lagrange space. The metric tensor,

$$g_{ij}(x, y^{(1)}, ..., y^{(k)}) = \frac{1}{2}\frac{\partial^2 L}{\partial y^{(k)i}\partial^{(k)j}}$$

has to be nondegenerate on T^kM. A central problem, about existence of regular Lagrangians of order k, arises in this case. The bundle of prolongations of order k, at T^kM, of a Riemannian space on M is an example for the Lagrange space of order k, [106].

We mention that the Euler–Lagrange equations given above are generated by the Craig–Synge covector

$$\overset{k-1}{E}{}_i(L) := \frac{\partial L}{\partial y^{(k-1)i}} - \frac{d}{dt}\frac{\partial L}{\partial y^{(k)i}}$$

that is used in the construction of the canonical semispray of $L^{(k)n}$. This is essential in defining the entire geometric mechanism of $L^{(k)n}$.

The geometric model of $L^{(k)n}$, is obtaining by lifting the whole construction to T^kM.

As a particular case, a Finsler space of order k is obtained if L is the square of a positive k–homogeneous function on the bundle of accelerations of order k. Also the class of generalized Lagrange spaces of order k may be considered.

Before starting to define the dual of $L^{(k)n}$, we should consider the geometrical entity $T^{*k}M$, having enough properties to deserve the name of dual of T^kM. The space $T^{*k}M$ should have the same dimension as T^kM, should carry a natural presymplectic structure and at least one Poisson structure. Although the subject was discussed in literature (see [85]) the above conditions are not full verified for the chosen duals.

Defining [110]:

$$T^{*k}M = T^{k-1}M \times_M T^*M,$$

then all the above conditions are satisfied. The two–form $\theta = dp_i \wedge dx^i$ defines a presymplectic structure and the Poisson brackets $\{f, g\} = \dfrac{\partial f}{\partial x^i}\dfrac{\partial g}{\partial p_i} - \dfrac{\partial g}{\partial x^i}\dfrac{\partial f}{\partial p_i}$ a Poisson structure.

The Legendre transformation is

$$Leg : (x, y^{(1)}, ..., y^{(k-1)}, y^{(k)}) \in T^kM \longrightarrow (x, y^{(1)}, ..., y^{(k-1)}, p) \in T^{*k}M$$

where $p_i = \dfrac{1}{2}\dfrac{\partial L}{\partial y^{(k)i}}$. It is a locall diffeomorphism.

Now, the geometry of a higher order regular Hamiltonians may be developed as we did for $k = 2$.

The book ends with a description of $C^{(2)n}$, the *Cartan spaces of order* 2, and $GH^{(2)n}$ the *Generalized Hamilton space or order* 2.

For the general case the extension seems to be more difficult since the L–duality process cannot be developed unless a nonlinear connection on T^{k-1} is given in advance.

We should add that this book naturally prolongates the main topics presented in the monographs: *The Geometry of Lagrange Spaces. Theory and Applications* (R. Miron and M. Anastasiei), Kluwer, FTPH no.59; *The Geometry of Higher Order Lagrange Spaces. Applications to Mechanics and Physics* (R. Miron), Kluwer, FTPH, nr.82.

This monograph was written as follows:

- Ch. 1,2,3 – H. Shimada and V.S. Sabău
- Ch. 4,5,6 – H. Shimada and R. Miron
- Ch. 7,8 – D. Hrimiuc
- Ch. 9,10,11,12,13 – R. Miron

The book is divided in two parts: Hamilton and Lagrange spaces and Hamilton space of higher order.

The readers can go in the heart of subject by studying the first part (Ch. 1–8). From this reason, the book is accessible for readers ranging from graduate students to researchers in Mathematics, Mechanics, Physics, Biology, Informatics etc.

Acknowledgements. We would like to express our gratitude to P.L. Antonelli, M. Anastasiei, M. Matsumoto for their continuous support, encouragement and numerous valuable suggestions. We owe special thanks to R.G. Beil, S.S. Chern, M. Crampin, R.S. Ingarten, D. Krupka, S. Kobayashi, R.M. Santilli, L. Tamassy, I. Vaisman for useful discussions and suggestions on the content of this book, to I. Bucătaru and M. Roman who gave the manuscript a meticulous reading. We are pleased to thank to Mrs. Elena Mocanu and Mrs. V. Spak who typeset our manuscript into its final excellent form.

Finally, we would like to thank the publishers for their co–operation and courtesy.

Chapter 1

The geometry of tangent bundle

The geometry of tangent bundle (TM, π, M) over a smooth, real, finite dimensional manifold M is one of the most important fields of the modern differential geometry. The tangent bundle TM carries some natural object fields, as: Liouville vector field $\mathbb{C}$, tangent structure J, the vertical distribution V. They allow to introduce the notion of semispray S, which is a tangent vector field of TM, having the property $J(S) = \mathbb{C}$. We will see that the geometry of the manifold TM can be constructed using only the notion of semispray.

The entire construction is basic for the introduction of the notion of Finsler space or Lagrange space [112], [113]. In the last twenty years this point of view was adopted by the authors of the present monograph in the development of the geometrical theory of the spaces which can be defined on the total space TM of tangent bundle. There exists a rich literature concerning this subject.

In this chapter all geometrical object fields and all mappings are considered of the class C^∞, expressed by the words "differentiable" or "smooth".

1.1 The manifold TM

Let M be a real differentiable manifold of dimension n. A point of M will be denoted by x and its local coordinate system by (U, φ), $\varphi(x) = (x^i)$. The indices $i, j, \ldots$ run over set $\{1, \ldots, n\}$ and Einstein convention of summarizing is adopted all over this book.

The tangent bundle (TM, π, M) of the manifold M can be identified with the 1-osculator bundle $(\mathrm{Osc}^1 M, \pi, M)$, see the definition below.

Indeed, let us consider two curves $\rho, \sigma : I \to M$, having images in a domain of local chart $U \subset M$. We say that ρ and σ have a "contact of order 1" or the "same tangent line" in the point $x_0 \in U$ if: $\rho(0) = \sigma(0) = x_0$, $(0 \in I)$, and for any function

$f \in \mathcal{F}(U)$:

$$\frac{d}{dt}(f \circ \rho)(t)\Big|_{t=0} = \frac{d}{dt}(f \circ \sigma)(t)\Big|_{t=0} \tag{1.1}$$

The relation "contact of order 1" is an equivalence on the set of smooth curves in M, which pass through the point x_0. Let $[\rho]_{x_0}$ be a class of equivalence. It will be called a "1–osculator space" in the point $x_0 \in M$. The set of 1–osculator spaces in the point $x_0 \in M$ will be denoted by $\mathrm{Osc}^1_{x_0}$, and we put

$$\mathrm{Osc}^1 M = \bigcup_{x_0 \in M} \mathrm{Osc}^1_{x_0}.$$

One considers the mapping $\pi : \mathrm{Osc}^1 M \to M$ defined by $\pi([\rho]_{x_0}) = x_0$. Clearly, π is a surjection.

The set $\mathrm{Osc}^1 M$ is endowed with a natural differentiable structure, induced by that of the manifold M, so that π is a differentiable mapping. It will be described below.

The curve $\rho : I \to M$, $(\mathrm{Im}\,\rho \subset U)$ is analyticaly represented in the local chart (U, φ) by $x^i = x^i(t)$, $t \in I$, $x_0 = (x^i_0 = x^i(0))$, taking the function f from (1.1), succesively equal to the coordinate functions x^i, then a representative of the class $[\rho]_{x_0}$ is given by

$$x^{*i} = x^i(0) + t\frac{dx^i}{dt}(0), \ t \in (-\varepsilon, \varepsilon) \subset I.$$

The previous polynomials are determined by the coefficients

$$x^i_0 = x^i(0), \ y^i_0 = \frac{dx^i}{dt}(0). \tag{1.2}$$

Hence, the pair $(\pi^{-1}(U), \phi)$, with $\phi([\rho]_{x_0}) = (x^i_0, y^i_0) \in R^{2n}$, $\forall\, [\rho]_{x_0} \in \pi^{-1}(U)$ is a local chart on $\mathrm{Osc}^1 M$. Thus a differentiable atlas $\mathcal{A}_M$ of the differentiable structure on the manifold M determines a differentiable atlas $\mathcal{A}_{\mathrm{Osc}^1 M}$ on $\mathrm{Osc}^1 M$ and therefore the triple $(\mathrm{Osc}^1 M, \pi, M)$ is a differentiable bundle.

Based on the equations (1.2) we can identify the point $[\rho]_{x_0} \in \mathrm{Osc}^1 M$ with the tangent vector $y^i_0 \in T_{x_0} M$. Consequently, we can indeed identify the 1–osculator bundle with the tangent bundle (TM, π, M).

By (1.2) a transformation of local coordinates $(x^i, y^i) \to (\widetilde{x}^i, \widetilde{y}^i)$ on the manifold TM is given by

$$\begin{cases} \widetilde{x}^i = \widetilde{x}^i(x^i, ..., x^n), \ \ \det\left(\dfrac{\partial \widetilde{x}^i}{\partial x^j}\right) \neq 0, \\ \widetilde{y}^i = \dfrac{\partial \widetilde{x}^i}{\partial x^j}\, y^j. \end{cases} \tag{1.3}$$

One can see that TM is of dimension $2n$ and is orientable.

Moreover, if M is a paracompact manifold, then TM is paracompact, too.

Let us present here some notations. A point $u \in TM$, whose projection by π is x, i.e. $\pi(u) = x$, will be denoted by (x, y), its local coordinates being (x^i, y^i). We put $\widetilde{TM} = TM \setminus \{0\}$, where $\{0\}$ means the null section of π.

The coordinate transformation (1.3) determines the transformation of the natural basis $\left(\dfrac{\partial}{\partial x^i}, \dfrac{\partial}{\partial y^i}\right)$, $(i = \overline{1,n})$, of the tangent space T_uTM at the point $u \in TM$ the following:

$$\frac{\partial}{\partial x^i} = \frac{\partial \widetilde{x}^j}{\partial x^i}\frac{\partial}{\partial \widetilde{x}^j} + \frac{\partial \widetilde{y}^j}{\partial x^i}\frac{\partial}{\partial \widetilde{y}^j}; \quad \frac{\partial}{\partial y^i} = \frac{\partial \widetilde{y}^j}{\partial y^i}\frac{\partial}{\partial \widetilde{y}^j}. \tag{1.4}$$

By means of (1.3) we obtain

$$\frac{\partial \widetilde{y}^i}{\partial y^j} = \frac{\partial \widetilde{x}^i}{\partial x^j}; \quad \frac{\partial \widetilde{y}^i}{\partial x^j} = \frac{\partial^2 \widetilde{x}^i}{\partial x^j \partial x^h}\, y^h. \tag{1.4$'$}$$

Looking at the formula (1.4) we remark the existence of some natural object fields on E.

First of all, the tangent space V_u to the fibre $\pi^{-1}(x)$ in the point $u \in TM$ is locally spanned by $\left\{\dfrac{\partial}{\partial y^1}, \dots, \dfrac{\partial}{\partial y^n}\right\}$. Therefore, the mapping $V : u \in TM \to V_u \subset T_uTM$ provides a regular distribution which is generated by the adapted basis $\left\{\dfrac{\partial}{\partial y^i}\right\}$, $(i = 1, ..., n)$. Consequently, V is an integrable distribution on TM. V is called the vertical distribution on TM.

Taking into account (1.3), (1.4), it follows that

$$\mathbb{C} = y^i \frac{\partial}{\partial y^i} \tag{1.5}$$

is a vertical vector field on TM, which does not vanish on the manifold $\widetilde{TM}$. It is called the *Liouville vector* field. The existence of the Liouville vector field is very important in the study of the geometry of the manifold TM.

Let us consider the $\mathcal{F}(TM)$–linear mapping $J : \mathcal{X}(TM) \to \mathcal{X}(TM)$,

$$J\left(\frac{\partial}{\partial x^i}\right) = \frac{\partial}{\partial y^i}, \; J\left(\frac{\partial}{\partial y^i}\right) = 0, \;\; (i = 1, .., n). \tag{1.6}$$

Theorem 1.1.1. *The following properties hold:*

1° *J is globally defined on TM.*

2° $J \circ J = 0$, $\operatorname{Im} J = \operatorname{Ker} J = V$, $\operatorname{rank}\|J\| = n$.

3° *J is an integrable structure on E.*

4° $J\mathbb{C} = 0$.

The proof can be found in [113].

We say that J is the *tangent* structure on E.

The previous geometrical notions are useful in the next sections of this book.

1.2 Homogeneity

The notion of homogeneity of functions $f \in \mathcal{F}(TM)$ with respect to the variables y^i is necessary in our considerations because some fundamental object fields on E have the homogeneous components.

In the osculator manifold $\mathrm{Osc}^1 M = TM$, a point $[\rho]_{x0}$ has a geometrical meaning, i.e. changing of parametrization of the curve $\rho : I \to M$ does not change the space $[\rho]_{x0}$. Taking into account the affine transformations of parameter

$$\bar{t} = at + b,\ t \in I,\ a \in R^* \tag{2.1}$$

we obtain the transformation of coordinates of $[\rho]_{x0}$ in the form

$$\bar{x}^i = x^i,\ \bar{y}^i = ay^i. \tag{2.2}$$

Therefore, the transformations of coordinates (1.3) on the manifold E preserve the transformations (2.2).

Let us consider

$$H = \{h_a : R \to R \mid a \in R^+\},$$

the group of homoteties of real numbers field R.

H acts as an uniparameter group of transformations on E as follows

$$H \times TM \to TM,\ (h_a, u) \to h_a(u),$$

where $\bar{u} = h_a(u),\ a \in R^+$, is the point $(\bar{x}, \bar{y}) = (x, ay),\ a \in R^+$. Consequently, H acts as a group of transformations on TM, with the preserving of the fibres.

The orbit of a point $u_0 = (x_0, y_0) \in E$ is given by

$$\begin{aligned} x^i &= x^i_0, \\ y^i &= ay^i_0,\ a \in R^+. \end{aligned}$$

The tangent vector to orbit in the point $u_0 = h_1(u_0)$ is given by

$$\mathbb{C}_{u_0} = y^i_0 \left(\frac{\partial}{\partial y^i}\right)_{u_0}.$$

This is the Liouville vector field $\mathbb{C}$ in the point u_0.

Now we can formulate:

Definition 1.2.1. A function $f : TM \to R$ differentiable on $\widetilde{TM}$ and continuous on the null section $0 : M \to TM$ is called homogeneous of degree r, $(r \in Z)$ on the fibres of TM, (briefly: r–homogeneous with respect to y^i) if:

$$f \circ h_a = a^r f, \ \forall\, a \in R^+. \tag{2.3}$$

The following Euler theorem holds [90], [106]:

Theorem 1.2.1. *A function $f \in \mathcal{F}(TM)$ differentiable on $\widetilde{TM}$ and continuous on the null sections is homogeneous of degree r on the fibres of TM if and only if we have*

$$\mathcal{L}_{\mathbb{C}} f = rf, \tag{2.4}$$

$\mathcal{L}_{\mathbb{C}}$ being the Lie derivative with respect to the Liouville vector field $\mathbb{C}$.

Remark. If we preserve Definition 1.2.1 and ask for $f : TM \to R$ to be differentiable on TM (inclusive on the null section), then the property of 1–homogeneity of f implies that f is a linear function in variables y^i.

The equality (2.4) is equivalent to

$$y^i \frac{\partial f}{\partial y^i} = rf. \tag{2.4$'$}$$

The following properties hold:

1° f_1, f_2-r-homogeneous$\Longrightarrow \lambda_1 f_1 + \lambda_2 f_2, \lambda_1, \lambda_2 \in R$ is r-homogeneous,

2° f_1-r-homogeneous, f_2 s-homogeneous$\Longrightarrow f_1 \cdot f_2$ is $(r+s)$-homogeneous,

3° f_1 r-homogeneous, $f_2 \neq 0$ s-homogeneous$\Longrightarrow \dfrac{f_1}{f_2}$ $(r-s)$ homogeneous.

Definition 1.2.2. A vector field $X \in \mathcal{X}(\widetilde{E})$ is r-homogeneous if

$$X \circ h_a = a^{r-1} h_a^* \circ X, \ \forall\, a \in R^+.$$

It follows:

Theorem 1.2.2. *A vector field $X \in \mathcal{X}(\widetilde{TM})$ is r-homogeneous if and only if we have*

$$\mathcal{L}_{\mathbb{C}} X = (r-1)X. \tag{2.5}$$

Of course, $\mathcal{L}_{\mathbb{C}} X = [\mathbb{C}, X]$is the Lie derivative of X with respect to $\mathbb{C}$.

Consequently, we can prove:

1° The vector fields $\frac{\partial}{\partial x^i}, \frac{\partial}{\partial y^i}$ are 1 and 0-homogeneous, respectively.

2° If $f \in \mathcal{F}(\widetilde{TM})$ is s-homogeneous and $X \in \mathcal{X}(\widetilde{TM})$ is r-homogeneous then fX is $s+r$-homogeneous.

3° A vector field on $\widetilde{TM}$:

$$X = X^{(0)i}\frac{\partial}{\partial x^i} + X^{(1)i}\frac{\partial}{\partial y^i}$$

is r–homogeneous if and only if $X^{(0)i}$ are functions $(r-1)$-homogeneous and $X^{(1)i}$ are functions r-homogeneous.

4° $X \in \mathcal{X}(\widetilde{TM})$ is r-homogeneous and $f \in \mathcal{F}(\widetilde{TM})$ is s-homogeneous, then $Xf \in \mathcal{F}(\widetilde{TM})$ is a $(r+s-1)$-homogeneous function.

5° The Liouville vector field $\mathbb{C}$ is 1-homogeneous.

6° If $f \in \mathcal{F}(\widetilde{TM})$ is an arbitrary s-homogeneous function, then $\frac{\partial f}{\partial y^i}$ is a $(s-1)$-homogeneous function and $\frac{\partial^2 f}{\partial y^i \partial y^j}$ is $(s-2)$-homogeneous function.

In the case of q-form we can give:

Definition 1.2.3. A q-form $\omega \in \Lambda^q(\widetilde{TM})$ is s-homogeneous if

$$\omega \circ h_a^* = a^s\omega, \forall\, a \in R^+.$$

It follows [106]:

Theorem 1.2.3. *A q-form $\omega \in \Lambda^q(\widetilde{TM})$ is s-homogeneous if and only if*

$$\mathcal{L}_{\mathbb{C}}\omega = s\omega. \tag{2.6}$$

Corollary 1.2.1. We have, [106]:

1° *$\omega \in \Lambda^q(\widetilde{TM})$ s-homogeneous and $\omega' \in \Lambda^{q'}(\widetilde{TM})$ is s'-homogeneous $\Longrightarrow \omega\wedge\omega'$ $(s+s')$-homogeneous.*

2° *$\omega \in \Lambda^q(\widetilde{TM})$ s-homogeneous, $\underset{1}{X}, ..., \underset{q}{X}$ -r-homogeneous$\Longrightarrow$ $\omega(\underset{1}{X}, ..., \underset{q}{X})$ is $r + (s-1)$-homogeneous.*

3° *dx^i $(i = 1, ..., n)$ are 0-homogeneous 1-forms.*
dy^i $(i = 1, ..., n)$ are 1-homogeneous 1-forms.

The applications of those properties in the geometry of Finsler space are numberless.

1.3 Semisprays on the manifold $\widetilde{TM}$

One of the most important notions in the geometry of tangent bundle is given in the following definition:

Definition 1.3.1. A *semispray* S on $\widetilde{TM}$ is a vector field $S \in \mathcal{X}(\widetilde{TM})$ with the property:

$$JS = \mathbb{C}. \tag{3.1}$$

If S is homogeneous, then S will be called a *spray*.

Of course, the notion of a local semispray can be formulated taking $S \in \mathcal{X}(\breve{U})$, $\breve{U}$ being an open set in the manifold $\widetilde{TM}$.

Theorem 1.3.1.

1° *A semispray S can be uniquely written in the form*

$$S = y^i \frac{\partial}{\partial x^i} - 2G^i(x,y)\frac{\partial}{\partial y^i}. \tag{3.2}$$

2° *The set of functions $G^i(x,y)$, $(i = 1, ..., n)$ are changed with respect to* (1.3) *as follows:*

$$2\widetilde{G}^i = 2\,\frac{\partial \widetilde{x}^i}{\partial x^j}\,G^j - \frac{\partial \widetilde{y}^i}{\partial x^j}\,y^j. \tag{3.3}$$

3° *If the set of functions G^i are a priori given on every domain of a local chart in $\widetilde{TM}$, so that* (3.3) *holds, then S from* (3.2) *is a semispray.*

Proof. 1° If a vector field $S = a^i(x,y)\,\dfrac{\partial}{\partial x^i} + b^i(x,y)\,\dfrac{\partial}{\partial y^i}$ is a semispray S, then $JS = \mathbb{C}$ implies $a^i = y^i$ and $b^i(x,y) = -2G^i(x,y)$.

So that G^i are uniquely determined and (3.2) holds.

2° The formula (3.3) followsfrom (1.3), (1.4) and the fact that S is a vector field on$\widetilde{TM}$, i.e. $S = y^i\dfrac{\partial}{\partial x^i} - 2G^i(x,y)\dfrac{\partial}{\partial y^i} = \widetilde{y}^i\dfrac{\partial}{\partial \widetilde{x}^i} - 2\widetilde{G}^i(\widetilde{x},\widetilde{y})\dfrac{\partial}{\partial \widetilde{y}^i}$.

3° Using the rule of transformation (3.3) of the set of functions G^i it follows that $S = y^i\dfrac{\partial}{\partial x^i} - 2G^i(x,y)\dfrac{\partial}{\partial y^i} = \widetilde{y}^i\dfrac{\partial}{\partial \widetilde{x}^i} - 2\widetilde{G}^i(\widetilde{x},\widetilde{y})\dfrac{\partial}{\partial \widetilde{y}^i}$ is a vector field which satisfies $JS = \mathbb{C}$. **q.e.d.**

From the previous theorem, it results that S is uniquely determined by $G^i(x,y)$ and conversely. Because of this reason, G^i are called *the coefficients* of the semispray S.

Theorem 1.3.2. *A semispray S is a spray if and only if its coefficients G^i are 2-homogeneous functions with respect to y^i.*

Proof. By means of 1° and 3° from the consequences of Theorem 2.2 it follows that $y^i \dfrac{\partial}{\partial x^i}$ is 2-homogeneous and $\dfrac{\partial}{\partial y^i}$ is 0–homogeneous vector fields. Hence, S is 2–homogeneous if and only if G^i are 2–homogeneous functions with respect to y^i.

The integral curves of the semispray S from (3.2) are given by

$$\frac{dx^i}{dt} = y^i, \ \frac{dy^i}{dt} = -2G^i(x,y). \tag{3.4}$$

It follows that, on M, these curves are expressed as solutions of the following differential equations

$$\frac{d^2x^i}{dt^2} + 2G^i\left(x, \frac{dx}{dt}\right) = 0. \tag{3.5}$$

The curves $c : t \in I \to (x^i(t)) \subset U \subset M$, solutions of (3.5), are called the paths of the semispray S. The differential equation (3.5) has geometrical meaning. Conversely, if the differential equation (3.5) is given on a domain of a local chart U of the manifold M, and this equation is preserved by the transformations of local coordinates on M, then coefficients $G^i(x,y)$, $\left(y^i = \dfrac{dx^i}{dt}\right)$ obey the transformations (3.3). Hence $G^i(x,y)$ are the coefficients of a semispray. Consequently:

Theorem 1.3.3. *A semispray S on $\widetilde{TM}$, with the coefficients $G^i(x,y)$ is characterized by a system of differential equations* (3.5), *which has a geometrical meaning.*

Now, we are able to prove

Theorem 1.3.4. *If the base manifold M is paracompact, then on $\widetilde{TM}$ there exist semisprays.*

Proof. M being paracompact, there is a Riemannian metric g on M. Consider $\gamma^i{}_{jk}(x)$ the Christoffel symbols of g. Then the set of functions

$$G^i(x,y) = \frac{1}{2}\, \gamma^i{}_{jk}(x) y^j y^k \tag{$*$}$$

is transformed, by means of a transformation (1.3), like in formula (3.3). Theorem 1.3.1 may be applied. It follows that the set of functions G^i are the coefficients of a semispray S. **q.e.d.**

Remarks.

1° $S = y^i \dfrac{\partial}{\partial x^i} - 2G^i \dfrac{\partial}{\partial y^i} is a spray, where\ G^i(x,y) = \dfrac{1}{2}\, \gamma^i{}_{jk}(x) y^j y^k$, whose differential equations (3.5) are

$$\frac{d^2x^i}{dt^2} + \gamma^i{}_{jk}(x)\frac{dx^j}{dt}\frac{dx^k}{dt} = 0.$$

So the paths of S in the canonical parametrization are the geodesics of the Riemann space (M, g).

2° $\dfrac{\partial G^i}{\partial y^j} = \gamma^i{}_{jk}(x) y^k$ is a remarkable geometrical object field on $\widetilde{TM}$ (called nonlinear connection).

Finally, in this section, taking into account the previous remark, we consider the functions determined by a semispray S:

$$N^i{}_j = \frac{\partial G^i}{\partial y^j}. \tag{3.6}$$

Using the rule of transformation (3.3) of the coefficients G^i we can prove, without difficulties:

Theorem 1.3.5. *If $G^i(x,y)$ are the coefficients of a semispray S, then the set of functions $N^i{}_j(x,y)$ from (3.6) has the following rule of transformationwith respect to (1.3):*

$$\widetilde{N}^i{}_m \frac{\partial \tilde{x}^m}{\partial x^j} = N^m{}_j \frac{\partial \tilde{x}^i}{\partial x^m} - \frac{\partial \tilde{y}^i}{\partial x^j}. \tag{3.7}$$

In the next section we shall prove that $N^i{}_j$ are the coefficients of a nonlinear connection on the manifold $E = TM$.

1.4 Nonlinear connections

The notion of nonlinear connection on the manifold $E = TM$ is essentially for study the geometry of TM. It is fundamental in the geometry of Finsler and Lagrange spaces [113].

Our approach will be two folded:

1° As a splitting in the exact sequence (4.1).

2° As a derivate notion from that of semispray.

Let us consider as previous the tangent bundle (TM, π, M) of the manifold M. It will be written in the form (E, π, M) with $E = TM$. The tangent bundle of the manifold E is $(TE, \pi^\top, E)$, where $\pi^\top$ is the tangent mapping of the projection π. As we know the kernel of $\pi^\top$ is the vertical subbundle (VE, π_V, E). Its fibres are the linear vertical spaces V_uE, $u \in E$.

A tangent vector vector field on E can be represented in the local natural frame $\left(\dfrac{\partial}{\partial x^i}, \dfrac{\partial}{\partial y^i}\right)$ on E by

$$\check{X} = X^i(x, y)\frac{\partial}{\partial x^i} + Y^i(x, y)\frac{\partial}{\partial y^i}.$$

It can be written in the form $\check{X} = (x^i, y^i, X^i, Y^i)$ or, shorter, $\check{X} = (x, y, X, Y)$. The mapping $\pi^\top : TE \to E$ has the local form

$$\pi^\top(x, y, X, Y) = (x, y).$$

The points of submanifold VE are of the form $(x, y, 0, Y)$. Hence, the fibres V_uE of the vertical bundle are isomorphic to the real vector space R^n.

Let us consider the pullback bundle

$$\pi^*E = E \times_\pi E = \{(u, v) \in E \times E \mid \pi(u) = \pi(v)\}.$$

The fibres of π^*E, i.e., π^*_uE are isomorphic to $T_{\pi(u)}M$. We can define the following morphism of vector bundles $\pi! : TE \to \pi^*E$, $\pi!(\check{X}_u) = (u, \pi^\top_u(\check{X}_u)$. It follows that

$$\operatorname{Ker} \pi! = \operatorname{Ker} \pi^\top = VE.$$

By means of these considerations one proves without difficulties that the following sequence is exact:

$$0 \longrightarrow VE \xrightarrow{i} TE \xrightarrow{\pi!} \pi^*E \longrightarrow 0 \tag{4.1}$$

Now, we can give:

Definition 1.4.1. A nonlinear connection on the manifold $E = TM$ is a left splitting of the exact sequence (4.1).

Therefore, a nonlinear connection on E is a vector bundle morphism $C : TE \to VE$, with the property $C \circ i = 1_{VE}$.

The kernel of the morphism C is a vector subbundle of the tangent bundle $(TE, \pi^\top, E)$, denoted by (HE, π_{HE}, E) and called the *horizontal* subbundle. Its fibres H_uE determine a distribution $u \in E \to H_uE \subset T_uE$, supplementary to the vertical distribution $u \in E \to V_uE \subset T_uE$. Therefore, a nonlinear connection N induces the following Whitney sum:

$$TE = HE \oplus VE. \tag{4.2}$$

The reciprocal property holds [112]. So we can formulate:

Theorem 1.4.1. *A nonlinear connection N on $E = TM$ is characterized by the existence of a subbundle (HE, π_{HE}, E) of tangent bundle of E such that the Whitney sum* (4.2) *holds.*

Consequences.

1° A nonlinear connection N on E is a distribution $H : u \in E \to H_u E \subset T_u E$ with the property

$$T_u E = H_u E \oplus V_u E, \ \forall\, u \in E, \tag{4.2'}$$

and conversely.

2° The restriction of the morphism $\pi! : TE \to \pi^* E$ to the HE is an isomorphism of vector bundles.

3° The component $\pi^\top : HE \to E$ of the mapping $\pi!$ is a morphism of vector bundles whose restrictions to fibres are isomorphisms. Hence for any vector field X on M there exists an horizontal vector field X^H on E such that $\pi^\top(X^H) = X$ is called *the horizontal lift* of the vector field X on M.

Using the inverse of the isomorphism $\pi|_{HE}$ we can define the morphism of vector bundles $D : \pi^* E \to TE$, such that $\pi! \circ D = \mathrm{id}|_{\pi^* E}$. In other words, D is a right splitting of the exact sequence (4.1). One can easy see that the bundle Im D coincides with the horizontal subbundle HE. The tangent bundle TE will decompose as Whitney sum of horizontal and vertical subbundle. We can define now the morphism $C : TE \to VE$ on fibres as being the identity on vertical vectors and zero on the horizontal vectors. It follows that C is a left splitting of the exact sequence (4.1). Moreover, the mapping C and D satisfy the relation:

$$i \circ C + D \circ \pi! = \mathrm{id}_{TE}.$$

So, we have

Theorem 1.4.2. *A nonlinear connection on the tangent bundle $E = (TM, \pi, M)$ is characterized by a right splitting of the exact sequence* (4.1), *$D : \pi^* E \to TE$, such that $\pi! \circ D = \mathrm{id}|_{\pi^* E}$.*

The set of isomorphisms $r_u : V_u E \to T_u E$, $u \in E$ defines a canonical isomorphism r between the vertical subbundle and the vector bundle $\pi^* E$.

Definition 1.4.2. The map $K : TE \to E$, given by $K = p_2 \circ r \circ C$ is called the *connection map* associated to the nonlinear connection C, where p_2 is the projection on the second factor of $\pi^* E$.

It follows that the connection map K is a morphism of vector bundles, whose kernel is the horizontal bundle HE. In general, the map K is not linear on the fibres of (E, π, M).

The local representation of the mapping K is

$$K(x, y, X, Y) = (x, Y^j + N^j_i(x, y)X^i). \tag{4.3}$$

Let us consider a nonlinear connection determined by C and K the connection map associated to C, with the local expression given by (4.3). Taking into account (4.3) and the definition of C, we get the local expression of the nonlinear connection:

$$C(x, y, X, Y) = (x, y, o, Y^j + N^j_i(x, y)X^i). \tag{4.3$'$}$$

The differential functions $(N^j_i(x, y))$, $i, j \in \{1, 2, ..., n\}$ defined on the domain of local charts on E are called *the coefficients of the nonlinear connection.* These functions characterize a nonlinear connection in the tangent bundle.

Proposition 1.4.1. *To give a nonlinear connection in the tangent bundle (TM, π, M) is equivalent to give a set of real functions $(N^i_j(x, y))$, $i, j \in \{1, 2, .., n\}$, on every coordinate neighbourhood of TM, which on the intersection of coordinate neighbourhoods satisfies the following transformation rule:*

$$\overline{N}^j_i \frac{\partial \bar{x}^i}{\partial x^k} = \frac{\partial \bar{x}^j}{\partial x^i} N^i_k - \frac{\partial^2 \bar{x}^j}{\partial x^k \partial x^i} y^i. \tag{4.4}$$

Proof. The formulae (4.4) are equivalent with the second components $Y^j + N^j_i(x, y)X^i$ of the connection map K from (4.3) under the overlap charts are changed as follows

$$\widetilde{Y}^j + \widetilde{N}^j_i(\tilde{x}, \tilde{y})\widetilde{X}^i = \frac{\partial \tilde{x}^j}{\partial x^k}(Y^k + N^k_i(x, y)X^i).$$

Applying Theorem 1.3.5 we get:

Theorem 1.4.3. *A semispray S on $\widetilde{TM}$, with the coefficients $G^i(x, y)$ determines a nonlinear connection N with the coefficients $N^i_j = \dfrac{\partial G^i}{\partial y^j}$.*

Conversely, if N^i_j are the coefficients of a nonlinear connection N, then

$$G^i(x, y) = N^i_j(x, y)y^j \tag{4.5}$$

are the coefficients of a semispray on $\widetilde{TM}$.

The nonlinear connection N, determined by the morphism C is called homogeneousand linear if the connection map K associated to C has this property, respectively.

Taking into account (4.3) and the local expression of the mapping $\pi^\top$ it follows that N is homogeneous iff its coefficients $N^i_j(x,y)$ are homogeneous.

Exactly as in Theorem 1.3.4, we can prove:

Theorem 1.4.4. *If the manifold M is pracompact, then there exists nonlinear connections on $\widetilde{TM}$.*

1.5 The structures $\mathbb{P}, \mathbb{F}$

Now, let $i_x : T_xM \to TM$ be the inclusion and for $u \in T_xM$ consider the usual identification $k_u : u \in T_xM \to u \in T_u(T_xM)$. We obtain a natural isomorphism

$$\ell^v_u = i^\top_{\pi(u)} \circ k_u : T_{\pi(u)}M \to V_uTM$$

called the *vertical lift*

In local coordinates, for any $z = z^i \dfrac{\partial}{\partial x^i} \in T_{\pi(u)}M$ it follows

$$\ell^v_u(z) = z^i \frac{\partial}{\partial y^i}.$$

The canonical isomorphism $\pi : VTM \to \pi^*TM$ is the inverse of the isomorphism $j : \pi^*TM \to VTM$ defined by $j(z_1, z_2) = \ell^v_{z_1}(z_2)$, $z_1, z_2 \in TM$, $\pi(z_1) = \pi(z_2)$. Explicitely, we have

$$r(X_u) = (u, (\ell^v_u)^{-1}(X_u)),\ u \in TM,\ X_u \in V_uTM.$$

Consequently, we can define $\mathcal{F}(\widetilde{TM})$–linear mapping $J : \mathcal{X}(TM) \to \mathcal{X}(TM)$ by

$$J_u = i_u \circ j_u \circ \pi!(u). \tag{5.1}$$

Proposition 1.5.1.

1° *The mapping* (5.1) *is the tangent structure J investigated in* §*1.*

2° *In the natural basis J is given by*

$$J = \frac{\partial}{\partial y^i} \otimes dx^i.$$

In the same manner we can introduce the notion of *almost product structure* $\mathbb{P}$ on TM.

Based on the fact that direct decomposition (4.2)′ holds when a nonlinear connection N is given, we consider the *vertical projector* $v : \mathcal{X}(TM) \to \mathcal{X}(TM)$ defined by

$$v(X) = C(X),\ \forall X \in \mathcal{X}(VTM),\ V(X) = 0,\ \forall X \in \mathcal{X}(HTM).$$

Of course, we have $v^2 = v$. The projector v coincides with the mapping C considered as morphism between modules of sections. So, N is characterized by a vertical projector v.

On the same way, a nonlinear connection on TM is characterized by a $\mathcal{F}(TM)$–linear mapping $h : \mathcal{X}(TM \to \mathcal{X}(TM)$ for which:

$$h^2 = h, \text{ Ker}\, h = \mathcal{X}(VTM).$$

The mapping h is called the *horizontal projector* determined by a nonlinear connection N.

We have $h + v = I$.

Finally, any vector field $X \in \mathcal{X}(TM)$ can be uniquely written as follows $X = hX + vX$. In the following we adopt the notations

$$hX = X^H, \; vX = X^V$$

and we say X^H is a *horizontal component* of vector field X, but X^V is the *vertical component.*

So, any $X \in \mathcal{X}(TM)$ can be uniquely written in the form

$$X = X^H + X^V. \tag{5.2}$$

Theorem 1.5.1. *A nonlinear connection N in the vector bundle (TM, π, M) is characterized by an almost product structure $\mathbb{P}$ on the manifold TM whose distribution of eigenspaces corresponding to the eigenvalue -1 coincides to the vertical distribution on TM.*

Proof. Given a nonlinear connection N, we consider the vertical projector v determined by N and set $\mathbb{P} = I - 2v$. It follows $\mathbb{P}^2 = I$. Hence $\mathbb{P}$ is an almost product structure on TM. We have

$$\mathbb{P}(X) = -X, \; \forall X \in \mathcal{X}(VTM). \tag{$*$}$$

Conversely, if an almost product structure $\mathbb{P}$ on TM is given, and $\mathbb{P}$ has the property $(*)$, we set $v := \dfrac{1}{2}(I - \mathbb{P})$. It results that v is a vertical projector and therefore it determines a nonlinear connection N.

The following relations hold:

$$\mathbb{P} = 2h - I, \; \mathbb{P} = h - v, \; \mathbb{P} = I - 2v. \tag{5.3}$$

Taking into account the properties of the tangent structure J and almost product structure $\mathbb{P}$ we obtain

$$J\mathbb{P} = J, \; \mathbb{P}J = -J. \tag{5.4}$$

Let us consider the horizontal lift determined by a nonlinear connection N with the local coefficients $N^i_j(x,y)$. Denote the horizontal lift of vector fields $\frac{\partial}{\partial x^i}$, $(i = 1, ..., n)$, by

$$\frac{\delta}{\delta x^i} = \left(\frac{\partial}{\partial x^i}\right)^H.$$

Remark that $\pi! : HTM \to \pi^*TM$ is an isomorphism of vector bundle. Then the horizontal lift induced by N is just the inverse map of $\pi!$ restricted to HTM. According to $(4.3)'$ we have

$$\frac{\delta}{\delta x^i} = \frac{\partial}{\partial x^i} - N^j_i(x,y)\frac{\partial}{\partial y^j}, \tag{5.5}$$

where $N^i{}_j$ are the coefficients of the nonlinear connection N.

Locally, if $X = X^i(x)\frac{\partial}{\partial x^i}$, then $X^H = X^i\left(\frac{\partial}{\partial x^i} - N^j_i(x,y)\frac{\partial}{\partial y^j}\right)$. Moreover, $\left(\frac{\delta}{\delta x^i}\right)$, $(i = 1, ..., n)$, is a local basis in the horizontal distribution HTM.

Consequently, it follows that $\left(\frac{\delta}{\delta x^i}, \frac{\partial}{\partial y^i}\right)$, $(i = 1, ..., n)$, is a *local basis adapted* to the horizontal distribution HTM and vertical distribution VTM.

Let $(dx^i, \delta y^i)$ the dual basis of the adapted basis $\left(\frac{\delta}{\delta x^i}, \frac{\partial}{\partial y^i}\right)$. It follows

$$\delta y^i = dy^i + N^i_j(x,y)dx^j. \tag{5.6}$$

Proposition 1.5.2. *The local adapted basis* $\left(\frac{\delta}{\delta x^i}, \frac{\partial}{\partial y^i}\right)$ *and its dual* $(dx^i, \delta y^i)$ *transform, under a transformation of coordinate* (1.3) *on* TM, *by*

$$\frac{\delta}{\delta x^i} = \frac{\partial \tilde{x}^j}{\partial x^i}\frac{\delta}{\delta \tilde{x}^j};\ \frac{\partial}{\partial y^i} = \frac{\partial \tilde{x}^j}{\partial x^i}\frac{\partial}{\partial \tilde{x}^j};\ d\tilde{x}^i = \frac{\partial \tilde{x}^i}{\partial x^j}dx^j;\ \delta\tilde{y}^i = \frac{\partial \tilde{x}^i}{\partial x^j}\delta y^j. \tag{5.7}$$

Indeed, the second formula is known from (1.4). The first one is a consequence of the formula $\frac{\delta}{\delta x^i} = \left(\frac{\partial}{\partial x^i}\right)^H$.

For the operators $h, v, \mathbb{P}$ we get:

$$\begin{cases} h\left(\frac{\delta}{\delta x^i}\right) = \frac{\delta}{\delta x^i};\ h\left(\frac{\partial}{\partial y^i}\right) = 0;\ v\left(\frac{\delta}{\delta x^i}\right) = 0,\ v\left(\frac{\partial}{\partial y^i}\right) = \frac{\partial}{\partial y^i}, \\ \mathbb{P}\left(\frac{\delta}{\delta x^i}\right) = \frac{\delta}{\delta x^i};\ \mathbb{P}\left(\frac{\partial}{\partial y^i}\right) = -\frac{\partial}{\partial y^i}. \end{cases} \tag{5.8}$$

Now, let us consider the $\mathcal{F}(TM)$–linear mapping $\mathbb{F} : \mathcal{X}(TM) \to \mathcal{X}(TM)$, defined by

$$\mathbb{F}\left(\frac{\delta}{\delta x^i}\right) = -\frac{\partial}{\partial y^i};\ \mathbb{F}\left(\frac{\partial}{\partial y^i}\right) = \frac{\delta}{\delta x^i},\ (i = 1, ..., n). \tag{5.9}$$

Theorem 1.5.2. *The mapping* $\mathbb{F}$ *has the properties:*

1° $\mathbb{F}$ *is globally defined on the manifold* TM.

2° $\mathbb{F}$ *is a tensor field of (1,1) type on* TM. *Locally it is given by*

$$\mathbb{F} = -\frac{\partial}{\partial y^i} \otimes dx^i + \frac{\delta}{\delta x^i} \otimes \delta y^i. \tag{5.10}$$

3° $\mathbb{F}$ *is an almost complex structure on* TM:

$$\mathbb{F} \circ \mathbb{F} = -I. \tag{5.11}$$

Proof. Since (5.9) and (5.10) are equivalent, it follows from (5.10) that $\mathbb{F}$ is globally defined on TM. From (5.9) we deduce (5.11). **q.e.d.**

By a straightforward calculation we deduce:

Lemma 1.5.1. *Lie brackets of the vector fields from adapted basis* $\left(\frac{\delta}{\delta x^i}, \frac{\partial}{\partial y^i}\right)$ *are given by*

$$\left[\frac{\delta}{\delta x^j}, \frac{\delta}{\delta x^k}\right] = R^i{}_{jk}\frac{\partial}{\partial y^i},\quad \left[\frac{\delta}{\delta x^j}, \frac{\partial}{\partial y^k}\right] = \frac{\partial N^i{}_j}{\partial y^k}\frac{\partial}{\partial y^i},\quad \left[\frac{\partial}{\partial y^j}, \frac{\partial}{\partial y^k}\right] = 0, \tag{5.12}$$

where

$$R^i{}_{jk} = \frac{\delta N^i{}_j}{\delta x^k} - \frac{\delta N^i{}_k}{\delta x^j}. \tag{5.13}$$

Let us consider the quantities

$$t^i{}_{jk} = \frac{\partial N^i{}_j}{\partial y^k} - \frac{\partial N^i{}_k}{\partial y^j}. \tag{5.14}$$

Also, by a direct calculation, we obtain:

Lemma 1.5.2. *Under a transformation of coordinates* (1.3) *on* TM, *we obtain*

$$(5.15)\qquad \begin{aligned} \widetilde{R}^i{}_{jk}(\widetilde{x},\widetilde{y}) &= \frac{\partial \widetilde{x}^i}{\partial x^r}\frac{\partial x^s}{\partial \widetilde{x}^j}\frac{\partial x^p}{\partial \widetilde{x}^k} R^r{}_{sp}(x,y) \\ \widetilde{t}^i{}_{jk}(\widetilde{x},\widetilde{y}) &= \frac{\partial \widetilde{x}^i}{\partial x^r}\frac{\partial x^s}{\partial \widetilde{x}^j}\frac{\partial x^p}{\partial \widetilde{x}^k} t^r{}_{sp}(x,y). \end{aligned}$$

Consequently, the tensorial equations $R^i{}_{jk}=0$, $t^i{}_{jk}=0$ have geometrical meaning.

Therefore, we get:

Lemma 1.5.3. *The horizontal distribution* HTM *is integrable if and only if we have on* TM:

$$R^i{}_{jk}(x,y)=0.$$

Indeed, from (5.12), the Lie brackets $\left[\frac{\delta}{\delta x^j},\frac{\delta}{\delta x^k}\right]$ give an horizontal vector fields if and only if $R^i{}_{jk}=0$.

The previous property allows to say that $R^i{}_{jk}$ is the curvature tensor field of the nonlinear connection N. We will say that $t^i{}_{jk}$ from (5.14) is the *torsion* of the nonlinear connection N.

Now, we can prove:

Theorem 1.5.3. *The almost complex structure* $\mathbb{F}$ *is integrable if and only if we have*

$$(5.16)\qquad R^i{}_{jk}=0,\ t^i{}_{jk}=0.$$

Proof. Applying Lemma 1.5.1, and taking into account the Nijenhuis tensor field of the structure $\mathbb{F}$ [113]:

$$\mathcal{N}_{\mathbb{F}}(X,Y) = -[X,Y]+[\mathbb{F}X,\mathbb{F}Y]-\mathbb{F}[\mathbb{F}X,Y]-\mathbb{F}[X,\mathbb{F}Y],$$

putting $X=\frac{\delta}{\delta x^i}$, $X=\frac{\partial}{\partial y^j}$ etc., we deduce

$$\begin{aligned} \mathcal{N}_{\mathbb{F}}\left(\frac{\delta}{\delta x^j},\frac{\delta}{\delta x^k}\right) &= R^i{}_{jk}\frac{\partial}{\partial y^i} - t^i{}_{jk}\frac{\delta}{\delta x^i} \\ \mathcal{N}_{\mathbb{F}}\left(\frac{\delta}{\delta x^j},\frac{\partial}{\partial y^k}\right) &= R^i{}_{jk}\frac{\delta}{\delta x^i} + t^i{}_{jk}\frac{\partial}{\partial y^i} \\ \mathcal{N}_{\mathbb{F}}\left(\frac{\partial}{\partial y^j},\frac{\partial}{\partial y^k}\right) &= -R^i{}_{jk}\frac{\partial}{\partial y^i} + t^i{}_{jk}\frac{\delta}{\delta x^i}. \end{aligned}$$

Now it follows that $\mathcal{N}_{\mathbb{F}}=0 \Longleftrightarrow \{R^i{}_{jk}=t^i{}_{jk}=0\}$. **q.e.d.**

1.6 d-tensor Algebra

Let N be a nonlinear connection on the manifold $E = TM$. We have the direct decomposition (4.2)$'$. We can write, uniquely a vector field $X \in \mathcal{X}(TM)$ in the form

$$X = X^H + X^V \tag{6.1}$$

where X^H belongs to the horizontal distribution HTM.

Taking the adapted basis $\left(\dfrac{\delta}{\delta x^i}, \dfrac{\partial}{\partial y^i}\right)$ to the direct decomposition (4.2)$'$ we can write:

$$X^H = X^i(x,y)\frac{\delta}{\delta x^i},\ X^V = \dot{X}^i(x,y)\frac{\partial}{\partial y^i}. \tag{6.1$'$}$$

With respect to (1.3) the components $X^i(x,y)$ and $\dot{X}^i(x,y)$ of X^H and X^V respectively obey the rules of transformation

$$\widetilde{X}^i = \frac{\partial \widetilde{x}^i}{\partial x^j}\, X^j,\quad \widetilde{\dot{X}}^i = \frac{\partial \widetilde{x}^i}{\partial x^j}\, \dot{X}^j. \tag{6.1$''$}$$

Also, a 1-form field $\omega \in \mathcal{X}^*(E)$ can be always set as follows

$$\omega = \omega^H + \omega^V, \tag{6.2}$$

where $\omega^H(X) = \omega(X^H),\ \omega^V(X) = \omega(X^V),\ \forall\, X \in \mathcal{X}(E)$.

Therefore in the adapted cobasis $(dx^i, \delta y^i) we have:$

$$\omega^H = \omega_j(x,y)dx^j,\ \omega^V = \dot{\omega}_j(x,y)\delta y^j. \tag{6.2$'$}$$

The changes of local coordinate on TM transform the components $\omega_j(x,y), \dot{\omega}_j(x,y)$ of the 1-form ω as the components of 1-forms on the base manifold M, i.e.:

$$\omega_j(x,y) = \frac{\partial \widetilde{x}^i}{\partial x^j}\, \widetilde{\omega}_i(\widetilde{x},\widetilde{y});\ \dot{\omega}_j(x,y) = \frac{\partial \widetilde{x}^i}{\partial x^j}\, \widetilde{\dot{\omega}}_i(\widetilde{x},\widetilde{y}). \tag{6.2$''$}$$

A curve $c : t \in I \to (x^i(t), y^i(t)) \in E$, has the tangent vector $\dfrac{dc}{dt} = \dot{c}$ given in the form (6.1), hence:

$$\dot{c} = \dot{c}^H + \dot{c}^V = \frac{dx^i}{dt}\frac{\delta}{\delta x^i} + \frac{\delta y^i}{dt}\frac{\partial}{\partial y^i}. \tag{6.3}$$

This is a *horizontal curve* if $\dfrac{\delta y^i}{dt} = 0,\ \forall\, t \in I$. So, if the functions $x^i = x^i(t),\ t \in I$ are given, then the curves $y^i = y^i(t),\ t \in I$, solutions of the system of differential equations $\dfrac{\delta y^i}{dt} + N^i{}_j(x,y)\dfrac{dx^j}{dt} = 0$, determine a horizontal curve c in $E = TM$.

A horizontal curve c with the property $y^i = \dfrac{dx^i}{dt}$ is said to be an *autoparallel* curve of the nonlinear connection N.

Proposition 1.6.1. *An autoparallel curve of the nonlinear connection N, with the coefficients $N^i{}_j(x,y)$, is characterized by the system of differential equations*

$$\frac{dy^i}{dt} + N^i{}_j(x,y)\,\frac{dx^j}{dt} = 0,\ \frac{dx^i}{dt} = y^i. \tag{6.4}$$

Now we study shortly the algebra of the distinguished tensor fields on the manifold $TM = E$.

Definition 1.6.1. A tensor field T of type (r,s) on the manifold E is called distinguished tensor field (briefly, a d-tensor) if it has the property

$$\begin{gathered} T(\overset{1}{\omega},...,\overset{r}{\omega},\underset{1}{X},...,\underset{s}{X}) = T(\overset{1}{\omega}{}^H,...,\overset{r}{\omega}{}^V,\underset{1}{X}{}^H,...,\underset{s}{X}{}^V), \\ \forall \underset{a}{\omega} \in \mathcal{X}^*(E),\ \forall \underset{b}{X} \in \mathcal{X}(E),\ (a=1,...,r;\ b=1,...,s). \end{gathered} \tag{6.5}$$

For instance, the components X^H and X^V from (6.1) of a vector field X are d–tensor fields. Also the components ω^H and ω^V of an 1-form ω, from (6.2) are d-1-form fields.

Clearly, the set $\mathcal{T}^r_s(E)$ of the d-tensor fields of type (r,s) is a $\mathcal{F}(E)$–module and the module $\mathcal{T}(E) = \bigoplus\limits_{r,s} \mathcal{T}^r_s$ is a tensor algebra. It is not difficult to see that any tensor field on E can be written as a sum of d-tensor fields.

We express a d-tensor field T from (6.5) in the adapted basis $\left(\dfrac{\delta}{\delta x^i}, \dfrac{\partial}{\partial y^i}\right)$ and adapted cobasis $(dx^i, \delta y^i)$. From (6.5) we get the components of T:

$$T^{i_1...i_r}_{j_1...j_s}(x,y) = T\left(dx^{i_1},...,\delta y^{i_r},\frac{\delta}{\delta x^{j_1}},\cdots,\frac{\partial}{\partial y^{j_s}}\right),\ (i_1,...,i_r,j_1,...,j_s=\overline{1,n}). \tag{6.5$'$}$$

So, T is expressed by

$$T = T^{i_1...i_r}_{j_1...j_s}(x,y)\frac{\delta}{\delta x^{i_1}} \otimes \cdots \otimes \frac{\partial}{\partial y^{i_r}} \otimes dx^{j_1} \otimes \cdots \otimes \delta y^{j_s}. \tag{6.6}$$

Taking into account the formulae (5.7) and (6.5)$'$, we obtain:

Proposition 1.6.2. *With respect to* (1.5) *the components $T^{i_1...i_r}_{j_1...j_s}(x,y)$ of a d-tensor field T of type (r,s) are transformed by the rules:*

$$\widetilde{T}^{i_1...i_r}_{j_1...j_s}(\widetilde{x},\widetilde{y}) = \frac{\partial \widetilde{x}^{i_1}}{\partial x^{h_1}}\cdots\frac{\partial \widetilde{x}^{i_r}}{\partial x^{h_r}}\,\frac{\partial \widetilde{x}^{k_1}}{\partial \widetilde{x}^{j_1}}\cdots\frac{\partial x^{k_s}}{\partial \widetilde{x}^{j_s}}\,T^{h_1...h_r}_{k_1...k_s}(x,y). \tag{6.7}$$

But (6.7) is just the classical law of transformation of the local coefficients of a tensor field on the base manifold M.

Of course, (6.7) characterizes the d-tensor fields of type (r, s) on the manifold $E = TM$ (up to the choice of the basis from (6.6)). Using the local expression (6.6) of a d-tensor field it follows that $\left\{1, \frac{\delta}{\delta x^i}, \frac{\partial}{\partial y^i}\right\}$, $(i = 1, ..., n)$, generate the d-tensor algebra $\mathcal{T}$ over the ring of functions $\mathcal{F}(E)$. Taking into account Lemma 1.5.2 it follows:

Proposition 1.6.3.

1° *$R^i{}_{jk}$ and $t^i{}_{jk}$ from (5.12), (5.13) are d-tensor fields of type (1, 2).*

2° *The Liouville vector field $\mathbb{C} = y^i \frac{\partial}{\partial y^i}$ is a d-vector field.*

1.7 N–linear connections

Let N be an a priori given nonlinear connection on the manifold $E = TM$.

The adapted basis to N and V is $\left(\frac{\delta}{\delta x^i}, \frac{\partial}{\partial y^i}\right)$ and adapted cobasis is its dual $(dx^i, \delta y^i)$.

Definition 1.7.1. A linear connection D (i.e. a Kozul connection or covariant derivative) on the manifold $E = TM$ is called an N–linear connection if:

1° D preserves by parallelism the horizontal distribution N.

2° The tangent structure J is absolute parallel with respect to D, that is $DJ = 0$.

Consequently, the following properties hold:

(7.1) $$(D_X Y^H)^V = 0,\ (D_X Y^V)^H = 0,$$

(7.1)′ $$D_X(JY^H) = J(D_X Y^H);\ D_X(JY^V) = J(D_X Y^V),$$

(7.1)″ $$D_X h = 0,\ D_X v = 0.$$

We will denote

(7.2) $$D^H_X Y = D_{X^H} Y,\ D^V_X Y = D_{X^V} Y.$$

Thus, we obtain the following expression of D:

(7.3) $$D_X = D^H_X + D^V_X,\ \forall X \in \mathcal{X}(E).$$

The operators D^H and D^V are special derivations in the algebra $\mathcal{T}$ of d-tensor fields on E. Of course, D^H, D^V are not covariant derivations, because $D^H_X f = X^H f \neq Xf$, $D^V_X f = X^V f \neq Xf$. However the operators D^H and D^V have similar properties to D. For instance, D^H and D^V satisfy the Leibniz rule with respect of tensorial product of d-tensor fields. It is important to remark that D^H and D^V applied to d-tensor fields give us the d-tensor fields, too. We can see these important properties on the local representtion of D^H and D^V in the adapted basis $\left(\dfrac{\delta}{\delta x^i}, \dfrac{\partial}{\partial y^i}\right)$. D^H and D^V will be called the h-covariant derivation and v-covariant derivation, respectively.

Remarking that $D_{\frac{\delta}{\delta x^i}} = D^H_{\frac{\delta}{\delta x^i}}$, $D_{\frac{\partial}{\partial y^i}} = D^V_{\frac{\partial}{\partial y^i}}$, we obtain:

Proposition 1.7.1. *In the adapted basis* $\left(\dfrac{\delta}{\delta x^i}, \dfrac{\partial}{\partial y^i}\right)$, *an N–linear connection D can be uniquely represented in the form:*

$$
(7.4)\qquad
\begin{aligned}
&D^H_{\frac{\delta}{\delta x^k}} \frac{\delta}{\delta x^j} = L^i{}_{jk}(x,y)\frac{\delta}{\delta x^i}, \quad D^H_{\frac{\delta}{\delta x^k}} \frac{\partial}{\partial y^j} = L^i{}_{jk}(x,y)\frac{\partial}{\partial y^i},\\
&D^V_{\frac{\partial}{\delta y^k}} \frac{\delta}{\delta x^j} = C^i{}_{jk}(x,y)\frac{\delta}{\delta x^i}, \quad D^V_{\frac{\partial}{\delta y^k}} \frac{\partial}{\partial y^j} = C^i{}_{jk}(x,y)\frac{\partial}{\partial y^i}.
\end{aligned}
$$

The system of functions $D\Gamma(N) = (L^i{}_{jk}(x,y), C^i{}_{jk}(x,y))$ gives us the coefficients of the h-covariant derivative D^H and of the v-covariant derivative D^V, respectively.

Proposition 1.7.2. *With respect to the changes of local coordinates on TM, the coefficients $L^i{}_{jk}(x,y)$, $C^i{}_{jk}(x,y)$ of an N–linear connection D are transformed as follows:*

$$
(7.5)\qquad
\begin{aligned}
\widetilde{L}^i{}_{jh}(\bar{x},\bar{y}) &= \frac{\partial \widetilde{x}^i}{\partial x^\ell}\frac{\partial x^s}{\partial \widetilde{x}^j}\frac{\partial x^r}{\partial \widetilde{x}^h} L^\ell{}_{sr} + \frac{\partial \widetilde{x}^i}{\partial x^r}\frac{\partial^2 x^r}{\partial \bar{x}^j \partial \bar{x}^h},\\
\widetilde{C}^i{}_{jh}(\bar{x},\bar{y}) &= \frac{\partial \widetilde{x}^i}{\partial x^\ell}\frac{\partial x^s}{\partial \widetilde{x}^j}\frac{\partial x^r}{\partial \widetilde{x}^h} C^\ell{}_{sr}.
\end{aligned}
$$

Indeed, the formulae (7.4) and (5.7) imply the rules transformation (7.5).

Remarks.

1° $C^i{}_{jk}(x,y)$ are the coordinates of a d-tensor field.

2° A reciprocal property of that expressed in the last proposition also holds.

Let us now consider a d-tensor field T in local adapted basis, given for simplicity by

$$T = T^{is}_{jh}\frac{\delta}{\delta x^i} \otimes \frac{\partial}{\partial y^s} \otimes dx^j \otimes \delta y^h.$$

Its covariant derivative with respect to $X = X^H + X^V = X^i \dfrac{\delta}{\delta x^i} + Y^j \dfrac{\partial}{\partial y^j}$ is given by

$$D_X T = \left(X^r T^{is}_{jh|r} + Y^\ell T^{is}_{jh}|_\ell\right) \frac{\delta}{\delta x^i} \otimes \frac{\partial}{\partial y^s} \otimes dx^j \otimes \delta y^h, \tag{7.6}$$

where we have the h–covariant derivative, $D^H_X T = X^r T^{is}_{jh|r} \dfrac{\delta}{\delta x^i} \otimes \dfrac{\partial}{\partial y^s} \otimes dx^j \otimes \delta y^h$. Its coefficients are

$$T^{is}_{jh|r} = \frac{\delta T^{is}_{jh}}{\delta x^r} + L^i_{\ell r} T^{\ell s}_{jh} + L^s_{\ell r} T^{i\ell}_{jh} - L^\ell_{jr} T^{is}_{\ell h} - L^\ell_{hr} T^{is}_{j\ell}. \tag{7.7}$$

Therefore, "$|$" is the operator of h-covariant derivative. Of course, $T^{is}_{jh|r}$ is a d-tensor field with one more index of a covariance.

The v-covariant derivative of T is $D^V_X T = Y^\ell T^{is}_{jh}|_\ell \dfrac{\delta}{\delta x^i} \otimes \dfrac{\partial}{\partial y^s} \otimes dx^j \otimes \delta y^h$, and the coefficients $T^{is}_{jh}|_\ell$ are as follows:

$$T^{is}_{jh}|_r = \frac{\partial T^{is}_{jh}}{\partial y^r} + C^i_{\ell r} T^{ls}_{jh} + C^s_{\ell r} T^{i\ell}_{jh} - C^\ell_{jr} T^{is}_{\ell h} - C^\ell_{hr} T^{is}_{j\ell}. \tag{7.8}$$

Here we denoted by "$|$" the operator of v-covariant derivative and remark that $T^{is}_{jh}|_r$ is a d-tensor field with one more index of a covariance.

The operators "$|$" and "$|$" have the known properties of a general covariant derivatives, applied to any d-tensor field T, taking into account the facts: $\dfrac{\delta f}{\delta x^i} = f_{|i}$, $\dfrac{\partial f}{\partial y^i} = f|_i$ for any function $f \in \mathcal{F}(E)$.

An important application can be done for the Liouville vector field $\mathbb{C} = y^i \dfrac{\partial}{\partial y^i}$.

The following d-tensor fields

$$D^i{}_j = y^i{}_{|j}, \; d^i{}_j = y^i|_j \tag{7.9}$$

are called the h- and v-deflection tensor fields of the N-linear connection D.

Proposition 1.7.3. *The deflection tensor fields are given by*

$$D^i{}_j = y^s L^i{}_{sj} - N^i{}_j, \; d^i{}_j = \delta^i_j + y^s C^i{}_{sj}. \tag{7.9$'$}$$

Indeed, applying the formulae (7.7), (7.8) we get the equalities (7.9)$'$.

The d-tensor of deflections are important in the geometry of tangent bundle.

A N–linear connection D is called of *Cartan type* if its tensor of deflection have the property:

$$D^i{}_j = 0,\ d^i{}_j = \delta^i{}_j.$$

From the last proposition, it follows

Proposition 1.7.4 *The N-linear connection $D\Gamma(N) = (L^i{}_{jk}, C^i{}_{jk})$ is of Cartan type if and only if we have*

$$N^i{}_j = y^s L^i{}_{sj},\ y^s C^i{}_{sj} = 0. \tag{7.10}$$

We will see that the canonical metrical connection in a Finsler space is of Cartan type.

We can prove [113]:

Theorem 1.7.1. *If M is a paracompact manifold then there exist N-linear connections on TM.*

1.8 Torsion and curvature

The *torsion* of a N–linear connection d is given by

$$\mathbb{T}(X,Y) = D_X Y - D_Y X - [X,Y],\ X,Y \in \mathcal{X}(TM). \tag{8.1}$$

Using the projectors, h, and v associated to the horizontal distribution N and to the vertical distribution V, we find

$$\mathbb{T}(X,Y) = \mathbb{T}(X^H,Y^H) + \mathbb{T}(X^H,Y^V) + \mathbb{T}(X^V,Y^H) + \mathbb{T}(X^V,Y^V).$$

Taking into account the property of skew-symmetry of $\mathbb{T}$ and the fact that $[X^V,Y^V]^H = 0$ we find

Theorem 1.8.1. *The torsion $\mathbb{T}$ of an N-linear connection is completely determined by the following d-tensor fields:*

$$\begin{aligned}
&h\mathbb{T}(X^H,Y^H) = D^H_X Y^H - D^H_Y X^H - [X^H,Y^H]^H,\\
&v\mathbb{T}(X^H,Y^H) = -[X^H,Y^H]^V,\\
&h\mathbb{T}(X^H,Y^V) = -D^V_Y X^H - [X^H,Y^V]^H,\\
&v\mathbb{T}(X^H,Y^V) = D^H_X Y^V - [X^H,Y^V]^V,\\
&v\mathbb{T}(X^V,Y^V) = D^V_X Y^V - D^V_Y X^V - [X^V,Y^V]^V,\ X,Y \in \mathcal{X}(TM).
\end{aligned} \tag{8.2}$$

Corollary 1.8.1. *The following properties hold:*

a) $v\mathbb{T}(X^H, Y^H) = 0 \Longleftrightarrow HTM$ *is an integrable distribution.*

b) $h\mathbb{T}(X^H, Y^V) = 0 \Longleftrightarrow vD^H_X Y^V = [X^H, Y^V]^V,\ X, Y \in \mathcal{X}(TM).$

We shall say that $h\mathbb{T}(X^H, Y^H)$ is $h(hh)$–torsion of D, that $vT(X^H, Y^H)$ is $v(hh)$–torsion, etc.

Since the Lie brackets of the vector field from the adapted basis are given by the formula (5.12), we obtain

Theorem 1.8.2. *The local components, in the adapted basis* $\left\{\dfrac{\delta}{\delta x^i}, \dfrac{\partial}{\partial y^i}\right\}$, *of the torsion* $\mathbb{T}$ *of an* N*–linear connection are as follows:*

$$
\begin{aligned}
&T^i{}_{jh} = L^i{}_{jh} - L^i{}_{hj},\ R^i{}_{jh},\ C^i{}_{jh},\\
(8.3)\qquad &P^i{}_{jh} = \frac{\partial N^i_j}{\partial y^h} - L^i{}_{jh},\ S^i{}_{jh} = C^i{}_{jh} - C^i{}_{hj}.
\end{aligned}
$$

Proof. These local coefficients are provided by the five formulae (8.2) if we consider instead of X and Y the components of the adapted basis $\left(\dfrac{\delta}{\delta x^i}, \dfrac{\partial}{\partial y^i}\right)$.

The curvature of a N–linear connection D is given by

$$
(8.4)\qquad \mathbb{R}(X, Y)Z = D_X D_Y Z - D_Y D_X Z - D_{[X,Y]}Z,\ \forall X, Y, Z \in \mathcal{X}(TM).
$$

It is not difficult to prove the following theorems:

Theorem 1.8.3. *The curvature tensor* $\mathbb{R}$ *of the* N*–linear connection* D *has the properties:*

$$
\begin{aligned}
&v\mathbb{R}(X, Y)Z^H = 0,\ h\mathbb{R}(X, Y)Z^V = 0,\\
(8.5)\qquad &\mathbb{R}(X, Y)Z = h\mathbb{R}(X, Y)Z^H + v\mathbb{R}(X, Y)Z^V,\ \forall X, Y, Z \in \mathcal{X}(TM).
\end{aligned}
$$

Theorem 1.8.4. *The curvature of an* N*–linear connection* D *on* TM *is completely determined by the following three d-tensor fields:*

$$
\begin{aligned}
&\mathbb{R}(X^H, Y^H)Z^H = D^H_X D^H_Y Z^H - D^H_Y D^H_X Z^H - D^H_{[X^H, Y^H]}Z^H - D^V_{[X^H, Y^H]}Z^H,\\
(8.6)\qquad &\mathbb{R}(X^V, Y^H)Z^H = D^V_X D^H_Y Z^H - D^H_Y D^V_X Z^H - D^H_{[X^V, Y^H]}Z^H - D^V_{[X^V, Y^H]}Z^H,\\
&\mathbb{R}(X^V, Y^V)Z^H = D^V_X D^V_Y Z^H - D^V_Y D^V_X Z^H - D^V_{[X^V, Y^V]}Z^H.
\end{aligned}
$$

Remark. The curvature $\mathbb{R}$ has six components. But the property $J(\mathbb{R}(X,Y)Z^H) = \mathbb{R}(X,Y)Z^V$ shows that only three components, namely the one in (8.6) are essential.

In the adapted basis, the local coefficients of the d-tensors of curvature are given by

$$\begin{aligned} \mathbb{R}\left(\frac{\delta}{\delta x^k}, \frac{\delta}{\delta x^j}\right)\frac{\delta}{\delta x^h} &= R_h{}^i{}_{jk}\frac{\delta}{\delta x^i}, \\ \mathbb{R}\left(\frac{\partial}{\partial y^k}, \frac{\delta}{\delta x^j}\right)\frac{\delta}{\delta x^h} &= P_h{}^i{}_{jk}\frac{\delta}{\delta x^i}, \\ \mathbb{R}\left(\frac{\partial}{\partial y^k}, \frac{\partial}{\partial y^j}\right)\frac{\delta}{\delta x^h} &= S_h{}^i{}_{jk}\frac{\delta}{\delta x^i}. \end{aligned} \tag{8.7}$$

Now, using Proposition 1.7.1, we obtain:

Theorem 1.8.5. *In the adapted basis the d-tensors of curvature $R_h{}^i{}_{jk}, P_h{}^i{}_{jk}$ and $S_h{}^i{}_{jk}$ of an N–linear connection $D\Gamma(N) = (L^i{}_{jk}, C^i{}_{jk})$ are as follows:*

$$\begin{aligned} R_h{}^i{}_{jk} &= \frac{\delta L^i_{hj}}{\delta x^k} - \frac{\delta L^i_{hk}}{\delta x^j} + L^m_{hj}L^i_{mk} - L^m_{hk}L^i_{mj} + C^i_{hm}R^m{}_{jk}, \\ P_h{}^i{}_{jk} &= \frac{\partial L^i_{hj}}{\partial y^k} - C^i_{hk|j} + C^i_{hm}P^m{}_{jk}, \\ S_h{}^i{}_{jk} &= \frac{\partial C^i_{hj}}{\partial x^k} - \frac{\partial L^i_{hk}}{\partial y^j} + C^m_{hj}C^i_{mk} - C^m_{hk}C^i_{mj}, \end{aligned} \tag{8.8}$$

where $_|$ denotes, as usual, the h-covariant derivative with respect to the N–linear connection $D\Gamma(N) = (L^i{}_{jk}, C^i{}_{jk})$.

The expressions (8.6) of the d-tensors of curvature $\mathbb{R}(X^H,Y^H)Z^H$, $\mathbb{R}(X^V,Y^H)Z^H$ and $\mathbb{R}(X^V,Y^V)Z^H$ in the adapted basis lead to the Ricci identities satisfied by an N–linear connection D.

Proposition 1.8.1. *The Ricci identities of the N–linear connection $D\Gamma(N) = (L^i{}_{jk}, C^i{}_{jk})$ are:*

$$\begin{aligned} X^i{}_{|k|h} - X^i{}_{|h|k} &= X^m R_m{}^i{}_{kh} - X^i{}_{|m}T^m{}_{kh} - X^i{}|_m R^m{}_{kh}, \\ X^i{}_{|k}|_h - X^i{}|_{h|k} &= X^m P_m{}^i{}_{kh} - X^i{}_{|m}C^m{}_{kh} - X^i{}|_m P^m{}_{kh}, \\ X^i{}|_k|_h - X^i{}|_h|_k &= X^m S_m{}^i{}_{kh} - X^i{}|_m S^m{}_{kh}, \end{aligned} \tag{8.9}$$

where X^i is an arbitrary d-vector field.

The Ricci identities for an arbitrary d-tensor field hold also.

For instance if $g_{ij}(x,y)$ is a d-tensor field, then the following formulae of the commutation of second h- and v-covariant derivative hold:

$$
(8.10)\qquad
\begin{aligned}
&g_{ij|k|h} - g_{ij|h|k} = -g_{sj}R_i{}^s{}_{kh} - g_{is}R_j{}^s{}_{kh} - g_{ij|s}T^s{}_{kh} - g_{ij}|_s R^s{}_{kh},\\
&g_{ij|k}|_h - g_{ij}|_{h|k} = -g_{sj}P_i{}^s{}_{kh} - g_{is}P_j{}^s{}_{kh} - g_{ij|s}C^s{}_{kh} - g_{ij}|_s P^s{}_{kh},\\
&g_{ij}|_k|_h - g_{ij}|_h|_k = -g_{sj}S_i{}^s{}_{kh} - g_{is}S_j{}^s{}_{kh} - g_{ij}|_s S^s{}_{kh}.
\end{aligned}
$$

Applying the Ricci identities(8.9) to the Liouville vector field $\mathbb{C} = y^i\dfrac{\partial}{\partial y^i}$ we deduce some fundamental identities in the theory of N–linear connections. Taking into account the h- and v-deflection tensors $D^i{}_j = y^i{}_{|j}$, $d^i{}_j = y^i|_j$ we have from (8.9):

Theorem 1.8.6. *For any N–linear connection $D\Gamma(N) = (L^i{}_{jk}, C^i{}_{jk})$ the following identities hold:*

$$
(8.11)\qquad
\begin{aligned}
&D^i{}_{k|h} - D^i{}_{h|k} = y^s R_s{}^i{}_{kh} - D^i{}_s T^s{}_{kh} - d^i{}_s R^s{}_{kh},\\
&D^i{}_k|_h - d^i{}_{h|k} = y^s P_s{}^i{}_{kh} - D^i{}_s C^s{}_{kh} - d^i{}_s P^s{}_{kh},\\
&d^i{}_k|_h - d^i{}_h|_k = y^s S_s{}^i{}_{kh} - d^i{}_s S^s{}_{kh}.
\end{aligned}
$$

Corollary 1.8.2. *If $D\Gamma(N)$ is an N–linear connection of Cartan type, then the following relations hold:*

$$
(8.12)\qquad y^s R_s{}^i{}_{kh} = R^i{}_{kh},\ y^s P_s{}^i{}_{kh} = P^i{}_{kh},\ y^s S_s{}^i{}_{kh} = S^i{}_{kh}.
$$

The d-torsions and d-curvature tensors of an N–linear connection $D\Gamma(N) = (L^i{}_{jk}, C^i{}_{jk})$ are not independent. They satisfy *the Bianchi* identities[113], obtained by writting in the adapted basis the following Bianchi identities, verified by the linear connection D:

$$
(8.13)\qquad
\begin{aligned}
&\Sigma[D_X\mathbb{T}(Y,Z) - \mathbb{R}(X,Y)Z + \mathbb{T}(\mathbb{T}(X,Y),Z)] = 0,\\
&\Sigma[(D_X\mathbb{R})(U,Y,Z) + \mathbb{R}(\mathbb{T}(X,Y),Z)U] = 0,
\end{aligned}
$$

where Σ means the cyclic sum over X, Y, Z.

1.9 Parallelism. Structure equations

Consider an N–linear connection D with the coefficients $D\Gamma(N) = (L^i{}_{jk}, C^i{}_{jk})$ in the adapted basis $\left(\dfrac{\delta}{\delta x^i}, \dfrac{\partial}{\partial y^i}\right)$.

If c is a parametrized curve in the manifold TM, $c{:}t{\in}I{\to}c(t){=}(x^i(t),y^i(t)){\in}TM$, with the property $\operatorname{Im} c \subset \pi^{-1}(U) \subset TM$, then its tangent vector field $\dot{c} = \dfrac{dc}{dt}$ can be written in the form (6.3), i.e.

$$\dot{c} = \dot{c}^H + \dot{c}^V = \frac{dx^i}{dt}\frac{\delta}{\delta x^i} + \frac{\delta y^i}{dt}\frac{\partial}{\partial y^i}. \tag{9.1}$$

The curve c is horizontal if $\dfrac{\delta y^i}{dt} = 0$ and it is an autoparallel curve of the nonlinear connection N if $\dfrac{\delta y^i}{dt} = 0$, $y^i = \dfrac{dx^i}{dt}$.

We denote

$$\frac{DX}{dt} = D_{\dot{c}}X,\ DX = \frac{DX}{dt}dt,\ \forall X \in \mathcal{X}(TM). \tag{9.2}$$

Here $\dfrac{DX}{dt}$ is the covariant differential along with the curve c of the N–linear connection D.

Setting $X = X^H + X^V$, $X^H = X^i\dfrac{\delta}{\delta x^i}$, $X^V = \dot{X}^i\dfrac{\partial}{\partial y^i}$ we have

$$\begin{aligned}\frac{DX}{dt} = \frac{DX^H}{dt} + \frac{DX^V}{dt} = & \left\{X^i{}_{|k}\frac{dx^k}{dt} + X^i|_k\frac{\delta y^k}{dt}\right\}\frac{\delta}{\delta x^i} + \\ & + \left\{\dot{X}^i{}_{|k}\frac{dx^k}{dt} + \dot{X}^i|_k\frac{\delta y^k}{dt}\right\}\frac{\partial}{\partial y^i}.\end{aligned} \tag{9.3}$$

Let us consider

$$\omega^i{}_j = L^i{}_{jk}dx^k + C^i{}_{jk}\delta y^k. \tag{9.4}$$

The objects $\omega^i{}_j$ are called the "1-forms connection" of D.

Then the equation (9.3) takes the form:

$$\frac{DX}{dt} = \left\{\frac{dX^i}{dt} + X^m\frac{\omega^i{}_m}{dt}\right\}\frac{\delta}{\delta x^i} + \left\{\frac{d\dot{X}^i}{dt} + \dot{X}^m\frac{\omega^i{}_m}{dt}\right\}\frac{\partial}{\partial y^i}. \tag{9.5}$$

The vector X on TM is said to be *parallel* along with the curve c, with respect to N–linear connection D if $\dfrac{DX}{dt} = 0$. A glance at (9.3) shows that the last equation is equivalent to $\dfrac{DX^H}{dt} = \dfrac{DX^V}{dt} = 0$. Using the formula (9.5) we find the following result:

Proposition 1.9.1. *The vector field $X = X^i\dfrac{\delta}{\delta x^i} + \dot{X}^i\dfrac{\partial}{\partial y^i}$ from $\mathcal{X}(TM)$ is parallel along the parametrized curve c in TM, with respect to the N–linear connection*

$D\Gamma(N) = (L^i{}_{jk}, C^i{}_{jk})$ if and only if its coefficients $X^i(x(t), y(t))$ and $\dot{X}^i(x(t), y(t))$ are solutions of the linear system of differential equations

$$\frac{dZ^i}{dt} + Z^m(x(t), y(t))\frac{\omega^i{}_m(x(t), y(t))}{dt} = 0.$$

A theorem of existence and uniqueness for the parallel vector fields along with a curve c on the manifold TM can be formulated.

A *horizontal path* of an N–linear connection D on TM is a horizontal parametrized curve $c : I \to TM$ with the property $D_{\dot{c}}\dot{c} = 0$.

Using (9.1) and (9.5), with $X^i = \dfrac{dx^i}{dt}$, $\dot{X}^i = \dfrac{\delta y^i}{dt} = 0$ we obtain the following theorem:

Theorem 1.9.1. *The horizontal paths of an N–linear connection $D\Gamma(N) = (L^i{}_{jk}, C^i{}_{jk})$ are characterized by the system of differential equations:*

$$\frac{d^2x^i}{dt^2} + L^i{}_{jk}(x, y)\frac{dx^j}{dt}\frac{dx^k}{dt} = 0, \quad \frac{dy^i}{dt} + N^i{}_j(x, y)\frac{dx^j}{dt} = 0. \tag{9.6}$$

Now we can consider a curve $c^v_{x_0}$ in the fibre $T_{x_0}M = \pi^{-1}(x_0)$. It can be represented by the equations

$$x^i = x^i_0, \; y^i = y^i(t), \; t \in I.$$

The above $c^v_{x_0}$ is called a *vertical curve* of TM in the point $x_0 \in M$.

A vertical curve $c^v_{x_0}$ is called a *vertical path* with respect to the N–linear connection D if $D_{\dot{c}^v_{x_0}}\dot{c}^v_{x_0} = 0$.

Again the formulae (9.1), (9.5) lead to:

Theorem 1.9.2. *The vertical paths in the point $x_0 \in M$, with respect to the N–linear connection $D\Gamma(N) = (L^i{}_{jk}, C^i{}_{jk})$ are characterized by the system of differential equations*

$$x^i = x^i_0, \; \frac{d^2y^i}{dt^2} + C^i{}_{jk}(x_0, y)\frac{dy^j}{dt}\frac{dy^k}{dt} = 0. \tag{9.7}$$

Obviously, the local existence and uniqueness of horizontal paths are assured if the initial conditions for (9.6) are given. The same consideration can be made for vertical paths, (9.7).

Considering the 1-form of connection $\omega^i{}_j$ from (9.4) and the exterior differential of the 1-forms from the adapted dual basis $(dx^i, \delta y^i)$ we can determine the structure equations of a N–linear connection D on the manifold TM.

Lemma 1.9.1. *The exterior differentials of* 1*-forms* $\delta y^i = dy^i + N^i{}_j dx^j$ *are given by*

$$d\delta y^i = \frac{1}{2} R^i{}_{jm} dx^m \wedge dx^j + B^i{}_{jm} \delta y^m \wedge dx^j \tag{9.8}$$

where

$$B^i{}_{jm} = \frac{\partial N^i{}_j}{\partial y^m}. \tag{9.8$'$}$$

Indeed, a straightforward calculus on the exterior differential $d\delta y^i$ leads to (9.8).

Remark. $B^i{}_{jm}$ from (9.8)$'$ are the h-coefficients of an N–linear connection, called the Berwald connection.

Lemma 1.9.2. *With respect to a changing of local coordinate on* TM*, the following* 2*-forms*

$$d(dx^i) - dx^m \wedge \omega^i{}_m$$
$$d(\delta y^i) - \delta y^m \wedge \omega^i{}_m$$

are transformed like a d-vector field and the 2*-forms*

$$d\omega^i{}_j - \omega^m{}_j \wedge \omega^i{}_m$$

are transformed like a d-tensor field of type $(1,1)$.

Indeed, taking into account Lemma 1.9.1 and the expression of 1-forms of connection $\omega^i{}_j$ the previous lemma can be proved.

Now, we can formulate the result:

Theorem 1.9.3. *The structure equations of an* N*–linear connection* $D\Gamma(N) = (L^i{}_{jk}, C^i{}_{jk})$ *on the manifold* TM *are given by*

$$\begin{aligned} d(dx^i) - dx^m \wedge \omega^i{}_m &= -\overset{(0)}{\Omega}{}^i \\ d(\delta y^i) - \delta y^m \wedge \omega^i{}_m &= -\overset{(1)}{\Omega}{}^i \\ d\omega^i{}_j - \omega^m{}_j \wedge \omega^i{}_m &= -\Omega^i{}_j \end{aligned} \tag{9.9}$$

where $\overset{(0)}{\Omega}{}^i$ *and* $\overset{(1)}{\Omega}{}^i$ *are the* 2*-forms of torsion*

$$\begin{aligned} \overset{(0)}{\Omega}{}^i &= \frac{1}{2} T^i{}_{jk} dx^j \wedge dx^k + C^i{}_{jk} dx^j \wedge \delta y^k \\ \overset{(1)}{\Omega}{}^i &= \frac{1}{2} R^i{}_{jk} dx^j \wedge dx^k + P^i{}_{jk} dx^j \wedge \delta y^k + \frac{1}{2} S^i{}_{jk} \delta y^j \wedge \delta y^k \end{aligned} \tag{9.10}$$

and the 2-forms of curvature $\Omega^i{}_j$ *are expressed by*

$$\Omega^i{}_j = \frac{1}{2} R_j{}^i{}_{kh} dx^k \wedge dx^h + P_j{}^i{}_{kh} dx^k \wedge \delta y^h + \frac{1}{2} S_j{}^i{}_{kh} \delta y^k \wedge \delta y^h. \tag{9.11}$$

Proof. By means of Lemma 1.9.2, the general structure equations of a linear connection on TM are particularized for an N–linear connection D in the form (9.9). Using 1-forms connection $\omega^i{}_j$ from (9.4) and the formula (9.8), we can calculate, without difficulties, the 2-forms of torsion $\overset{(0)}{\Omega}{}^i$, $\overset{(1)}{\Omega}{}^i$ and the 2-forms of curvature $\Omega^i{}_j$, obtaining the expressions (9.10) and (9.11).

The geometrical theory on the manifold TM of tangent bundle will be used in next chapters for studying the geometries of Finsler and Lagrange spaces.

Chapter 2

Finsler spaces

The notion of general metric space appeared for the first time in the disertation of B. Riemann in 1854. After sixty five years, P. Finsler in his Ph.D. thesis introduced the concept of general metric function, which can be studied by means of variational calculus. Later, L. Berwald, J.L. Synge and E. Cartan precisely gave the correct definition of a Finsler space.

During eighty years, famous geometers studied the Finsler geometry in connection with variational problem, geometrical theory of tangent bundle and for its applications in Mechanics, Physics or Biology. In the last 40 years, some remarkable books on Finsler geometry were published by H. Rund, M. Matsumoto, R. Miron and M. Anastasiei, A. Bejancu, Abate–Patrizio, D. Bao, S.S. Chern and Z.Shen, P. Antonelli, R.Ingarden and M. Matsumoto.

In the present chapter we made a brief introduction in the geometry of Finsler spaces in order to study the relationships between these spaces and the dual notion of Cartan spaces.

In the following we will study: Finsler metrics, Cartan nonlinear connection, canonical metrical connections and their structure equations.

Some special classes of Finsler manifolds as (α, β)–metrics, Berwald spaces will be pointed out. We underline the important role which the Sasaki lift plays for almost Kählerian model of a Finsler manifold, as well as the new notion of homogeneous lift of the Finsler metrics in the framework of this theory.

2.1 Finsler metrics

At the begining we define the notion of Finsler metric and Finsler manifold.

Definition 2.1.1. A Finsler manifold (or Finsler space) is a pair $F^n = (M, F(x, y))$ where M is a real n-dimensional differentiable manifold and $F : TM \to \mathbb{R}$ a scalar function which satisfy the following axioms:

1° F is a differentiable function on the manifold $\widetilde{TM} = TM\setminus\{0\}$ and F is continuous on the null section of the projection $\pi : TM \to M$.

2° F is a positive function.

3° F is positively 1-homogeneous on the fibres of tangent bundle TM.

4° The Hessian of F^2 with elements

$$g_{ij}(x,y) = \frac{1}{2}\frac{\partial^2 F^2}{\partial y^i \partial y^j} \tag{1.1}$$

is positively defined on $\widetilde{TM}$.

Proposition 2.1.1. *The set of functions $g_{ij}(x,y)$ from* (1.1) *is transformed, with respect to* (1.3) *in Ch.1, by the rule*

$$\widetilde{g}_{ij}(\widetilde{x},\widetilde{y}) = \frac{\partial x^r}{\partial \widetilde{x}^i}\frac{\partial x^s}{\partial \widetilde{x}^j}\, g_{rs}(x,y). \tag{1.2}$$

Indeed, in virtue of (1.4) in Ch.1 we have:

$$\frac{\partial \widetilde{F}^2}{\partial \widetilde{y}^i} = \frac{\partial x^r}{\partial \widetilde{x}^i}\frac{\partial F^2}{\partial y^r} \text{ and } \frac{1}{2}\frac{\partial^2 \widetilde{F}^2}{\partial \widetilde{y}^i \partial \widetilde{y}^j} = \frac{\partial x^r}{\partial \widetilde{x}^i}\frac{\partial x^s}{\partial \widetilde{x}^j}\frac{1}{2}\frac{\partial^2 F^2}{\partial y^r \partial y^s}.$$

Consequently (1.2) holds.

Because of (1.2) we say that g_{ij} is a distinguished tensor field (briefly d-tensor field). Of course, g_{ij} is a covariant symmetric of order 2 d-tensor field defined on the manifold $\widetilde{TM}$.

The function $F(x,y)$ is called *fundamental function* and the d-tensor field g_{ij} is called *fundamental* (or *metric*) tensor of the Finsler space $F^n = (M, F(x,y))$.

Examples.

1° A Riemannian manifold $(M, \gamma_{ij}(x))$ determines a Finsler manifold $F^n = (M, F(x,y))$, where

$$F(x,y) = \sqrt{\gamma_{ij}(x) y^i y^j}. \tag{1.3}$$

The fundamental tensor of this Finsler space is coincident to the metric tensor $\gamma_{ij}(x)$ of the Riemann space $(M, \gamma_{ij}(x))$.

2° Let us consider the function

$$F(x,y) = \sqrt[4]{(y^1)^4 + \cdots + (y^n)^4} \tag{1.4}$$

defined in a preferential local system of coordinate on $\widetilde{TM}$.

The pair $F^n = (M, F(x,y))$, with F defined in (1.4) satisfies the axioms 1-4 from Definition 2.1.1. So, F^n is a Finsler space. The fundamental tensor field g_{ij} can be easy calculated.

Remark. This was the first example of Finsler space from the literature of the subject. It was given by B. Riemann in 1854.

3° Antonelli–Shimada's ecological metric is given, in a preferential local system of coordinate on $\widetilde{TM}$, by

$$F(x,y) = e^{\phi} L, \ \phi = \alpha_i x^i \ (\alpha_i \text{ are positive constants}),$$

where $L = \{(y^1)^m + (y^2)^m + \cdots + (y^n)^m\}^{1/m}$, $m \geq 3$, m being even.

4° Randers metric. Let us consider the function of a Finsler space F^n :

$$F(x,y) = \alpha(x,y) + \beta(x,y),$$

where $\alpha^2 := a_{ij}(x) y^i y^j$ is a Riemannian metric and $\beta(x,y) := b_i(x) y^i$ is a differential linear function in y^i. This metric is called a *Randers metric* and was introduced by the paper [135]. The fundamental tensor of the Randers space is given by [89]:

$$g_{ij} = \frac{\alpha+\beta}{\alpha} \overset{0}{h}_{ij} + d_i d_j, \ \overset{0}{h}_{ij} := a_{ij} - \overset{0}{\ell}_i \overset{0}{\ell}_j, \ d_i := b_i + \overset{0}{\ell}_i, \ \overset{0}{\ell}_i := \frac{\partial \alpha}{\partial y^i}$$

and one can prove that the fundamental tensor field g_{ij} is positive definite under the condition $b^2 = a^{ij} b_i b_j < 1$ (see the book [24]).

The first example motivates the following theorem:

Theorem 2.1.1. *If the base manifold M is paracompact, then there exist functions $F : TM \to R$ which are fundamental functions for Finsler manifolds.*

Regarding the axioms 1-4 formulated in Definition 2.1.1, we can prove without difficulties.

Theorem 2.1.2. *The system of axioms of a Finsler space is minimal.*

However, the axiom 4° of this system is sometimes too strong in applications of Finsler geometry in construction of geometrical models in other scientific disciplines, for instance in theoretical physics.

Let us consider the fibre $T_{x0}M$ in a point $x_0 \in M$ of the tangent bundle (TM, π, M). It is known that $T_{x0}M$ is a real n-dimensional vector space. The set of points

$$I(x_0) = \{u \in T_{x0}M \mid F(u) \leq 1\} \subset T_{x_0}M \tag{1.5}$$

is called the *indicatrix* of Finsler space F^n in the point $x_0 \in M$.

The restriction $F(x_0, y)$ of the fundamental function F to the fibre $T_{x_0}M$ determines a Riemann metric $g_{ij}(x_0, y)$ in the submanifold $T_{x_0}M$ immersed in the manifold TM. Since $g_{ij}(x_0, y)$ is positively defined, the following property holds, [139]:

Theorem 2.1.3. *If a Finsler manifold $F^n = (M, F(x, y))$ has the property: in every point $x_0 \in M$, the indicatrix $I(x_0)$ is strictly convex, then the axiom $4°$ is satisfied.*

If one retains the axioms $1°$, $2°$, $3°$ from the Definition 2.1.1 of a Finsler manifold and add the following axiom $4'$:

$4'$ a. rank$||g_{ij}(x, y)|| = n$ on $\widetilde{TM}$.
   b. signature of d-tensor field $g_{ij}(x, y)$ is constant,

it results the notion of Finsler manifold with semidefinite metric. If the axioms $1°$, $2°$, $3°$, $4'$ are satisfied on an open set $\pi^{-1}(U) \subset TM$ we will say that we have a Finsler space with the semidefinite metric on the open set $\pi^{-1}(U)$.

In general, Randers metric $\alpha + \beta$ (without the condition $||b|| < 1$) give rise to Finsler spaces with semi–definite Finsler metrics.

In the following sections of chapters of the present monograph we refer to the Finsler spaces with the definite or semidefinite metric without mention the difference between them.

2.2 Geometric objects of the space F^n

The property of 1-homogeneity of the fundamental function $F(x, y)$ of the Finsler space F^n induces properties of homogeneity of the geometrical objects derived from it.

Theorem 2.2.1. *On a Finsler manifold F^n the following properties hold:*

$1°$ *The components of the fundamental tensor field g_{ij} are 0–homogeneous, i.e.,*

$$y^i \frac{\partial g_{jk}}{\partial y^i} = 0. \tag{2.1}$$

$2°$ *The functions*

$$p_i = \frac{1}{2}\frac{\partial F^2}{\partial y^i} \tag{2.2}$$

are 1–homogeneous.

3° *The functions*

$$C_{ijk} = \frac{1}{4}\frac{\partial^3 F^2}{\partial y^i \partial y^j \partial y^k} \tag{2.3}$$

are (-1)–homogeneous.

In the same time we have some natural object fields:

Proposition 2.2.1.

1° *The quantity p_i from (2.2) is a d–covector field.*

2° *The set of functions C_{ijk} from (2.3) is a covariant of order 3 symmetric d–tensor field.*

3° $\|X\|^2 := g_{ij}(x,y)X^iX^j$ *is a scalar field, if X^i is a d-vector field.*

4° $\langle X, Y\rangle := g_{ij}(x,y)X^iY^j$ *is a scalar field, X^i, Y^i being d-vector fields.*

The proof of previous properties is elementary. $\|X\|^2$ is called the square of norm of vector field X and $\langle X, Y\rangle$ is the scalar product (calculated in a point $u \in \widetilde{TM}$).

Assuming $\|X\|_u \neq 0$, $\|Y\|_u \neq 0$ the angle $\varphi = \sphericalangle(X, Y)$, in a point $u = (x, y) \in \widetilde{TM}$, between vectors $X_u = (X^i(u))$, $Y_u = (Y^i(u))$ are given by the solution φ of the trigonometric equation

$$\cos\varphi = \frac{< X, Y > (u)}{\|X\|_u \|Y\|_u}. \tag{2.4}$$

The number φ is uniquely determined, because in a Finsler space with the definite metric $\cos\varphi$ from (2.4) satisfy the condition $-1 \leq \cos\varphi \leq 1$.

Other properties are given by

Proposition 2.2.2. *In a Finsler manifold $F^n = (M, F)$ the following identities hold:*

1° $p_i y^i = F^2$

2° $y_i := g_{ij}y^j = p_i$

3° $C_{0jh} = y^i C_{ijh} = 0,\ C_{j0h} = C_{jh0} = 0$

4° $F^2(x, y) = g_{ij}(x, y)y^iy^j$.

Some natural object fields are introduced in the following:

Theorem 2.2.2. *In a Finsler manifold $F^n = (M, F)$ we have the following natural object fields:*

1° *The Liouville vector field*

$$\mathbb{C} = y^i \frac{\partial}{\partial y^i}. \tag{2.5}$$

2° *The Hamilton 1-form*

$$\omega = p_i dx^i. \tag{2.6}$$

3° *The symplectic structure*

$$\theta = d\omega = dp_i \wedge dx^i. \tag{2.7}$$

Proof. 1° The Liouville vector field $\mathbb{C}$ exists on the manifold TM independently of the metrical function F of the space F^n (cf. §1, Ch.1).

2° By means of Proposition 2.2.1 and on the fact that with respect to (1.1) it follows that ω does not depend on the changing of local coordinates.

3° θ is a closed 2-form on $\widetilde{TM}$ and

$$\text{rank}||\theta|| = 2n = \dim \text{ of } \widetilde{TM}.$$

q.e.d.

Definition 2.2.2. A Finsler space $F^n = (M, F(x, y))$ is called reducible to a Riemannian space if its fundamental tensor field does not depend on the directional variables y^i.

The previous definition has a geometrical meaning, since the equation $\dfrac{\partial g_{ij}}{\partial y^k} = 2C_{ijk} = 0$ does not depend on the changing of local coordinates.

Theorem 2.2.3. *A Finsler space F^n is reducible to a Riemannian space iff the tensor C_{ijk} is vanishing on the manifold $\widetilde{TM}$.*

Proof. If g_{ij} does not depend on y^i, thus from the identity

$$\frac{\partial g_{ij}}{\partial y^k} = 2C_{ijk} \tag{2.8}$$

the condition $\dfrac{\partial g_{ij}}{\partial y^k} = 0$ implies $C_{ijk} = 0$. Conversely, $C_{ijk} = 0$ on $\widetilde{TM}$ and (2.8) implies $\dfrac{\partial g_{ij}}{\partial y^k} = 0$ on $\widetilde{TM}$.

Let us consider another geometrical notion: the arc length of a smooth curve in a Finsler manifold $F^n = (M, F(x,y))$.

Let c be a parametrized curve in the manifold M:

$$c : t \in [0,1] \longrightarrow c(t) \in U \subset M \tag{2.9}$$

U being a domain of a local chart in M.

The curve c has an analytical expression of form:

$$x^i = x^i(t), \ t \in [0,1]. \tag{2.7)'$$

The extension $\widetilde{c}$ of c to $\widetilde{TM}$ is defined by the equations

$$x^i = x^i(t), \ y^i = \frac{dx^i}{dt}(t), \ t \in [0,1]. \tag{2.7)''$$

Thus the restriction of the fundamental function $F(x,y)$ to $\widetilde{c}$ is

$$F\left(x(t), \frac{dx}{dt}(t)\right), \ t \in [0,1].$$

We define the "length" of curve c with extremities $c(0), c(1)$ by the number

$$L(c) = \int_0^1 F\left(x(t), \frac{dx}{dt}(t)\right) dt. \tag{2.10}$$

The number $L(c)$ does not depend by the changing of coordinates on $\widetilde{TM}$ and, by means of 1-homogeneity of the fundamental function F, $L(c)$ does not depend on the parametrization of the curve c. So $L(c)$ depends on c, only.

We can fix a canonical parameter on the curve c, given by the arclength of c.

Indeed, the function $s = s(t)$, $t \in [0,1]$, given by

$$s(t) = \int_{t_0}^t F\left(x(\tau), \frac{dx}{dt}(\tau)\right) d\tau, \ t_0, t \in [0,1]$$

is derivable, having the derivative:

$$\frac{ds}{dt} = F\left(x(\tau), \frac{dx}{dt}(\tau)\right) > 0, \ t \in (0,1).$$

So the function $s = s(t)$, $t \in [0,1]$, is invertible. Let $t = t(s)$ be its inverse. The change of parameter $t \to s$, given by $s = s(t)$, has the property

$$F\left(x(s), \frac{dx}{ds}(s)\right) = 1. \tag{2.11}$$

So, we have:

Theorem 2.2.4. *In a Finsler space $F^n = (M, F)$, for any smooth curve $c: t \in [0,1] \to c(t) \in U \subset M$ exists a canonical parameter s with the property* (2.11).

2.3 Geodesics

Let us consider a smooth parametrized curve $c : t \in [0,1] \to c(t) \in U \subset M$ having the ending points $c(0)$ and $c(1)$. Its length is given by the formula (2.10). We will formulate the variational problem for the functional $L(c)$. Consider a vector field $V^i(x(t))$ along the curve c with the properties $V^i(c(0)) = V^i(c(1)) = 0$. Let c_ε be a set of smooth curve given by the mappings

$$c_\varepsilon : t \in [0,1] \to c_\varepsilon(t) \in U,$$

such that the analytical expression of $c_\varepsilon(t)$ being

$$c_\varepsilon(t) : x^i(t) + \varepsilon V^i(x(t))$$

with $|\varepsilon|$ small, ε being real number. So, the curves $c_\varepsilon(t)$ have the same end points $c(0), c(1)$ and same tangent vectors in this points with curve c. The length of the curve c_ε is given by

$$L(c_\varepsilon) = \int_0^1 F\left(x(t) + \varepsilon V(t), \frac{dx}{dt} + \varepsilon \frac{dV}{dt}\right) dt.$$

The necessary condition for $L(c)$ to be extremal value of $L(c_\varepsilon)$ is as follows

$$\left.\frac{dL(c_\varepsilon)}{d\varepsilon}\right|_{\varepsilon=0} = 0.$$

This equality is equivalent to

$$\int_0^1 \left(\frac{\partial F}{\partial x^i} V^i + \frac{\partial F}{\partial V^i}\frac{dV^i}{dt}\right) dt = 0.$$

Integrating by parts the second term one obtains

$$\int_0^1 V^i \left[\frac{\partial F}{\partial x^i} - \frac{d}{dt}\left(\frac{\partial F}{\partial y^i}\right)\right] dt = 0, \; y^i = \frac{dx^i}{dt}\cdot$$

Since V^i is an arbitrary d-vector field we get from the previous equation the following Euler–Lagrange equations

$$(3.1)\qquad \frac{\partial F}{\partial x^i} - \frac{d}{dt}\left(\frac{\partial F}{\partial y^i}\right) = 0,\ y^i = \frac{dx^i}{dt}.$$

Definition 2.3.1. The curves $c = \{(x^i(t)),\ t \in [0,1]\}$ solutions of the Euler–Lagrange equations (3.1) are called *geodesics* of theFinsler space F^n.

Since $F(x,y) = \sqrt{g_{ij}(x,y)y^iy^j}$ the equations (3.1) are equivalent to the system of equations

$$(3.1)'\qquad \frac{d}{dt}\left(\frac{\partial F^2}{\partial y^i}\right) - \frac{\partial F^2}{\partial x^i} = 2\,\frac{dF}{dt}\,\frac{\partial F}{\partial y^i},\ y^i = \frac{dx^i}{dt}.$$

Substituting $F^2 = g_{ij}y^iy^j$ we get the following form of the previous Euler–Lagrange equations

$$(3.2)\qquad \frac{d^2x^i}{dt^2} + 2G^i\left(x, \frac{dx}{dt}\right) = 2\frac{dF}{dt}\,\frac{\partial F}{\partial y^i},\ y^i = \frac{dx^i}{dt},$$

where

$$(3.3)\qquad G^i = \frac{1}{2}\,\gamma^i{}_{jk}(x,y)y^jy^k,$$

the functions $\gamma^i{}_{jk}$ being the Christoffel symbols of the fundamental tensor field g_{ij}. This is

$$(3.4)\qquad \gamma^i{}_{jk}(x,y) = \frac{1}{2}\,g^{ir}\left(\frac{\partial g_{rk}}{\partial x^j} + \frac{\partial g_{jr}}{\partial x^k} - \frac{\partial g_{jk}}{\partial x^r}\right).$$

Changing now to the canonical parameter s, we have $F\left(x, \dfrac{dx}{ds}\right) = 1$. The equations (3.2) become

$$(3.5)\qquad \frac{d^2x^i}{ds^2} + \gamma^i{}_{jk}\left(x, \frac{dx}{ds}\right)\frac{dx^j}{ds}\,\frac{dx^k}{ds} = 0.$$

Theorem 2.3.1. *Geodesics in a Finsler space F^n in the canonical parametrization are given by the differential equations* (3.5).

A theorem of existence and uniqueness of the solutions of differential equations (3.5) can be formulated.

2.4 Canonical spray. Cartan nonlinear connection

For a Finsler space $F^n = (M, F)$, we can define a canonical spray S and an important nonlinear connection. Noticing that the function F^2 is a regular Lagrangian, we introduce its energy by:

$$L = y^i \frac{\partial F^2}{\partial y^i} - F^2 = F^2.$$

Thus, integral action of the Lagrangian $L(x, y)$ along with smooth parametrized curve $c : [0, 1] \to U \subset M$ is given by the functional

$$\varepsilon(c) = \int_0^1 F^2 \left(x, \frac{dx}{dt} \right) dt. \tag{4.1}$$

The Euler–Lagrange equations are given by:

$$\overset{0}{E}_i (F^2) := \frac{\partial F^2}{\partial x^i} - \frac{d}{dt} \left(\frac{\partial F^2}{\partial y^i} \right) = 0, \; y^i = \frac{dx^i}{dt}\cdot \tag{4.2}$$

Remark. Starting from the property that $L(x, y) = F^2(x, y)$ is energy function of the regular Lagrangian F^2, we will prove in the next chapter the following two properties.

Theorem A. *Along with the integral curves of the Euler–Lagrange equations* $\overset{0}{E}_i (F^2) = 0$, $y^i = \dfrac{dx^i}{dt}$ *we have:*

$$\frac{dF^2}{dt} = 0\cdot$$

Theorem B. (Noether) *For any infinitesimal symmetry* $x'^i = x^i + \varepsilon V^i(x, t)$, $t' = t + \varepsilon \tau(x, t)$ ($\varepsilon =$ *const.*) *of the regular Lagrangian* $F^2(x, y)$ *and for any* C^∞*–function* $\phi(x)$ *the following function*

$$\mathcal{F}(F^2, \phi) := V^i \frac{\partial F^2}{\partial y^i} - \tau F^2 - \phi(x)$$

is conserved along with the integral curves of the Euler–Lagrange equations $\overset{0}{E}_i (F^2) = 0$, $y^i = \dfrac{dx^i}{dt}\cdot$

Proposition 2.4.1. *The Euler–Lagrange equations* (4.1) *can be expressed in the form*

$$\frac{d^2x^i}{dt^2} + 2G^i\left(x, \frac{dx}{dt}\right) = 0, \tag{4.3}$$

where G^i *is given by*

$$G^i = \frac{1}{2}\,\gamma^i{}_{jk}(x,y)y^jy^k. \tag{4.4}$$

Proof. The same calculus as in the previous section, taking $F^2 = g_{ij}(x,y)y^iy^j$ shows us that the equations (4.2) are equivalent to (4.3).

Remarking that the equations (4.3), (4.4) give the integral curve of the spray

$$S = y^i \frac{\partial}{\partial x^i} - 2G^i\,(x,y)\,\frac{\partial}{\partial y^i}\cdot \tag{4.5}$$

The vector field S is called canonical spray of the space.

Proposition 2.4.2. *The spray* S *of* (4.5) *is determined only by the fundamental function* $F(x,y)$. *Its integral curve are given by the equation* (4.3), (4.4).

Consequently, we have:

Proposition 2.4.3. *In a Finsler space* $F^n = (M,F)$ *the integral curves of the canonical spray are the geodesics in canonical parametrization.*

Indeed, the equations (3.5) of geodesics in the canonical parametrization are coincident with the equations (4.3), (4.4).

Now, applying the theory from the section 4, ch.1, one can derive from the canonical spray S the notion of the nonlinear connection for the Finsler space $F^n = (M,F)$.

Definition 2.4.1. The nonlinear connection determined by the canonical spray S of the Finsler space F^n is called *Cartan nonlinear connection* of the Finsler space $F^n = (M,F)$.

Theorem 2.4.1. *The Cartan nonlinear connection* N *has the coefficients*

$$N^i{}_j = \frac{1}{2}\,\frac{\partial}{\partial y^j}\left(\gamma^i{}_{rs}(x,y)y^ry^s\right). \tag{4.5$'$}$$

It is *globally defined on the manifold* $\widetilde{TM}$ *and depends only on the fundamental function* $F(x,y)$.

From now on we will use only the adapted basis $\left(\frac{\delta}{\delta x^i}, \frac{\partial}{\partial y^i}\right)$ to the distributions N and V, determined only by the Cartan nonlinear connection.

The coefficients of $N^i{}_j$ from (4.5)' are 1–homogeneous functions with respect to y^i and have the properties

$$N^i{}_j y^j = \frac{\partial G^i}{\partial y^i} = 2G^i = \gamma^i{}_{00}, \tag{4.6}$$

where $\gamma^i{}_{00} := \gamma^i{}_{jk} y^j y^k$.

We have

Theorem 2.4.2.

1) *The horizontal curves $\gamma : I \to \widetilde{TM}$ with respect to Cartan nonlinear connection are characterized by the following system of differential equations:*

$$x^i = x^i(t),\ t \in I,\ \frac{\delta y^i}{dt} = \frac{dy^i}{dt} + N^i{}_j(x(t), y(t))\frac{dx^j}{dt} = 0. \tag{4.7}$$

2) *The autoparallel curves of the Cartan nonlinear connection are characterized by the system of differential equations:*

$$\frac{dx^i}{dt} = y^i,\ \frac{\delta y^i}{dt} = \frac{dy^i}{dt} + N^i{}_j(x(t), y(t))\frac{dx^j}{dt} = 0. \tag{4.8}$$

2.5 Metrical Cartan connection

The famous metrical Cartan connection in a Finsler space $F^n = (M, F)$ can be defined as an N–linear connection metrical with respect to the fundamental tensor field g_{ij} and with h- and v-torsions vanish, N being Cartan nonlinear connection. Indeed, we have:

Theorem 2.5.1. *The following properties hold:*

1) *There exists a unique N–linear connection D on $\widetilde{TM}$ with coefficients $D\Gamma(N) = (L^i{}_{jk}, C^i{}_{jk})$ satisfying the following axioms:*

A1 *D is h-metrical, i.e.* $g_{ij|h} = 0$.
A2 *D is v-metrical, i.e.* $g_{ij}|_k = 0$.
A3 *D is h-torsion free, i.e.* $T^i{}_{jk} := L^i{}_{jk} - L^i{}_{kj} = 0$.
A4 *D is v-torsion free, i.e.* $S^i{}_{jk} := C^i{}_{jk} - C^i{}_{kj} = 0$.

2) *The coefficients* $D\Gamma(N) = (L^i{}_{jk}, C^i{}_{jk})$ *are given by the generalized Christoffel symbols:*

$$(5.1)\qquad \begin{aligned} L^i{}_{jh} &= \frac{1}{2}g^{im}\left(\frac{\delta g_{mh}}{\delta x^j} + \frac{\delta g_{jm}}{\delta x^h} - \frac{\delta g_{jh}}{\delta x^m}\right), \\ C^i{}_{jh} &= \frac{1}{2}g^{im}\left(\frac{\partial g_{mh}}{\partial y^j} + \frac{\partial g_{jm}}{\partial y^h} - \frac{\partial g_{jh}}{\partial y^m}\right). \end{aligned}$$

3) D *depends only on the fundamental function* F *of the Finsler space* F^n.

The proof of the theorem is made by using the known techniques. It was initiated by M. Matsumoto [88].

The connection $C\Gamma(N)$ from the previous theorem will be called the *canonical metrical Cartan connection.*

Taking into account this theorem one can demonstrate without difficulties the following properties of the Finsler spaces endowed with the canonical Cartan nonlinear connection and the canonical metrical Cartan connection.

Proposition 2.5.1. *The deflection tensor field of the Cartan metrical connection* $C\Gamma(N)$ *satisfies the following equations:*

$$(5.2)\qquad D^i{}_j = y^i{}_{|j} = 0,\ d^i{}_j = y^i{}\big|_j = \delta^i{}_j.$$

Remark. If we consider the following Matsumoto's system of axioms A1–A4 and the axiom

A5 $\quad D^i{}_j = 0$

we obtain the system of axioms which uniquely determined the Cartan metrical connection $C\Gamma(N)$. Miron, Aikou, Hashiguchi proved the following result:

The Matsumoto's axioms A1–A5 *of the Cartan metrical connection* $C\Gamma(N)$ *are independent.*

Proposition 2.5.2. *The following properties hold with respect to Cartan metrical connection:*

1. $F_{|k} = 0, \quad F\big|_k = \frac{1}{F}y_k,$
2. $F^2{}_{|k} = 0, \quad F^2\big|_k = 2y_k,$
3. $y_{i|k} = 0, \quad y_i\big|_k = g_{ik}.$

Proposition 2.5.3. *The Ricci identities of the metrical Cartan connection $C\Gamma(N)$ are:*

$$(5.3)\qquad \begin{aligned} &X^i{}_{|k|h} - X^i{}_{|h|k} = X^r R_r{}^i{}_{kh} - X^i{}_{|_r} R^r{}_{kh},\\ &X^i{}_{|k}\big|_h - X^i\big|_{h|k} = X^r P_r{}^i{}_{kh} - X^i{}_{|r} C_k{}^r{}_h - X^i\big|_r P^r{}_{kh},\\ &X^i\big|_k\big|_h - X^i\big|_h\big|_k = X^r S_r{}^i{}_{kh}. \end{aligned}$$

where the torsion tensors are:

$$(5.4)\qquad R^i{}_{jk} = \frac{\delta N^i{}_j}{\delta x^k} - \frac{\delta N^i{}_k}{\delta x^j},\ C_j{}^i{}_k,\ P^i{}_{jk} = \frac{\partial N^i{}_j}{\partial y^k} - F^i{}_{kj},$$

and the curvature d-tensors are:

$$(5.5)\qquad \begin{aligned} R_h{}^i{}_{jk} &= \frac{\delta F^i_{hj}}{\delta x^k} - \frac{\delta F^i{}_{hk}}{\delta x^j} + F^s{}_{hj}F^i{}_{sk} - F^s{}_{hk}F^i{}_{sj} + C_h{}^i{}_s R^s{}_{jk},\\ P_h{}^i{}_{jk} &= \frac{\partial F^i{}_{hj}}{\partial y^k} - C^i{}_{hk|j} + C_h{}^i{}_s P^s{}_{jk},\\ S_h{}^i{}_{jk} &= \frac{\partial C_h{}^i{}_j}{\partial y^k} - \frac{\partial C_h{}^i{}_k}{\partial y^j} + C_h{}^s{}_j C_s{}^i{}_k - C_h{}^s{}_k C_s{}^i{}_j. \end{aligned}$$

Hereafter we denote the metrical Cartan connection $D\Gamma(N)$ by $C\Gamma(N)$ or by $C\Gamma$.

Let us consider the covariant d-tensors of curvature

$$R_{ijkh} = g_{js}R_i{}^s{}_{kh},\ P_{ijkh} = g_{js}P_i{}^s{}_{kh},\ S_{ijkh} = g_{js}S_i{}^s{}_{kh}.$$

Proposition 2.5.4. *The covariant d-tensors of curvature satisfy the following identities:*

$$(5.5)'\qquad \begin{aligned} &R_{ijkh} + R_{jikh} = 0,\quad P_{ijkh} + P_{jikh} = 0,\quad S_{ijkh} + S_{jikh} = 0,\\ &R_{ijkh} + R_{ijhk} = 0,\quad S_{ijkh} + S_{ijhk} = 0. \end{aligned}$$

Indeed, the last two identities are evident.

Applying the Ricci identities to the fundamental tensor g_{ij} and taking into account Theorem 2.5.1, we get the first three identities.

Proposition 2.5.5. *The Cartan connection $C\Gamma$ has the following properties:*

$$(5.6)\qquad \begin{aligned} &R_0{}^i{}_{hk} = R^i{}_{hk},\quad P_0{}^i{}_{hk} = P^i{}_{hk},\qquad S_0{}^i{}_{hk} = 0,\\ &P_{ijk} = C_{ijk|0},\quad (P_{ijk} := g_{is}P^s{}_{jk})\\ &\underset{(ijk)}{\sigma}(R_{ijk}) = 0,\quad (R_{ijk} := g_{im}R^m{}_{jk}) \end{aligned}$$

where $\underset{(ijk)}{\sigma}$ *means the cyclic sum in the indices* i, j, k..

Indeed, by applying the Ricci identities to the Liouville d-vector field y^i and looking at the tensor of deflection $D^i{}_j = y^i{}_{|j} = 0$ and $d^i{}_j = \delta^i{}_j$, we get the first identity (5.6). For the other identities, we will write the symplectic structure $\theta = dp_i \wedge dx^i$ in the form $\theta = \delta p_i \wedge dx^i$ and write that its exterior differential vanishes, $d\theta = 0$.

2.6 Parallelism. Structure equations

Let $C\Gamma$ be a metrical Cartan connection of the Finsler space F^n. The coefficients $(L^i{}_{jk}, C^i{}_{jk})$ of $C\Gamma$ are given by the formula (5.1). As usually, the adapted basis $\left(\frac{\delta}{\delta x^i}, \frac{\partial}{\partial y^i}\right)$ of Cartan nonlinear connection N and vertical connection V and the dual adapted basis $(dx^i, \delta y^i)$ we can study the notion of parallelism of the vector fields in Finsler geometry.

Let $\gamma : [0, 1] \to TM$, $t \mapsto \gamma(t)$ be a parametrized curve of the manifold TM and $\frac{d\gamma}{dt}$ be the tangent vector field along with the curve γ. Then, we can write

$$\frac{d\gamma}{dt} = \frac{dx^i}{dt}\frac{\delta}{\delta x^i} + \frac{\delta y^i}{dt}\frac{\partial}{\partial y^i}. \tag{6.1}$$

As we know γ is horizontal curve with respect to the nonlinear connection N if $\frac{\delta y^i}{dt} = 0$. Also, γ is autoparallel curve of the nonlinear connection N if $\frac{\delta y^i}{dt} = 0$, $y^i = \frac{dx^i}{dt}$. We denote the tangent vector field along with γ by $\dot{\gamma} = \frac{d\gamma}{dt}$ and taking into account (6.1), we can set for the vector field X along with γ:

$$\frac{DX}{dt} = D_{\dot{\gamma}}X, \; DX = \frac{DX}{dt} \cdot dt, \; \forall X \in \mathcal{X}(TM). \tag{6.2}$$

$\frac{DX}{dt}$ is called the *covariant differential* along with the curve γ.

Setting $X = X^H + X^V$, $X^H = X^i \frac{\delta}{\delta x^i}$, $X^V = \dot{X}^i \frac{\partial}{\partial y^i}$, we get

$$\begin{aligned} \frac{DX}{dt} = \frac{DX^H}{dt} + \frac{DX^V}{dt} = &\left\{X^i{}_{|k}\frac{dx^k}{dt} + X^i{}_{|k}\frac{\delta y^k}{dt}\right\}\frac{\delta}{\delta x^i} + \\ &+ \left\{\dot{X}^i{}_{|k}\frac{dx^k}{dt} + \dot{X}^i{}_{|k}\frac{\delta y^k}{dt}\right\}\frac{\partial}{\partial y^i}. \end{aligned} \tag{6.3}$$

Let us consider

$$\omega^i{}_j = L^i{}_{jk}dx^k + C^i{}_{jk}\delta y^k. \tag{6.4}$$

where $\omega^i{}_j$ are the 1-forms connection of D. Then the equation (6.3) can be written in the form:

$$\frac{DX^i}{dt} = \left\{\frac{dX^i}{dt} + X^m\frac{\omega^i{}_m}{dt}\right\}\frac{\delta}{\delta x^i} + \left\{\frac{d\dot{X}^i}{dt} + \dot{X}^m\frac{\omega^i{}_m}{dt}\right\}\frac{\partial}{\partial y^i}. \tag{6.5}$$

Definition 2.6.1. We say that the vector field X on TM is said to be *parallel* along with the curve γ with respect to Cartan mectrical connection $C\Gamma$, if $\dfrac{DX}{dt} = 0$.

By means of (6.3), the equation $\dfrac{DX}{dt} = 0$ is equivalent with equations $\dfrac{DX^H}{dt} = 0$, $\dfrac{DX^V}{dt} = 0$.

From the formula (6.5), one obtains the following result:

Theorem 2.6.1. *The vector field* $X = X^i\dfrac{\delta}{\delta x^i} + \dot{X}^i\dfrac{\partial}{\partial y^i}$ *is parallel along with the parametrized curve* γ *with respect to the metrical Cartan connection if and only if its coefficients* $X^i(x,y)$, $\dot{X}^i(x,y)$ *are solutions of the linear system of the differential equations*

$$\frac{dZ^i}{dt} + Z^m(x(t),y(t))\frac{\omega^i{}_m(x(t),y(t))}{dt} = 0.$$

A theorem of the existence and uniqueness for the parallel vector field along with a given curve on TM can be formulated.

A *horizontal path* of the metrical Cartan connection D on TM is a horizontal parametrized curve γ with the property $D_{\dot{\gamma}}\dot{\gamma} = 0$.

Using (6.5) for $X^i = \dfrac{dx^i}{dt}$ and taking into account the previous theorem, we get:

Theorem 2.6.2. *The horizontal paths of Cartan metrical connection in Finsler space* F^n *are characterized by the system of differential equations*

$$\frac{d^2x^i}{dt^2} + L^i{}_{jk}(x,y)\frac{dx^j}{dt}\frac{dx^k}{dt} = 0, \quad \frac{dy^i}{dt} + N^i{}_j(x,y)\frac{dx^j}{dt} = 0. \tag{6.6}$$

If we describe the initial conditions of the previous system, we obtain the existence and uniqueness of the horizontal paths in the Finsler space F^n.

Now let us consider a curve γ^v_{x0} in the fibre $\pi^{-1}(x_0)$ of TM. It can be represented the equations:

$$x^i = x^i_0,\ y^i = y^i(t),\ t \in I.$$

The curve γ^v_{x0} is called a *vertical curve in the point* $x_0 \in M$.

A vertical curve γ^v_{x0} is called a *vertical path* with respect to metrical Cartan connection $C\Gamma$ if $D_{\dot\gamma^v_{x0}}\dot\gamma^v_{x0} = 0$. Now, applying equation (6.5), we have

Theorem 2.6.3. *In the Finsler space F^n the vertical paths in the point $x_0 \in M$ with respect to metrical Cartan connection $C\Gamma$ are characterized by the system of differential equations*

$$x^i = x^i_0,\ \frac{d^2y^i}{dt^2} + C^i{}_{jk}(x_0, y)\frac{dy^j}{dt}\frac{dy^k}{dt} = 0. \tag{6.7}$$

Now, taking into account the theory of structure equations of N–linear connection given in the section of chapter 1, we can apply it to the case of metrical Cartan connection $C\Gamma$. We get the following result:

Theorem 2.6.4. *The structure equations of the Cartan metrical connection $C\Gamma(N)$ are given by*

$$\begin{aligned} d(dx^i) - dx^m \wedge \omega^i{}_m &= -\overset{(0)}{\Omega}{}^i, \\ d(\delta y^i) - \delta y^m \wedge \omega^i{}_m &= -\overset{(1)}{\Omega}{}^i, \\ d\omega^i{}_j - \omega^m{}_j \wedge \omega^i{}_m &= -\Omega^i{}_j, \end{aligned} \tag{6.8}$$

where the 2-forms of torsion $\overset{(0)}{\Omega}{}^i$, $\overset{(1)}{\Omega}{}^i$ and 2-form of curvature $\Omega^i{}_j$ are as follows:

$$\begin{aligned} \overset{(0)}{\Omega}{}^i &= C^i{}_{jk} dx^j \wedge \delta y^h, \\ \overset{(1)}{\Omega}{}^i &= \frac{1}{2} R^i{}_{jh} dx^j \wedge dx^h + P^i{}_{jh} dx^j \wedge \delta y^h, \\ \Omega^i{}_j &= \frac{1}{2} R_j{}^i{}_{hm} dx^h \wedge dx^m + P_j{}^i{}_{hm} dx^h \wedge \delta y^m + \frac{1}{2} S_j{}^i{}_{hm} \delta y^h \wedge \delta y^m. \end{aligned} \tag{6.9}$$

Now, the Bianchi identities of the metrical Cartan connection $C\Gamma(N)$ can be obtained from the system of exterior equations (6.8) by calculating the exterior differential of (6.8), modulo of the same system (6.8) and using the exterior differential of 2-forms $\overset{(0)}{\Omega}{}^i$, $\overset{(1)}{\Omega}{}^i$ and of 2-form of curvature $\Omega^i{}_j$.

We obtain:

Theorem 2.6.5. *The Bianchi identities of the Cartan metrical connection $C\Gamma(N)$ of Finsler spaces F^n are as follows:*

$$
\begin{aligned}
&\underset{(jk\ell)}{\sigma}\{R_j{}^i{}_{k\ell} - R^h{}_{jk}C_h{}^i{}_\ell\} = 0,\\
(6.10)\qquad &\underset{(jk\ell)}{\sigma}\{R_j{}^i{}_{k|\ell} - R^h{}_{jk}P^i{}_{h\ell}\} = 0,\\
&\underset{(jk\ell)}{\sigma}\{R_s{}^i{}_{jk|\ell} - R^h{}_{jk}P_s{}^i{}_{\ell h}\} = 0,
\end{aligned}
$$

$$
\begin{aligned}
&\underset{(jk\ell)}{\sigma}\{S_j{}^i{}_{k\ell}\} = 0,\\
(6.11)\qquad &\underset{(jk\ell)}{\sigma}\{S_h{}^i{}_{jk}|_\ell\} = 0,
\end{aligned}
$$

$$
\begin{aligned}
&\underset{(jk)}{\mathcal{M}}\{C_k{}^i{}_{\ell|j} + C_j{}^i{}_h P^h{}_{k\ell} - P_j{}^i{}_{k\ell}\} = 0,\\
(6.12)\qquad &\underset{(jk)}{\mathcal{M}}\{R^i{}_{jh}C_k{}^h{}_\ell + P^i{}_{jh}P^h{}_{h\ell} + P^i{}_{k\ell|j}\} = R_\ell{}^i{}_{jk} - R^i{}_{jk|\ell},\\
&\underset{(jk)}{\mathcal{M}}\{R_s{}^i{}_{jh}C_k{}^h{}_\ell + P_s{}^i{}_{jh}P^h{}_{k\ell} + P^i{}_{sk|j\}} = -S_s{}^i{}_{\ell h}R^h{}_{jk} - R_s{}^i{}_{jk}|_\ell,
\end{aligned}
$$

$$
\begin{aligned}
&\underset{(jk)}{\mathcal{M}}\left\{\frac{\partial C_s{}^i{}_j}{\partial y^k} + C_s{}^h{}_j\, C_h{}^i{}_k\right\} = S_s{}^i{}_{jk},\\
(6.13)\qquad &\underset{(jk)}{\mathcal{M}}\left\{P_j{}^i{}_{\ell k} + P^i{}_{hj}C_\ell{}^h{}_k - P^i{}_{\ell k|j}\right\} = 0,\\
&\underset{(jk)}{\mathcal{M}}\left\{P_s{}^i{}_{hj}C_\ell{}^h{}_k - S_s{}^i{}_{jh}P^h{}_{\ell k} - P_s{}^h{}_{\ell k}|_j\right\} = -S_s{}^i{}_{jk|\ell},
\end{aligned}
$$

where $\underset{(jk)}{\mathcal{M}}$ *means the interchange of the indices j, k and subtraction and* $\underset{(jk\ell)}{\sigma}$ *means the cyclic permutation of indices j, k, l and summation.*

Remark. The structure equations given in Theorem 2.6.4 are extremely useful in the theory of submanifolds of the Finsler manifold F^n.

2.7 Remarkable connections of Finsler spaces

Let us consider an N–linear connection D with the coefficients $D\Gamma(N) = (L^i{}_{jk}, C^i{}_{jk})$. To these coefficients we add the coefficients $N^i{}_j$ of the nonlinear connection and we write D with the coefficients $D\Gamma(N^i{}_j, L^i{}_{jk}, C^i{}_{jk})$. For metrical Cartan connection $C\Gamma(N)$ we have the coefficients $C\Gamma(N^i{}_j, L^i{}_{jk}, C^i{}_{jk})$ given by the formulae (4.5)' and (5.1).

To the metrical Cartan connection $C\Gamma(N^i{}_j, L^i{}_{jk}, C^i{}_{jk})$ we associate the following N–linear connections:

1° Berwald connection $B\Gamma\left(N^i{}_j, \dfrac{\partial N^i{}_j}{\partial y^k}, 0\right)$.

2° Chern–Rund connection $R\Gamma\left(N^i{}_j, L^i{}_{jk}, 0\right)$.

3° Hashiguchi connection $H\Gamma\left(N^i{}_j, \dfrac{\partial N^i{}_j}{\partial y^k}, C^i{}_{jk}\right)$.

These remarkable connections satisfy a commutative diagram:

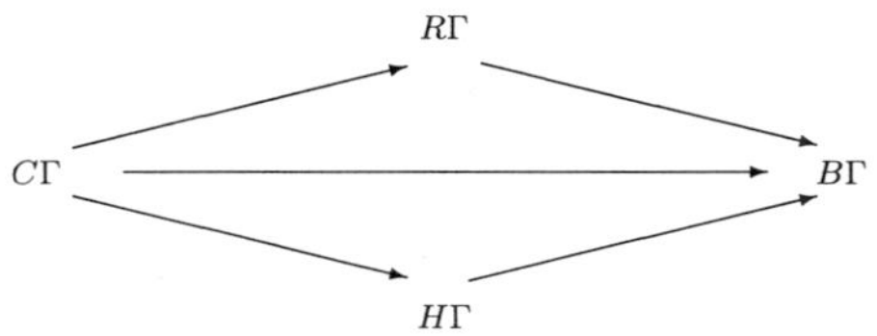

obtained by means of connection transformations [113].

The properties of metrizability of those connections can be expressed by the following table:

$C\Gamma(N)$	h – metrical	v – metrical
$B\Gamma(N)$	$g_{ij\|k}^{(h)} = -2C_{ijk\|0}$	$g_{ij}\|_k^{(v)} = 2C_{ijk}$
$R\Gamma(N)$	h – metrical	$g_{ij}\|_k^{(v)} = 2C_{ijk}$
$H\Gamma(N)$	$g_{ij\|k}^{(h)} = -2C_{ijk\|0}$	v – metrical

Remark. It is shown that the Chern connection (introduced in [25], [42]) can be identified with the Rund connection (cf. M. Anastasiei [7]).

2.8 Special Finsler manifolds

Berwald space is a class of Finsler spaces with geometrical properties similar to those Riemann spaces. Based on the holonomy group of Berwald connection Z. Szabo made a first classification of Berwald spaces. Other important classes of special Finsler spaces are Landsberg spaces and locally Minkowski Finsler spaces.

In this section we briefly describe some of the main properties of these spaces.

Definition 2.8.1. A Finsler space is called Berwald space if the connection coefficients $G^i{}_{jk} = \dfrac{\partial N^i{}_j}{\partial y^k}$ of the Berwald connection $B\Gamma$ are function of position alone, i.e.

$$\frac{\partial G^i{}_{jk}}{\partial y^h} = 0.$$

We denote by $B^n = (M, F(x,y))$ a Berwald space.

The space B^n can be characterized by the following tensor equation.

Theorem 2.8.1. *A Finsler space is Berwald space if and only if*

$$C_{ijk|\ell} = 0. \tag{8.1}$$

Proof. It is not difficult to prove that the Cartan connection $C\Gamma = (N^i{}_j, L^i{}_{jk}, C^i{}_{jk})$ and Berwald connection $B\Gamma = (G^i{}_j, G^i{}_{jk}, 0)$ are related by the formulae:

$$\begin{aligned} G^i{}_{jk} &= L^i{}_{jk} + C^i{}_{jk|0} \\ G^i{}_j &= N^i{}_j = G^i{}_{j0}. \end{aligned} \tag{8.2}$$

From here we obtain:

$$G_h{}^i{}_{jk} = F_h{}^i{}_{jk} + \frac{\partial C^i{}_{hj|0}}{\partial y^k},$$

where

$$G_h{}^i{}_{jk} := \frac{\partial G^i{}_{jh}}{\partial y^k}, F_h{}^i{}_{jk} := \frac{\partial L^i{}_{jh}}{\partial y^k}.$$

From the second expression of (5.5), we have

$$F_h{}^i{}_{jk} = P_h{}^i{}_{jk} + C_h{}^i{}_{k|j} - C^i{}_{hr} P^r{}_{jk}. \tag{8.3}$$

It follows that the condition $\dfrac{\partial G^i{}_{jh}}{\partial y^k} = 0$ is equivalent to

$$F_h{}^i{}_{jk} = -\frac{\partial C_j{}^i{}_{h|0}}{\partial y^k}. \tag{8.4}$$

Eliminating the term $F_h{}^i{}_{jk}$ from (8.3) and (8.4), we find

$$P_{i\ell jk} + C_{i\ell k|j} - C_{i\ell r} P^r{}_{jk} + \frac{\partial P_{i\ell j}}{\partial y^k} - 2C_{\ell kr} P^r{}_{ij} = 0, \tag{8.5}$$

where we used $g_{\ell h} \dfrac{\partial P^h{}_{ij}}{\partial y^k} = \dfrac{\partial P_{i\ell j}}{\partial y^k} - 2C_{\ell kh} P^h{}_{ij}$.

Permutating now the indices $(i\ell j)$ in (8.5) and taking into account the identity $\underset{(hkj)}{\sigma}\{P_{hkji}\} = 0$, one obtain:

$$3\frac{\partial P_{i\ell j}}{\partial y^k} + \underset{(i\ell j)}{\sigma}\{C_{i\ell k|j} - C_{i\ell r}P^r{}_{jk} - 2C_{\ell kr}P^r{}_{ij}\} = 0.$$

Contracting this by y^ℓ, we obtain

$$\text{(8.6)} \qquad 3\frac{\partial P_{i\ell j}}{\partial y^k}y^\ell + C_{jik|0} = -3P_{ikj} + C_{jik|0} = -2C_{jik|0} = 0,$$

which means $P_{ikj} = 0$. So the equation (8.5) reduces to

$$\text{(8.7)} \qquad P_{i\ell jk} + C_{i\ell k|j} = 0.$$

On the other hand, in general, the hv–curvature tensor P_{hkji} of $C\Gamma$ can be written as, [88]:

$$\text{(8.8)} \qquad P_{hkji} + C_{kij|h} - C_{hij|k} + C_{hjr}P^r{}_{ki} - C_{kjr}P^r{}_{hi}.$$

Taking into account of $P_{ikj} = 0$, from (8.8) we obtain:

$$\text{(8.9)} \qquad P_{hkji} = C_{kij|h} - C_{hij|k}.$$

Hence, from (8.7), (8.9) we have $C_{ijk|\ell} = 0$.

Conversely, from $C_{ijk|\ell} = 0$ we obtain $P_{ijk} = 0$, $P_{hkij} = 0$. Therefore, (8.3) gives $F_h{}^i{}_{jk} = 0$, and then $G_h{}^i{}_{jk} = 0$, i.e. $G^i{}_{hj} = G^i{}_{hj}(x)$. **q.e.d.**

Corollary 2.8.1. *A Finsler space is Berwald space if and only if the hv–curvature tensor* $G_h{}^i{}_{jk}\left(:= \dfrac{\partial G^i{}_{jk}}{\partial y^h}\right)$ *of* $B\Gamma$ *vanishes identically.*

Another important class of Finsler space is given by the Landsberg spaces [88]:

Definition 2.8.2. A Finsler space is called *Landsberg space* if its Berwald connection $B\Gamma$ is h–metrical, i.e.

$$\text{(8.10)} \qquad \underset{ij\,|\,k}{g}{}^{(b)} = -2C_{ijk|0} = 0,$$

where the index 0 means contraction by y^i.

Theorem 2.8.2. *A Finsler space is Landsberg space if and only if the hv–curvature tensor* P_{hijk} *of Cartan connection* $C\Gamma(N)$ *vanishes identically.*

Proof. From Proposition 2.5.6, a Landsberg space is characterized by $P_{ijk} = 0$. In general, from the equation $P_{ijk} = C_{ijk|0}$ we obtain

$$\begin{aligned} P_{ijk|h} - P_{hjk|i} &= \left(C_{ijk|r|h} - C_{hjk|r|i} \right) y^r + C_{ijk|\ell} - C_{hjk|\ell} \\ &= \left(C_{ijk|h} - C_{hjk|i} \right)_{|r} y^r + C_{ijk|h} - C_{hjk|i} \\ &= C_{ijk|h} - C_{hjk|i}, \end{aligned}$$

where we used the Ricci identity (5.3) and $C_{ijk|h} - C_{hjk|i} = 0$.

Taking into account equation (8.8) and the above equation we obtain:

$$P_{hijk} = \frac{\partial P_{ijk}}{\partial y^h} - \frac{\partial P_{hjk}}{\partial y^i} + C^r_{ik} P_{rhj} - C^r_{hk} P_{rij}. \tag{8.11}$$

From (8.11), if $P_{ijk} = 0$ hold good, then we obtain $P_{hijk} = 0$.

Conversely, from the relation $P_0{}^i{}_{hk} = P^i{}_{hk}$ (cf. (5.6)), we can conclude the assertion. **q.e.d.**

The following result is now immediate.

Corollary 2.8.2. *If a Finsler space is Berwald space, then it is a Landsberg space.*

The locally Minkowski Finsler space are introduced by the following definition.

Definition 2.8.3. A Finsler space $F^n = (M, F(x, y))$ is called *locally Minkowski* if in every point $x \in M$ there is a coordinate system (x^i, U) such that on $\pi^{-1}(U) \subset TM$ its fundamental function $F(x, y)$ depends only on directional variable y^i. Such a coordinate system (x^i, U) is called *adapted* to a locally Minkowski space.

Theorem 2.8.3. *A Finsler space is locally Minkowski if and only if the covariant tensor of curvature R_{ijhk} of the Cartan connection $C\Gamma$ vanishes and the tensorial equation $C_{ijk|h} = 0$ holds.*

Proof. In an adapted coordinate system, first two coefficients of the Cartan connection $C\Gamma$ are given by $N^i{}_j = 0$, $L^i{}_{jk} = 0$. Hence from the definitions of $R^i{}_{jk}$ and $R_h{}^i{}_{jk}$, taking into account of (5.4), (5.5), respectively, we obtain $R_{hijk} = 0$.

Next, we see easily that under the adapted coordinate system

$$C_{hij|k} = \frac{\partial C_{hij}}{\partial x^k} = \frac{1}{2} \frac{\partial}{\partial y^j} \left(\frac{\partial g_{hi}}{\partial x^h} \right) = 0.$$

Conversely, if $R_{hijk} = 0$ and $C_{ijk|\ell} = 0$, from (5.6) and (8.8) we obtain $P_{ijk} = P_{hijk} = 0$. Hence from (8.8) we obtain $F_h{}^i{}_{jk} = \left(:= \frac{\partial F^i{}_{jh}}{\partial y^k} \right) = 0$. Hence, $L^i{}_{hj} = L^i{}_{hj}(x)$. So, the first equation of (5.5) reduces to

$$\mathop{\mathcal{M}}_{(jk)} \left\{ \frac{\partial L^i{}_{hj}}{\partial x^k} + L^r_{hj} L^i_{rk} \right\} = 0.$$

This implies Riemannian flatness. Hence there exists a coordinate system $(\bar{x}^a)$ such that $\overline{L}^a_{bc} = 0$, from which $\overline{N}^a_b = 0$.

Consequently, from the axiom of h-metrizability, we obtain

$$\frac{\partial \bar{g}_{ab}}{\partial \bar{x}^c} = \frac{\partial \bar{g}_{ab}}{\partial \bar{y}^d} \overline{N}^d_c + \overline{L}_{abc} + \overline{L}_{bac} = 0, \ \overline{L}_{abc} := \bar{g}_{bd} \overline{L}^d_{ac},$$

from which we obtain $\bar{g}_{ab} = \bar{g}_{ab}(\bar{y})$. **q.e.d.**

Since we have an example given by Antonelli of a locally Minkowski space [11], that means:

On the paracompact manifolds there exists locally Minkowski Finsler spaces.

So, we obtain the following sequences of inclusions of special Finsler spaces:

$$\mathcal{L} \supset \mathcal{B} \supset \mathcal{M} \supset \mathcal{M} \cap \mathcal{R},$$

where $\mathcal{L}$ is the class of Landsberg spaces, $\mathcal{B}$ the class of the Berwald spaces, $\mathcal{M}$ the class of the locally Minkowski spaces, and $\mathcal{R}$ the class of Riemannian spaces.

We remark that $\mathcal{M} \cap \mathcal{R}$ is the class of flat Riemannian spaces.

In 1978, Y. Ichijyo has shown the geometrical meaning of the vanishing hv–curvature tensor using the holonomy mapping as follows:

Theorem 2.8.4. (Ichijyo) *Let us assume that (M, F) is a connected Finsler space with the Cartan connection $C\Gamma$. Let p and q be two arbitrary points of M, and let ℓ be any piecewise differentiable curve joining p and q. In order that the holonomy mapping from T_pM to T_qM along with ℓ, with respect to the nonlinear connection N, be always a C–affine mapping, it is necessary and sufficient that the hv–curvature tensor $P_h{}^i{}_{jk}$ vanishes identically.*

It still remain an open problem: If there exists Landsberg space with vanishing hv–curvature tensor.

In [68], Y. Ichijyo introduced the following interesting fundamental function

$$\overline{F}(x, dx) = F(a^\alpha), \ a^\alpha = a^\alpha_i(x) dx^i, \tag{8.12}$$

where $\overline{F}$ is a fundamental Finsler function, the function $F(a^\alpha)$ is 1–positively homogeneous in a^α, $(\alpha = 1, 2, ..., n)$ and $a^\alpha_i(x)dx^i$ are linearly independent differentiable 1–forms.

Definition 2.8.4. The Finsler metric (8.12) is called 1-form Finsler metric and the space (M, F) is called 1-form Finsler space.

There are some special 1-form metrics:

1. Berwald–Mòor metric: $F = (y^1y^2...y^n)^{\frac{1}{n}}$ which is a typical Minkowski metric in a local coordinat system. G.S.Asanov introduced a more general form: $F = (a^1a^2...a^n)^{\frac{1}{n}}$, where a^α $(\alpha = 1, 2, ..., n)$ are linearly independent 1-forms.
2. m–th root metric: $F = \{(a^1)^m + (a^2)^m + \cdots + (a^n)^m\}^{\frac{1}{m}}$ which was studied by Shimada [152] and Antonelli and Shimada [19] for the case $n = 2$.
3. A special Randers metric: $F = \{(a^1)^2 + (a^2)^2 + \cdots + (a^n)^2\}^{\frac{1}{2}} + ka^1$, where k is a constant. This example was given by Y. Ichijyo.

Using the Cartan connection, Matsumoto and Shimada proved the following result for a 2-dimensional 1-form Finsler space [91]:

Theorem 2.8.5. *If a 2-dimensional 1-form Finsler space is a Landsberg space, then it is a Berwald space.*

In order to introduce the notion of Douglas space, let us observe first that the geodesics of a Finsler space F^n can be written in the form:

$$\ddot{x}^i\dot{x}^j - \ddot{x}^j\dot{x}^i + 2D^{ij}(x, \dot{x}) = 0, \tag{8.13}$$

where $D^{ij}(x, \dot{x}) := G^i\dot{x}^j - G^j\dot{x}^i$.

Definition 2.8.5. A Finsler space F^n is called Douglas space if the functions $D^{ij}(x, \dot{x})$ are homogeneous polynomials in $\dot{x}$ of degree three.

This is equivalent to the fact that the Douglas tensors of F^n vanishes. M. Matsumoto and S. Bacso [23] proved:

Theorem 2.8.6. *If a Landsberg space is a Douglas space, then it is a Berwald space.*

Theorem 2.8.7. *A Randers space $F^n = (M, F = \alpha + \beta)$ is a Douglas space if and only if the differential 1-form $\beta$is a closed form.*

We will describe in the sequel the Finsler spaces with constant curvature.

Definition 2.8.6. The quantity $K(x, y, X)$ given by

$$K(x, y, X) = \frac{H_{hijk}y^hX^iy^jX^k}{(g_{hj}g_{ik} - g_{hk}g_{ij})y^hX^iy^jX^k}, \ \forall X \in T_xM, \ X \neq 0, \ \forall\, (x, y) \in \widetilde{TM},$$

is called the scalar curvature at (x, y) with respect to X, where H_{hijk} is the h-curvature tensor of the Berwald connection $B\Gamma$ [88].

One can remark that the h-curvature tensor of the Berwald connection $B\Gamma$ is ([88][p.118])

$$H_h{}^i{}_{jk} = K_h{}^i{}_{jk} + \mathop{\mathcal{M}}_{(jk)} \{C_j{}^i{}_{h|0|k} + C_h{}^i{}_{r|0} C_j{}^r{}_{h|0}\}.$$

From Definition 2.8.6 it can be said that the scalar curvature $K(x, y, X)$ is defined as the sectional curvature of a 2-section spanned by y and X, with $g_{ij} = g_{ij}(x, y)$ and $H_{hijk} = H_{hijk}(x, y)$, in general.

Definition 2.8.7. If the scalar curvature K for a Finsler space of scalar curvature is a constant, the Finsler space is called a space of constant curvature K.

The following theorem is known ([88]):

Theorem 2.8.8. *A Finsler space is of scalar curvature if and only if*

$$R_{iok} = KF^2 h_{ik}, \tag{8.14}$$

where R_{ijk} is the torsion tensor of the Cartan connection $C\Gamma$.

The left hand of (8.14) is also called flag curvature:

$$R_k{}^\ell{}_{ij} y^k y^j = y^j \left\{ \frac{\delta N^\ell{}_j}{\delta x^i} - \frac{\delta N^\ell{}_i}{\delta x^j} \right\}.$$

The flag curvature is one of the important numerical invariants because it lies in the second variation formula of arc length and takes the place of sectional curvature from the Riemannian case.

In 1975 the following interesting result concerning scalar curvature was obtained:

Theorem 2.8.9. (Numata [129]) *Let F^n, $n \geq 3$, be a Berwald space of scalar curvature K. Then F^n is a Riemannian space of constant curvature or a locally Minkowski space, according $K \neq 0$ or $K = 0$, respectively.*

Lastly in this chapter we remark that Finsler spaces with (α, β)–metric were studied in the paper [89]. And using the invariants of a Finsler space, was made a classification of some Finsler spaces with (α, β)–metric, namely Randers class, Kropina class, Matsumoto class, etc. The classes are providing new concrete examples of (α, β)–metrics.

2.9 Almost Kählerian model of a Finsler manifold

A Finsler space $F^n = (M, F)$ can be thought as an almost Kähler space on the manifold $\widetilde{TM} = TM\setminus\{0\}$, called the geometrical model of the Finsler space F^n.

In this section we present the Sasaki–Matsumoto lift of the metric tensor g_{ij} of the space F^n and a new lift, given by R. Miron [109], which is homogeneous and allows us to study problems concerning the global properties of the Finsler spaces.

In this way the theory of Finsler spaces gets more geometrical consistence.

If we consider the Cartan nonlinear connection $N^i{}_j$ of the Finsler space $F^n = (M, F)$, then we can define an almost complex structure $\mathbb{F}$ on TM by:

$$\mathbb{F}\left(\frac{\delta}{\delta x^i}\right) = -\frac{\partial}{\partial y^i},\ \mathbb{F}\left(\frac{\partial}{\partial y^i}\right) = \frac{\delta}{\delta x^i}. \tag{9.1}$$

It is easy to see that $\mathbb{F}$ is well defined on $\widetilde{TM}$, $\mathbb{F}^2 = -I$ and it is determined only by the fundamental function F of the Finsler space F^n.

Theorem 2.9.1. *The almost complex structure $\mathbb{F}$ is integrable if and only if the h-coefficients $R^i{}_{jk}$ of the torsion of $C\Gamma$ vanishes.*

Let $(dx^i, \delta y^i)$ be the dual basis of the adapted basis $\left(\frac{\delta}{\delta x^i}, \frac{\partial}{\partial y^i}\right)$. Then, the Sasaki–Matsumoto lift of the fundamental tensor g_{ij} can be introduced as follows:

$$G = g_{ij}dx^i \otimes dx^j + g_{ij}\delta y^i \otimes \delta y^j. \tag{9.2}$$

Consequently, G is a Riemannian metric on $\widetilde{TM}$ determined only by the fundamental function F of the Finsler space F^n and the horizontal and vertical distributions are orthogonal with respect to it.

The following results can be proved without difficulties, [113]:

Theorem 2.9.2.

(i) *The pair $(G, \mathbb{F})$ is an almost Hermitian structure on $\widetilde{TM}$.*

(ii) *The almost symplectic 2-form associated to the almost Hermitian structure $(G, \mathbb{F})$ is*

$$\theta = g_{ij}(x, y)\delta y^i \otimes dx^j. \tag{9.3}$$

(iii) *The space $H^{2n} = (\widetilde{TM}; G, \mathbb{F})$ is an almost Kählerian space, constructed only by means of the fundamental function F of the Finsler space F^n.*

The space $H^{2n} = (\widetilde{TM}; G, \mathbb{F})$ is called the *almost Kählerian model* of the Finsler space F^n.

Theorem 2.9.3. *The N-linear connection D with the coefficients $C\Gamma(N) = (F^i{}_{jk}, C^i{}_{jk})$ of the Cartan connection is an almost Kählerian connection, i.e.:*

$$D_X G = 0,\ D_X \mathbb{F} = 0,\ \forall X \in \mathcal{X}(TM). \tag{9.4}$$

Hence, the geometry of the almost Kählerian model H^{2n} can be studied by means of Cartan connection of the Finsler space F^n. For instance, the Einstein equations in F^n are given by the Einstein equations in the previous model H^{2n}. One can find it in [113].

Remarking that the Sasaki–Matsumoto lift (9.2) is not homogeneous with respect to y^i, a homogeneous lift to $\widetilde{TM}$ of the fundamental tensor field $g_{ij}(x,y)$ was introduced by R. Miron in the paper [109].

We describe here this new lift for its theoretical and applicative interest.

Definition 2.9.1. We call the following tensor field on $\widetilde{TM}$:

$$\overset{0}{\mathbb{G}}(x,y) = g_{ij}(x,y)dx^i \otimes dx^j + \frac{a^2}{\|y\|^2}\, g_{ij}(x,y)\delta y^i \otimes \delta y^j,\ \forall\,(x,y) \in \widetilde{TM}, \tag{9.5}$$

the homogeneous lift to $\widetilde{TM}$ of the fundamental tensor field g_{ij} of a Finsler space F^n, where $a > 0$ is a constant, imposed by applications (in order to preserve the physical dimensions of the components of $\overset{0}{\mathbb{G}}$) and where $\|y\|^2$ is the square of the norm of the Liouville vector field:

$$\|y\|^2 = g_{ij}(x,y)y^i y^j = y_i y^i = F^2(x,y), \tag{9.6}$$

with $y_i = g_{ij}y^j = \dfrac{1}{2}\dfrac{\partial F^2}{\partial y^i}$.

We obtain, without difficulties:

Theorem 2.9.4.

1° *The pair $(\widetilde{TM}, \overset{0}{\mathbb{G}})$ is a Riemannian space.*

2° *$\overset{0}{\mathbb{G}}$ is 0–homogeneous on the fibers of TM.*

3° *$\overset{0}{\mathbb{G}}$ depends only on the fundamental function $F(x,y)$ of the Finsler space F^n.*

4° *The distributions N and V are orthogonal with respect to $\overset{0}{\mathbb{G}}$.*

We shall write $\overset{0}{\mathbb{G}}$ in the form

$$\overset{0}{\mathbb{G}} = \overset{0}{\mathbb{G}}{}^H + \overset{0}{\mathbb{G}}{}^V,\quad \overset{0}{\mathbb{G}}{}^H = g_{ij}(x,y)dx^i \otimes dx^j,\quad \overset{0}{\mathbb{G}}{}^V = h_{ij}(x,y)\delta y^i \otimes \delta y^j \tag{9.7}$$

where

$$h_{ij} = \frac{a^2}{\|y\|^2}\, g_{ij}(x,y). \tag{9.8}$$

Consequently, we can apply the theory of the (h, v)–Riemannian metric on TM investigated by R. Miron and M. Anastasiei in the book [113].

The equation $F(x_0, y) = a$ determines the so called indicatrix of the Finsler space F^n in the point $x_0 \in M$.

Therefore, we have:

Theorem 2.9.5. *The homogeneous lift* $\overset{0}{\mathbb{G}}$ *of the metric tensor* $g_{ij}(x, y)$ *coincides with the Sasaki–Matsumoto lift of* $g_{ij}(x, y)$ *on the indicatrix* $F(x_0, y) = a$*, for every point* $x_0 \in M$.

A linear connection D on $\widetilde{TM}$ is called a metrical N–connection, with respect to $\overset{0}{\mathbb{G}}$, if $D\overset{0}{\mathbb{G}} = 0$ and D preserves by parallelism the horizontal distribution N.

As we know,[113] there exist the metrical N–connection on $\widetilde{TM}$.

We represent a linear connection D, in the adapted basis, in the following form:

$$
\begin{aligned}
&D_{\frac{\delta}{\delta x^k}} \frac{\delta}{\delta x^j} = \overset{H}{L}{}^i{}_{jk} \frac{\delta}{\delta x^i} + \tilde{L}{}^i{}_{jk} \frac{\partial}{\partial y^i}, \quad D_{\frac{\delta}{\delta x^k}} \frac{\partial}{\partial y^j} = \overset{\approx}{L}{}^i{}_{jk} \frac{\delta}{\delta x^i} + \overset{V}{L}{}^i{}_{jk} \frac{\partial}{\partial y^i} \\
&D_{\frac{\partial}{\partial y^k}} \frac{\delta}{\delta x^j} = \overset{H}{C}{}^i{}_{jk} \frac{\delta}{\delta x^i} + \tilde{C}{}^i{}_{jk} \frac{\partial}{\partial y^i}, \quad D_{\frac{\partial}{\partial y^k}} \frac{\partial}{\partial y^j} = \overset{\approx}{C}{}^i{}_{jk} \frac{\delta}{\delta x^i} + \overset{V}{C}{}^i{}_{jk} \frac{\partial}{\partial y^i},
\end{aligned}
\tag{9.9}
$$

where $(\overset{H}{L}{}^i{}_{jk}, \tilde{L}{}^i{}_{jk}, \overset{\approx}{L}{}^i{}_{jk}, \overset{V}{L}{}^i{}_{jk}, \overset{H}{C}{}^i{}_{jk}, \tilde{C}{}^i{}_{jk}, \overset{\approx}{C}{}^i{}_{jk}, \overset{V}{C}{}^i{}_{jk})$ are the coefficients of D.

Theorem 2.9.6. *There exist the metrical* N*–connections* D *on* $\widetilde{TM}$*, with respect to* $\overset{0}{\mathbb{G}}$*, which depend only on the fundamental function* $F(x, y)$ *of the Finsler space* F^n*. One of them has the following coefficients*

$$
\begin{aligned}
&\tilde{L}{}^i{}_{jk} = \overset{\approx}{L}{}^i{}_{jk} = \tilde{C}{}^i{}_{jk} = \overset{\approx}{C}{}^i{}_{jk} = 0, \\
&\overset{H}{L}{}^i{}_{jk} = F^i{}_{jk} = \frac{1}{2} g^{is} \left(\frac{\delta g_{sk}}{\delta x^j} + \frac{\delta g_{js}}{\delta x^k} - \frac{\delta g_{jk}}{\delta x^s} \right); \ \overset{V}{L}{}^i{}_{jk} = B^i{}_{jk} + \frac{1}{2} h^{is} h_{sj\|k}, \\
&\overset{V}{C}{}^i{}_{jk} = \frac{1}{2} h^{is} \left(\frac{\partial h_{sk}}{\partial y^j} + \frac{\partial h_{js}}{\partial y^k} - \frac{\partial h_{jk}}{\partial y^s} \right); \ \overset{H}{C}{}^i{}_{jk} = C^i{}_{jk},
\end{aligned}
\tag{9.10}
$$

where $C\Gamma(N) = (F^i{}_{jk}, C^i{}_{jk})$ *is the Cartan connection of the Finsler space* F^n, $B\Gamma(N) = (B^i{}_{jk}, 0)$ *is the Berwald connection and* $\|$ *means the* h*–covariant derivation with respect to* $B\Gamma(N)$.

Of course, the structure equations of the previous connection can be written as in the books [112], [113].

In order to study the Riemannian space $(\widetilde{TM}, \overset{0}{\mathbb{G}})$ it is important to express the coefficients

$$(\overset{H}{L}{}^i{}_{jk}, \tilde{L}{}^i{}_{jk}, \overset{\approx}{L}{}^i{}_{jk}, \overset{V}{L}{}^i{}_{jk}, \overset{H}{C}{}^i{}_{jk}, \tilde{C}{}^i{}_{jk}, \overset{\approx}{C}{}^i{}_{jk}, \overset{V}{C}{}^i{}_{jk}).$$

To this aim, expressing in the adapted basis the conditions:

$$X \overset{0}{\mathbb{G}} (Y,Z) - \overset{0}{\mathbb{G}} (D_X Y, Z) - \overset{0}{\mathbb{G}} (Y, D_X Z) = 0,$$
$$D_X Y - D_Y X - [X,Y] = 0, \ \forall X,Y,Z \in \mathcal{X}(TM)$$

and using the torsions $R^i{}_{jk}$ and $C^i{}_{jk}$ of the Cartan connection $C\Gamma(N)$, we get by a direct calculus:

Theorem 2.9.7. *The Levi–Civita connection of the Riemannian metric* $\overset{0}{\mathbb{G}}$ *have in an adapted basis the following coefficients*

(9.11)
$$\begin{aligned}
\overset{H}{L}{}^i{}_{jk} &= \frac{1}{2} g^{im} \left(\frac{\delta g_{mj}}{\delta x^k} + \frac{\delta g_{mk}}{\delta x^i} - \frac{\delta g_{jk}}{\delta x^m} \right), \\
\overset{V}{C}{}^i{}_{jk} &= \frac{1}{2} h^{im} \left(\frac{\partial h_{mj}}{\partial y^k} + \frac{\partial h_{mk}}{\partial y^j} - \frac{\partial h_{jk}}{\partial y^m} \right), \\
\overset{H}{C}{}^i{}_{kj} &= \overset{\approx}{L}{}^i{}_{kj} = C^i{}_{jk} + \frac{1}{2} g^{is} h_{mj} R^m{}_{sk}, \\
\overset{V}{L}{}^i{}_{jk} &= B^i{}_{kj} + \frac{1}{2} h^{is} h_{sj\|k} \\
\tilde{L}^i{}_{jk} &= -h^{is} C_{skj} - \frac{1}{2} R^i{}_{jk}, \\
\tilde{C}^i{}_{jk} &= \frac{1}{2} h^{is} h_{sk\|j}, \quad \overset{\approx}{C}{}^i{}_{jk} = -\frac{1}{2} g_{is} h^{is} h_{jk\|s}.
\end{aligned}$$

The structure equations of the Levi–Civita connection (9.11) can be written in the usual way.

Let us prove that the almost complex structure $\mathbb{F}$, defined by (9.1) does not preserve the property of homogeneity of the vector fields. Indeed, it applies the 1–homogeneous vector fields $\frac{\delta}{\delta x^i}$, $(i = 1, ..., n)$ onto the 0–homogeneous vector fields $\frac{\partial}{\partial y^i}$, $(i = 1, ..., n)$.

We can eliminate this incovenient by defining a new kind of almost complex structure $\overset{0}{\mathbb{F}}: \mathcal{X}(\widetilde{TM}) \to \mathcal{X}(\widetilde{TM})$, setting:

(9.12)
$$\overset{0}{\mathbb{F}} \left(\frac{\delta}{\delta x^i} \right) = -\frac{\|y\|}{a} \frac{\partial}{\partial y^i}, \ \overset{0}{\mathbb{F}} \left(\frac{\partial}{\partial y^i} \right) = \frac{a}{\|y\|} \frac{\delta}{\delta x^i}, \ (i = 1, ..., n).$$

Taking into account that the norm of the Liouville vector field $\|y\|$ and the Cartan nonlinear connection N are defined on $\widetilde{TM}$, it is not difficult to prove:

Theorem 2.9.8. *The following properties hold:*

1° $\overset{0}{\mathbb{F}}$ *is a tensor field of type* (1.1) *on* $\widetilde{TM}$.

2° $\overset{0}{\mathbb{F}} \circ \overset{0}{\mathbb{F}} = -I$.

3° $\overset{0}{\mathbb{F}}$ *depends only on the fundamental function* F *of the Finsler space* F^n.

4° *The* $\mathcal{F}(\widetilde{TM})$*–linear mapping* $\overset{0}{\mathbb{F}}: \mathcal{X}(\widetilde{TM}) \to \mathcal{X}(\widetilde{TM})$ *preserves the property of homogeneity of the vector fields form* $\mathcal{X}(\widetilde{TM})$.

It is important to know when $\overset{0}{\mathbb{F}}$ is a complex structure.

Theorem 2.9.9. $\overset{0}{\mathbb{F}}$ *is a complex structure on* $\widetilde{TM}$ *if and only if the Finsler space* F^n *has the following property:*

$$R^h{}_{ij} = \frac{1}{a^2}\left(y_i\delta^h{}_j - y_j\delta^h{}_i\right). \tag{9.13}$$

Proof. The Nijenhuis tensor $\mathcal{N}_{\overset{0}{\mathbb{F}}}$:

$$\mathcal{N}_{\overset{0}{\mathbb{F}}}(X,Y) = \overset{0}{\mathbb{F}}{}^2[X,Y] + [\overset{0}{\mathbb{F}}X, \overset{0}{\mathbb{F}}Y] - \overset{0}{\mathbb{F}}[\overset{0}{\mathbb{F}}X, Y] - \overset{0}{\mathbb{F}}[X, \overset{0}{\mathbb{F}}Y],\ \ X, Y \in \mathcal{X}(\widetilde{TM}),$$

vanishes if and only if the previous equations hold.

Remark. If F^n is a Riemann space, the equation (9.13) is a necessary and sufficient condition that it to be of constant sectional curvature.

The pair $(\overset{0}{\mathbb{G}}, \overset{0}{\mathbb{F}})$ has remarkable properties:

Theorem 2.9.10. *We have*

1° $(\overset{0}{\mathbb{G}}, \overset{0}{\mathbb{F}})$ *is an almost Hermitian structure on* $\widetilde{TM}$ *and depend only on the fundamental function* F *of the Finsler space* F^n.

2° *The associated almost symplectic structure* $\overset{0}{\theta}$ *has the expression*

$$\overset{0}{\theta} = \frac{a}{\|y\|}\,\theta, \tag{9.14}$$

where θ *is the symplectic structure* (9.3).

3° *The following formula holds:*

$$d\overset{0}{\theta} = d\frac{a}{\|y\|} \wedge \theta. \tag{9.15}$$

4° *Consequently,* $(\overset{0}{\mathbb{G}}, \overset{0}{\mathbb{F}})$ *is a conformal almost Kählerian structure and we have*

$$d\overset{0}{\theta} = 0 \text{ (modulo } \overset{0}{\theta}).$$

Remarks.

1° The previous theorem shows that $(\overset{0}{\mathbb{G}}, \overset{0}{\mathbb{F}})$ is a special almost Hermitian structure.

2° There exist the linear connection compatible with the conformal almost Kählerian structure $(\overset{0}{\mathbb{G}}, \overset{0}{\mathbb{F}})$.

The conformal almost Kählerian space $(\widetilde{TM}, \overset{0}{\mathbb{G}}, \overset{0}{\mathbb{F}})$ is another geometrical model of the Finsler space F^n. It is based on the homogeneous lift $\overset{0}{\mathbb{G}}$, (9.5).

The previous considerations are important for study the Finslerian gauge theory, [22], [177], and in general in the Geometry of Finsler space F^n. The importance of such kind of lift was emphasized by G.S. Asanov [22]. Namely he proved that some (h, v) metrics on $\widetilde{TM}$ satisfy the principle of the Post Newtonian calculus. The metric $\overset{0}{\mathbb{G}}$ belongs to this category, while Sasaki–Matsumoto lift has not this property.

The theory of subspaces of Finsler spaces can be found in the books [112], [113].

Chapter 3

Lagrange spaces

The notion of Lagrange space was introduced and studied by J. Kern [76] and R. Miron [96]. It was widely developed by the first author of the present monograph [106]. Since this notion includes that of Finsler space it is expected that the geometry of these spaces to be more rich and applications in Mechanics or Physics to be more important.

We will develop the geometry of Lagrange spaces, using the fundamental notions from Analytical Mechanics as: integral of action, Euler–Lagrange equations, the law of conservation of energy, Noether symmetries, etc. Remarking that the Euler–Lagrange equations determine the canonical spray of the space, we can construct all geometry of Lagrange space by means of its canonical spray, following the methods given in the Chapter 1. So the geometry of Lagrange space is a direct and natural extension of the geometry of Finsler space.

At the end of this chapter, we emphasize the notion of generalized Lagrange spaces, useful in the geometrical models for the Relativistic Optics.

3.1 The notion of Lagrange space

At the begining we define the notions of differentiable Lagrangian using the manifolds TM and $\widetilde{TM}$, where M is a differentiable real manifold of dimension n.

Definition 3.1.1. A differentiable Lagrangian is a mapping $L : (x, y) \in TM \to L(x, y) \in R$, of class C^∞ on manifold $\widetilde{TM}$ and continuous on the null section $0 : M \to TM$ of the projection $\pi : TM \to M$.

The Hessian of a differentiable Lagrangian L, with respect to y^i, has the elements:

$$(1.1) \qquad g_{ij}(x, y) = \frac{1}{2}\frac{\partial^2 L(x, y)}{\partial y^i \partial y^j}, \text{ on } \widetilde{TM}.$$

Exactly as in the Finsler case we prove that $g_{ij}(x, y)$ is a d–tensor field, covariant of order 2, symmetric.

Definition 3.1.2. A differentiable Lagrangian $L(x, y)$ is called *regular* if the following condition holds:

$$\text{rank}\|g_{ij}(x, y)\| = n, \quad \text{on } \widetilde{TM}. \tag{1.2}$$

Now we can give:

Definition 3.1.3. A Lagrange space is a pair $L^n = (M, L(x, y))$ formed by a smooth real n-dimensional manifold M and a regular Lagrangian $L(x, y)$ for which d-tensor g_{ij} has a constant signature over the manifold $\widetilde{TM}$.

For the Lagrange space $L^n = (M, L(x, y))$ we say that $L(x, y)$ is fundamental function and $g_{ij}(x, y)$ is fundamental (or metric) tensor. We will denote, as usually, by g^{ij} the contravariant of the tensor g_{ij}.

Examples.

1° The following Lagrangian from electrodynamics [112], [113]

$$L(x, y) = mc\gamma_{ij}(x)y^i y^j + \frac{2e}{m} A_i(x)y^i + U(x) \tag{1.3}$$

where $\gamma_{ij}(x)$ is a pseudo–Riemannian metric, $A_i(x)$ a covector field and $U(x)$ a smooth function, m, c, e being the known constants from Physics, determine a Lagrange space L^n.

More general:

2° The Lagrangian

$$L(x, y) = F^2(x, y) + A_i(x)y^i + U(x) \tag{1.4}$$

where $F(x, y)$ is the fundamental function of a Finsler space $F^n = (M, F(x, y))$, $A_i(x)$ is a covector field and $U(x)$ a smooth function gives rise to a remarkable Lagrange space, called the Almost Finsler–Lagrange space (shortly AFL–space).

In particular, ($A_i(x) = 0$, $U(x) = 0$) the pair $L^n = (M, F^2(x, y))$ is a Lagrange space.

In other words,

Proposition 3.1.1. *Any Finsler space $F^n = (M, F(x, y))$ is a Lagrange space $L^n = (M, F^2)$. Conversely, any Lagrange space $L^n = (M, L(x, y))$ for which fundamental function $L(x, y)$ is positive and 2-homogeneous with respect to y^i and $\dfrac{1}{2}\dfrac{\partial^2 L(x, y)}{\partial y^i \partial y^j}$ is positive definite determine a Finsler space $F^n = (M, \sqrt{L(x, y)})$.*

The previous example proves that:

Theorem 3.1.1. *If the base manifold M is paracompat then there exist regular Lagrangians $L(x, y)$ such that the pair $L^n = (M, L(x, y))$ is a Lagrange space.*

3.2 Variational problem Euler–Lagrange equations

The variational problem can be formulated for differentiable Lagrangians and can be solved in the case when we consider the parametrized curves, even if the integral of action depends on the parametrization of the considered curve.

Let $L : TM \to R$ be a differentiable Lagrangian and $c : t \in [0, 1] \to (x^i(t)) \in U \subset M$ a curve (with a fixed parametrization) having the image in the domain of a chart U on the manifold M. The curve c can be extended to $\pi^{-1}(U) \subset \widetilde{TM}$ as

$$c^* : t \in [0, 1] \longrightarrow \left(x^i(t), \frac{dx^i}{dt}(t)\right) \in \pi^{-1}(U).$$

Since the vector field $\dfrac{dx^i}{dt}(t)$, $t \in [0, 1]$, vanishes nowhere, the image of the mapping c^* belongs to $\widetilde{TM}$.

The integral of action of the Lagrangian L on the curve c is given by the functional

$$I(c) = \int_0^1 L\left(x, \frac{dx}{dt}\right) dt. \tag{2.1}$$

Consider the curves

$$c_\varepsilon : t \in [0, 1] \longrightarrow (x^i(t) + \varepsilon V^i(t)) \in M \tag{2.2}$$

which have the same end points $x^i(0)$, $x^i(x)$ as the curve c, $V^i(t) = V^i(x(t))$ being a regular vector field on the curve c, with the property $V^i(0) = V^i(1) = 0$ and ε a real number, sufficiently small in absolute value, so that $\operatorname{Im} c_\varepsilon \subset U$.

The extension of curves c_ε to $\widetilde{TM}$ is given by

$$c^*_\varepsilon : t \in [0, 1] \longrightarrow \left(x^i(t) + \varepsilon V^i(t), \frac{dx^i}{dt} + \varepsilon\frac{dV^i}{dt}\right) \in \pi^{-1}(U).$$

The integral of action of the Lagrangian L on the curve c_ε is given by

$$I(c_\varepsilon) = \int_0^1 L\left(x + \varepsilon V, \frac{dx}{dt} + \varepsilon\frac{dV}{dt}\right) dt. \tag{2.2$'$}$$

A necessary condition for $I(c)$ to be an extremal value of $I(c_\varepsilon)$ is

$$\left.\frac{dI(c_\varepsilon)}{d\varepsilon}\right|_{\varepsilon=0} = 0. \tag{2.3}$$

Under our condition of differentiability, the operator $\dfrac{d}{d\varepsilon}$ is permuting with operator of integration.

From $(2.2)'$ we obtain

$$\frac{dI(\varepsilon)}{dt} = \int_0^1 \frac{d}{d\varepsilon} L\left(x+\varepsilon V, \frac{dx}{dt}+\varepsilon\frac{dV}{dt}\right) dt. \tag{2.4}$$

A straightforward calculus leads to:

$$\left.\frac{d}{d\varepsilon} L\left(x+\varepsilon V, \frac{dx}{dt}+\varepsilon\frac{dV}{dt}\right)\right|_{\varepsilon=0} = \frac{\partial L}{\partial x^i}V^i + \frac{\partial L}{\partial y^i}\frac{dV^i}{dt} =$$
$$= \left\{\frac{\partial L}{\partial x^i} - \frac{d}{dt}\frac{\partial L}{\partial y^i}\right\}V^i + \frac{d}{dt}\left\{\frac{\partial L}{\partial y^i}V^i\right\},\ y^i = \frac{dx^i}{dt}.$$

Substituting in (2.4) and taking into account the fact that $V^i(x(t))$ is arbitrary, we obtain the following.

Theorem 3.2.1. *In order that the functional $I(c)$ be an extremal value of $I(c_\varepsilon)$ it is necessary that c be the solution of the Euler–Lagrange equations:*

$$E_i(L) := \frac{\partial L}{\partial x^i} - \frac{d}{dt}\frac{\partial L}{\partial y^i} = 0,\ y^i = \frac{dx^i}{dt}. \tag{2.5}$$

Some important properties of the Euler–Lagrange equations can be done.

Introducing the notion of energy of the Lagrangian $L(x,y)$, by:

$$E_L := y^i\frac{\partial L}{\partial y^i} - L \tag{2.6}$$

we can prove the so called theorem of conservation of energy:

Theorem 3.2.2. *The energy E_L of the Lagrangian L is conserved along to every integral curve c of the Euler–Lagrange equation $E_i(L) = 0$, $y^i = \dfrac{dx^i}{dt}$.*

Indeed, along to the integral curve of the equations $E_i(L) = 0$, $\dfrac{dx^i}{dt} = y^i$, we have:

$$\frac{d}{dt}E_L = \frac{dy^i}{dt}\frac{\partial L}{\partial y^i} + y^i\frac{d}{dt}\left(\frac{\partial L}{\partial y^i}\right) - y^i\frac{\partial L}{\partial x^i} - \frac{dy^i}{dt}\frac{\partial L}{\partial y^i} =$$
$$= -y^i\left(\frac{\partial L}{\partial x^i} - \frac{d}{dt}\frac{\partial L}{\partial y^i}\right) = 0.$$

Consequently, we have $\dfrac{dE_L}{dt} = 0.$ **q.e.d.**

Remark. A theorem of Noether type for the Lagrangian $L(x, y)$ can be found in the book [106].

3.3 Canonical semispray. Nonlinear connection

A Lagrange space $L^n = (M, L(x, y))$ determines an important nonlinear connection which depends only on the fundamental function $L(x, y)$. Remarking that $E_i(L)$, from (2.5) is a d–covector and that the fundamental tensor of the space, g_{ij}, is nondegenerate we can establish:

Theorem 3.3.1. *If $L^n = (M, L)$ is a Lagrange space then the system of differential equations*

$$g^{ij}E_j(L) = 0, \ y^i = \frac{dx^i}{dt} \tag{3.1}$$

can be written in the form:

$$\frac{d^2x^i}{dt^2} + 2G^i\left(x, \frac{dx}{dt}\right) = 0 \tag{3.1'}$$

where

$$2G^i(x, y) = \frac{1}{2}g^{ij}\left\{\frac{\partial^2 L}{\partial y^j \partial x^k}y^k - \frac{\partial L}{\partial x^j}\right\}. \tag{3.2}$$

Indeed, the formula

$$E_i(L) = \frac{\partial L}{\partial x^i} - \frac{d}{dt}\frac{\partial L}{\partial y^i} = \frac{\partial L}{\partial x^i} - \left\{\frac{\partial^2 L}{\partial y^j \partial x^h}y^h + 2g_{ij}\frac{dy^h}{dt}\right\}, \ \frac{dx^i}{dt} = y^i$$

holds. Hence (3.1) is equivalent to (3.1)′, G^i being expressed in (3.2).

Theorem 3.3.2. *The differential equation* (3.1)′ *gives the integral curves of the semi–spray*

$$S = y^i \frac{\partial}{\partial x^i} - 2G^i(x, y)\frac{\partial}{\partial y^i}, \tag{3.3}$$

where $G^i(x, y)$ are expressed in (3.2).

Indeed, the differential equations $(3.1)'$ do have a geometrical meaning and therefore it follows that G^i are the coefficients of a semispray S from (3.3). The integral curves of S are given by

$$\frac{dx^i}{dt} = y^i, \ \frac{dy^i}{dt} = -2G^i(x,y),$$

hence the differential equations $(3.1)'$ are satisfied. **q.e.d.**

The previous semispray is determined only by the fundamental function $L(x,y)$ of the space L^n. It will be called *canonical semispray* of Lagrange space L^n.

Corollary 3.3.1. *The integral curves of the Euler–Lagrange equations* $E_i(L) = 0$, $\dfrac{dx^i}{dt} = y^i$ *are the integral curves of the canonical semispray* S *from* (3.3).

Indeed, we can apply Theorems 3.3.1 and 3.3.2 to get the announced property.

As we know, (Ch.1), a semispray S determines a nonlinear connection. Applying Theorem 1.4.3, we obtain:

Theorem 3.3.3. *In a Lagrange space* $L^n = (M, L)$ *there exists the nonlinear connections which depend on the fundamental function* L*. One of them has the coefficients*

$$N^i{}_j = \frac{\partial G^i}{\partial y^j} = \frac{1}{4}\frac{\partial}{\partial y^j}\left\{g^{ik}\left(\frac{\partial^2 L}{\partial y^k \partial x^h}\, y^h - \frac{\partial L}{\partial x^k}\right)\right\}. \tag{3.4}$$

Proposition 3.3.1. *The nonlinear connection* N *with coefficients* $N^i{}_j$ (3.4) *is invariant with respect to the Carathéodory transformations*

$$L'(x,y) = L(x,y) + \frac{\partial \varphi(x)}{\partial x^i}\, y^i \tag{3.5}$$

where $\varphi(x)$ *is an arbitrary smooth function.*

Indeed, we have $E_i(L'(x,y)) = E_i\left(L(x,y) + \dfrac{d\varphi}{dt}\right) = E_i(L(x,y))$. So, $E_i(L'(x,y)) = 0$ determine the same canonical spray with $E_i(L(x,y)) = 0$. Thus, the previous theorem shows that the Carathéodory transformation (3.5) does not change the nonlinear connection N.

Because the coefficients of N are expressed by means of the fundamental function L, we say that N is a *canonical nonlinear connection* of the Lagrange space L^n.

Example. The Lagrange space of electrodynamics, $L^n = (M, L(x,y))$, $L(x,y)$ being given by (1.3), with $U(x) = 0$, has the coefficients $G^i(x,y)$ of the canonical semispray S, of the form:

$$G^i(x,y) = \frac{1}{2}\gamma^i{}_{jk}(x)y^jy^k - g^{ij}F_{jk}(x)y^k, \tag{3.6}$$

where $\gamma^i{}_{jk}(x)$ are the Christoffel symbols of the metric tensor $g_{ij}(x) = mc\gamma_{ij}(x)$ of the space L^n and F_{jk} is the electromagnetic tensor

$$F_{jk}(x,y) = \frac{c}{2m}\left(\frac{\partial A_k}{\partial x^j} - \frac{\partial A_j}{\partial x^k}\right). \tag{3.7}$$

Therefore, the integral curves of the Euler–Lagrange equation are given by solution curves of the *Lorentz equations*:

$$\frac{d^2x^i}{dt^2} + \gamma^i{}_{jk}(x)\frac{dx^j}{dt}\frac{dx^k}{dt} = g^{ij}(x)F_{jk}(x)\frac{dx^k}{dt}. \tag{3.8}$$

The canonical nonlinear connection of L^n has the coefficients (3.4) of the form

$$N^i{}_j(x,y) = \gamma^i{}_{jk}(x)y^k - g^{ik}(x)F_{kj}(x). \tag{3.9}$$

It is remarkable that the coefficients $N^i{}_j$ of the canonical nonlinear connection N of the Lagrange spaces of electrodynamics are linear with respect to y^i. This fact has some consequences:

1° The Berwald connection of the space, $B\Gamma(N)$, has the coefficients $\gamma^i{}_{jk}(x)$.

2° The solution curves of the Euler–Lagrange equation and the autoparallel curves of the canonical nonlinear connection N are given by the Lorentz equation (3.8).

In the end part of this section, we underline the following theorem:

Theorem 3.3.4. *The autoparallel curves of the canonical nonlinear connection N are given by the following system of differential equations:*

$$\frac{d^2x^i}{dt^2} + N^i{}_j\left(x, \frac{dx}{dt}\right)\frac{dx^j}{dt} = 0, \tag{3.10}$$

where $N^i{}_j$ *is expressed in* (3.4).

This results from Section 3, Ch.1.

3.4 Hamilton–Jacobi equations

Let us consider a Lagrange space $L^n = (M, L(x,y))$ and $N(N^i{}_j)$ its canonical non-linear connection. The adapted basis $\left(\frac{\delta}{\delta x^i}, \frac{\partial}{\partial y^i}\right)$ to the horizontal distribution N and the vertical distribution V has the vector fields $\frac{\delta}{\delta x^i}$:

$$(4.1) \qquad \frac{\delta}{\delta x^i} = \frac{\partial}{\partial x^i} - N^j{}_i \frac{\partial}{\partial y^j}.$$

Its dual is $(dx^i, \delta y^i)$, with

$$(4.1)' \qquad \delta y^i = dy^i + N^i{}_j\, dx^j.$$

The momenta p_i of the space L^n can be defined by

$$(4.2) \qquad p_i = \frac{1}{2}\frac{\partial L}{\partial y^i}.$$

Thus p_i is a d–covector field. We can consider the following forms

$$(4.3) \qquad \omega = p_i\, dx^i,$$

$$(4.4) \qquad \theta = g_{ij}(x,y)\delta y^i \wedge dx^j.$$

Proposition 3.4.1. *The forms ω and θ are globally defined on $\widetilde{TM}$, and we have*

$$(4.5) \qquad \theta = d\omega.$$

Proof. 1° Indeed, $p_i dx^i$, and θ from (4.4) do not depend on the transformation of local coordinates on $\widetilde{TM}$.

2° $d\omega = dp_i \wedge dx^i = \frac{1}{2} d\frac{\partial L}{\partial y^i}\wedge dx^i = \frac{1}{2}\left(\frac{\delta}{\delta x^m}\frac{\partial L}{\partial y^i}\, dx^m + \frac{\partial^2 L}{\partial y^m \partial y^i}\,\delta y^m\right)\wedge dx^i =$
$= \frac{1}{4}\left(\frac{\delta}{\delta x^m}\frac{\partial L}{\partial y^i} - \frac{\delta}{\delta x^i}\frac{\partial L}{\partial y^m}\right) dx^m \wedge dx^i + g_{im}\delta y^m \wedge dx^i = g_m \delta y^m \wedge dx^i.$

Because of a direct calculus shows that

$$\frac{\delta}{\delta x^m}\frac{\partial L}{\partial y^i} - \frac{\delta}{\delta x^i}\frac{\partial L}{\partial y^m} = 0.$$

Theorem 3.4.1. *The 2–form θ, from (4.4), determines on $\widetilde{TM}$ a symplectic structure, which depends only on the fundamental function $L(x,y)$ of the space L^n.*

Proof. Using the previous proposition results that θ is integrable, i.e. $d\theta = 0$, and $\text{rank}\|\theta\| = \dim \widetilde{TM}$. **q.e.d.**

Corollary 3.4.1. *The triple $(\widetilde{TM}, \theta, L)$ is a Lagrangian system.*

The energy E_L of the space L^n is given by (2.6), Ch.3. Denoting $\mathcal{H} = \frac{1}{2} E_L$, $\mathcal{L} = \frac{1}{2} L$, we get from (2.6), Ch.3:

$$\mathcal{H} = p_i y^i - \mathcal{L}(x, y). \tag{4.6}$$

But, along the integral curve of the Euler–Lagrange equations we have

$$\frac{\partial \mathcal{H}}{\partial x^i} = -\frac{\partial \mathcal{L}}{\partial x^i} = -\frac{d}{dt}\frac{\partial \mathcal{L}}{\partial y^i} = -\frac{dp_i}{dt}.$$

And from (4.6), we have also

$$\frac{\partial \mathcal{H}}{\partial p_i} = y^i = \frac{dx^i}{dt}.$$

So, we obtain:

Theorem 3.4.2. *Along to the integral curves of Euler–Lagrange equations we have the Hamilton–Jacobi equations*

$$\frac{dx^i}{dt} = \frac{\partial \mathcal{H}}{\partial p_i}; \quad \frac{dp_i}{dt} = -\frac{\partial \mathcal{H}}{\partial x^i}. \tag{4.7}$$

Corollary 3.4.2. *The energy E_L is conserved along to every integral curve of the Hamilton–Jacobi equations.*

3.5 The structures $\mathbb{P}$ and $\mathbb{F}$ of the Lagrange space L^n

The canonical nonlinear connection N determines some global structures on the manifold $\widetilde{TM}$. One of them is the almost product structure $\mathbb{P}$. It is given by the difference of the projectors h and v

$$\mathbb{P} = h - v. \tag{5.1}$$

It follows

$$\mathbb{P}^2 = I. \tag{5.2}$$

Theorem 3.5.1. *The canonical nonlinear connection N of the Lagrange space $L^n = (M, L(x,y))$ determines an almost product structures $\mathbb{P}$, which depends only on the Lagrangian $L(x,y)$. The eigensubspace of $\mathbb{P}$ corresponding to the eigenvalue -1 is the vertical distribution V and eigensubspace of $\mathbb{P}$ corresponding to the eigenvalue $+1$ is the horizontal distribution N.*

Conform to a general result, we get from Ch.1.

Theorem 3.5.2. *The almost product structure $\mathbb{P}$ determined by the canonical nonlinear connection N is integrable if and only if the horizontal distribution N is integrable.*

This condition is expressed by the equations

$$R^i{}_{jk}(x,y) = 0 \text{ on } \widetilde{TM}, \tag{5.3}$$

where

$$R^i{}_{jk} = \frac{\delta N^i{}_j}{\delta x^k} - \frac{\delta N^i{}_k}{\delta x^j}. \tag{5.4}$$

Proof. It is not difficult to see that the Nijenhuis tensor of the structure $\mathbb{P}$ vanishes if and only if $[X^H, Y^H]^H = 0$. Taking X^H, Y^H in the adapted basis we have that $\left[\dfrac{\delta}{\delta x^i}, \dfrac{\delta}{\delta x^j}\right] = R^h{}_{ij}\dfrac{\partial}{\partial y^h} = 0$ hold iff $R^h{}_{ij} = 0$. But $R^h{}_{ij}$ is a d-tensor field, so the condition $R^h{}_{ij} = 0$ is verified on the manifold $\widetilde{TM}$. **q.e.d.**

On the manifold $\widetilde{TM}$ there exists another important structure defined by the canonical nonlinear connection N: It is the almost complex structure, given by the $\mathcal{F}(\widetilde{TM})$-linear mapping $\mathbb{F} : \mathcal{X}(\widetilde{TM}) \longrightarrow \mathcal{X}(\widetilde{TM})$

$$\mathbb{F}\left(\frac{\delta}{\delta x^i}\right) = -\frac{\partial}{\partial y^i};\ \mathbb{F}\left(\frac{\partial}{\partial y^i}\right) = \frac{\delta}{\delta x^i},\quad (i = 1, ..., n), \tag{5.5}$$

or by tensor field

$$\mathbb{F} = -\frac{\partial}{\partial y^i}\otimes dx^i + \frac{\delta}{\delta x^i}\otimes \delta y^i. \tag{5.5$'$}$$

It follows:

Theorem 3.5.3. *We have:*

1° *$\mathbb{F}$ is an almost complex structure globally defined on the manifold $\widetilde{TM}$.*

2° *$\mathbb{F}$ is determined only by the fundamental function $L(x,y)$ of the Lagrange space L^n.*

Indeed, 1° The form (5.5)′ of $\mathbb{F}$ shows that the $\mathbb{F}$ is a tensor field defined on $\widetilde{TM}$. From (5.5) we prove that $\mathbb{F}$ is an almost complex structure. This is that we have

$$\mathbb{F} \circ \mathbb{F} = -I. \tag{5.6}$$

This can be proved by means of (5.5), using the adapted basis $\left(\dfrac{\delta}{\delta x^i}, \dfrac{\partial}{\partial y^i}\right)$.

2° Clearly, $\mathbb{F}$ is determined only by the canonical nonlinear connection, which depends only on $L(x,y)$. **q.e.d.**

Theorem 3.5.4. *The almost complex structure $\mathbb{F}$ is a complex structure (i.e. $\mathbb{F}$ is integrable) if and only if the canonical nonlinear connection N is integrable.*

Indeed, the Nijenhuis tensor $\mathcal{N}_{\mathbb{F}}$ vanishes if and only if

1° the distribution N is integrable, i.e., $R^i{}_{jk} = 0$;

2° the d–tensor $t^i{}_{jk} = \dfrac{\partial N^i{}_j}{\partial y^k} - \dfrac{\partial N^i{}_k}{\partial y^j} = 0$.

But $N^i{}_j = \dfrac{\partial G^i}{\partial y^j}\cdot$ So, it follows $t^i{}_{jk} = 0$. **q.e.d.**

3.6 The almost Kählerian model of the space L^n

Following the construction of the almost Kählerian model from geometry of Finsler space, we extend for Lagrange spaces the almost Hermitian structure $(\mathbb{G}, \mathbb{F})$ determined by the lift of Sasaki type $\mathbb{G}$ of the fundamental tensor field g_{ij} and by the almost complex structure $\mathbb{F}$.

The metric tensor $g_{ij}(x,y)$ of the space $L^n = (M, L(x,y))$ and its canonical nonlinear connection $N(N^i_j)$ allows to introduce a pseudo–Riemannian structure $\mathbb{G}$ on the manifold $\widetilde{TM}$, given by the following lift of Sasaki type:

$$\mathbb{G}(x,y) = g_{ij}(x,y)dx^i \otimes dx^j + g_{ij}(x,y)\delta y^i \otimes \delta y^j \tag{6.1}$$

We have:

Theorem 3.6.1.

1° *$\mathbb{G}$ is a pseudo–Riemannian structure on the manifold $\widetilde{TM}$ determined only by the fundamental function $L(x,y)$.*

2° *The distributions N and V are orthogonal with respect to $\mathbb{G}$.*

Indeed,

1° The tensorial character of g_{ij}, dx^i and δy^i shows that $\mathbb{G}$ does not depend on the transformations of local coordinates on $\widetilde{TM}$.

2° $\mathbb{G}\left(\frac{\delta}{\delta x^i}, \frac{\partial}{\partial y^j}\right) = 0$, $(i, j = 1, ..., n)$.

q.e.d.

For the pair $(\mathbb{G}, \mathbb{F})$ we get:

Theorem 3.6.2.

1° *The pair* $(\mathbb{G}, \mathbb{F})$ *is an almost Hermitian structure on* $\widetilde{TM}$*, determined only by the fundamental function* $L(x, y)$.

2° *The almost symplectic structure associated to the pair* $(\mathbb{G}, \mathbb{F})$ *is given by* (4.4), *i.e.*

$$\theta = g_{ij}\delta y^i \wedge dx^j. \tag{6.2}$$

3° *The space* $(\widetilde{TM}, \mathbb{G}, \mathbb{F})$ *is almost Kählerian.*

Proof. 1° is evident. $\mathbb{G}$ from (6.1) and $\mathbb{F}$ from (5.5) depend only on $L(x, y)$ and we have, $\mathbb{G}(\mathbb{F}X, \mathbb{F}Y) = \mathbb{G}(X, Y)$, $\forall X, Y \in \mathcal{X}(\widetilde{TM})$ hold.

2° Calculating in the adapted basis $\theta(X, Y) = \mathbb{G}(\mathbb{F}X, Y)$ we obtain (6.2).

3° Taking into account Theorem 3.4.1, it follows that θ is a symplectic structure. **q.e.d.**

The space $K^{2n} = (\widetilde{TM}, \mathbb{G}, \mathbb{F})$ is called almost Kählerian model of the Lagrange space $L^n = (M, L(x, y))$.

We can use it to study the geometry of Lagrange space L^n. For instance, the Einstein equations of the Riemannian space $(\widetilde{TM}, \mathbb{G})$ can be considered as "the Einstein equations" of the space L^n.

G.S. Asanov showed [22] that the metric $\mathbb{G}$ given by the lift (6.1) does not satisfy the principle of the Post–Newtonian calculus. This fact is because the two terms of $\mathbb{G}$ has not the same phyisical dimensions. This is the reason to introduce a new lift [109] which can be used in a gauge theory.

Let us consider the scalar field:

$$\|y\|^2 = g_{ij}(x, y)y^i y^j. \tag{6.2'}$$

It is determined only by $L(x, y)$. We assume $\|y\|^2 > 0$. As in the case of Finsler geometry (cf. Ch.2), the following lift of the fundamental tensor field $g_{ij}(x, y)$:

$$\overset{0}{\mathbb{G}}(x, y) = g_{ij}(x, y)dx^i \otimes dx^j + \frac{a^2}{\|y\|^2} g_{ij}(x, y)\delta y^i \otimes \delta y^j \tag{6.3}$$

where $a > 0$ is a constant, imposed by applications in theoretical Physics. This is to preserve the physical dimensions of the both members of $\overset{0}{\mathbb{G}}$.

Let us consider also the tensor field on $\widetilde{TM}$:

$$(6.4)\qquad \overset{0}{\mathbb{F}} = -\frac{\|y\|}{a}\frac{\partial}{\partial y^i}\otimes dx^j + \frac{a}{\|y\|}\frac{\delta}{\delta x^i}\otimes \delta y^i,$$

and 2–form

$$(6.5)\qquad \overset{0}{\theta} = \frac{a}{\|y\|}\theta,$$

where θ is given by (6.2).

As in the Finslerian case we can prove:

Theorem 3.6.3. *We have*

1° *The pair* $(\overset{0}{\mathbb{G}}, \overset{0}{\mathbb{F}})$ *is an almost Hermitian structure on* $\widetilde{TM}$*, depending only on the fundamental function* $L(x,y)$.

2° *The almost symplectic structure* $\overset{0}{\theta}$ *associated to the structure* $(\mathbb{G}, \mathbb{F})$ *is given by the formula* (6.5).

3° $\overset{0}{\theta}$ *being conformal to symplectic structure* θ*, the pair* $(\overset{0}{\mathbb{G}}, \overset{0}{\mathbb{F}})$ *is conformal almost Kählerian structure.*

We can remark now that the conformal almost Kählerian space $\overset{0}{K}{}^{2n} = (\widetilde{TM}, \overset{0}{\mathbb{G}}, \overset{0}{\mathbb{F}})$ can be used for applications in gauge theories which implies the notion of the regular Lagrangian.

3.7 Metrical N–linear connections

Now applying the methods exposed in the first chapter, we will determine some metrical connections compatible with the Riemannian metric $\mathbb{G}$ determined by the formula (6.1). Such kind of metrical connection will give the metrical N–linear connections for the Lagrange space L^n. These connections depend only on the fundamental function $L(x,y)$ and this is the reason for the N–metrical connection to be called *canonical.*

Applying the theory of N–linear connection from Chapter 1, one proves without difficulties the following theorem:

Theorem 3.7.1. *On the manifold* $\widetilde{TM}$ *there exists the linear connection* D *which satisfy the axioms:*

1° *D is metrical connection with respect to $\mathbb{G}$ i.e.*

$$D\mathbb{G} = 0. \tag{7.1}$$

2° *D preserves by parallelism the horizontal distribution of the canonical nonlinear connection N.*

3° *The almost tangent structure J is absolute parallel with respect to D, i.e., $DJ = 0$.*

From this theorem it follows that we have $D = D^H + D^V$ and the h–component D^H and v–component D^V have the properties

$$D^H_X\mathbb{G} = 0,\ D^V_X\mathbb{G} = 0,\ \ \forall X \in \mathcal{X}(\widetilde{TM}). \tag{7.2}$$

Moreover,

$$\begin{cases} (D^H_X Y^H)^V = 0, & (D^H_X Y^V)^H = 0, \\ (D^V_X Y^H)^V = 0, & (D^V_X Y^V)^H = 0, \quad \forall X \in \mathcal{X}(\widetilde{TM}). \\ D^H_X J = 0, & D^V_X J = 0. \end{cases} \tag{7.3}$$

Consequently, D is an N–linear connection and has two coefficients $D\Gamma(N) = (L^i{}_{jk}, C^i{}_{jk})$ which verify the following tensorial equations:

$$g_{ij|k} = 0,\ g_{ij}|_k = 0 \tag{7.4}$$

where $_|$ ($|$) is the h (v)–covariant derivative with respect to $D\Gamma(N)$ respectively. And conversely, if an N–linear connection with the coefficients $D\mathbb{G} = (L^i{}_{jk}, C^i{}_{jk})$ verifies the properties (7.3), at (7.4) then it is metrical with respect to $\mathbb{G}$, i.e. the equations (7.2) are verified.

But if (7.2) are verified then the equation $D\mathbb{G} = 0$ is verified also. So we can refer the (7.4). We shall determine the general solution $(L^i{}_{jk}, C^i{}_{jk})$ of the tensorial equations (7.4). First of all we prove

Theorem 3.7.2.

1° *There exists only one N–linear connection $D\Gamma(N) = (L^i{}_{jk}, C^i{}_{jk})$ which verifies the following axioms:*

> A_1 *N is canonical nonlinear connection of the space L^n.*
> A_2 $g_{ij|k} = 0$ *(it is h-metrical).*
> A_3 $g_{ij}|_k = 0$ *(it is v-metrical).*
> A_4 $T^i{}_{jk} = 0$ *(it is h-torsion free).*
> A_5 $S^i{}_{jk} = 0$ *(it is v-torsion free).*

2° *The coefficients* $L^i{}_{jk}$ *and* $C^i{}_{jk}$ *are expressed by the following generalized Christoffel symbols:*

$$(7.5)\qquad \begin{cases} L^i{}_{jk} = \dfrac{1}{2}\, g^{ir}\left(\dfrac{\delta g_{rk}}{\delta x^j} + \dfrac{\delta g_{rj}}{\delta x^k} - \dfrac{\delta g_{jk}}{\delta x^r}\right) \\ C^i{}_{jk} = \dfrac{1}{2}\, g^{ir}\left(\dfrac{\partial g_{rk}}{\partial y^j} + \dfrac{\partial g_{rj}}{\partial y^k} - \dfrac{\partial g_{jk}}{\partial y^r}\right) \end{cases}$$

3° *This connection depends only on the fundamental function* $L(x,y)$ *if the Lagrange space* L^n.

This theorem can be provided in the usual way (see [113]). And this metrical N–linear connection will be called canonical and will be denoted by $C\Gamma(N)$.

Similar as in the case of Finsler spaces one can determine all N–linear connections which satisfy only the axioms A_1, A_2, A_3 from the previous theorem.

Now we can study the geometry of Lagrange space L^n by means of canonical metrical connection or by the general metrical connections which satisfy the axioms A_1, A_2, A_3. In this respect we can determine as in the Finsler geometry, the structure equations of the metrical N–linear connection $D\Gamma(N)$. Moreover, Ricci identities and Bianchi identities can be written in the usual manner.

Therefore, by means of §.9, Ch. 1, the connection 1-forms $\omega^i{}_j$ of the canonical metrical N–connection $C\Gamma(N)$ are

$$(7.6)\qquad \omega^i{}_j = L^i{}_{jk}\, dx^k + C^i{}_{jk}\, \delta y^k,$$

where $L^i{}_{jk}, C^i{}_{jk}$ are given in (7.5).

Theorem 3.7.3. *The connection* 1*-forms* $\omega^i{}_j$ *of the canonical metrical* N*–connection* $C\Gamma(N)$ *satisfy the following structure equations*

$$(7.7)\qquad \begin{aligned} d(dx^i) - dx^k \wedge \omega^i{}_k &= -\overset{(0)}{\Omega}{}^i \\ d(\delta y^i) - \delta y^k \wedge \omega^i{}_k &= -\overset{(1)}{\Omega}{}^i \end{aligned}$$

and

$$(7.8)\qquad d\omega^i{}_j - \omega^k{}_j \wedge \omega^i{}_k = -\Omega^i{}_j$$

where the 2*-forms of torsion* $\overset{(0)}{\Omega}{}^i, \overset{(1)}{\Omega}{}^i$ *are as follows*

$$(7.9)\qquad \begin{cases} \overset{(0)}{\Omega}{}^i = C^i{}_{jk}\, dx^j \wedge \delta y^k \\ \overset{(1)}{\Omega}{}^i = \dfrac{1}{2}\, R^i{}_{jk} dx^j \wedge dx^k + P^i{}_{jk} dx^j \wedge \delta y^k \end{cases}$$

and the 2-forms of curvature $\Omega^i{}_j$ *are*

$$\Omega^i{}_j = \frac{1}{2}\, R_j{}^i{}_{kh} dx^k \wedge dx^h + P_j{}^i{}_{kh} dx^k \wedge \delta y^h + \frac{1}{2} S_j{}^i{}_{kh} \delta y^k \wedge \delta y^h \tag{7.10}$$

the d–tensors of curvature $R_j{}^i{}_{kh}, P_j{}^i{}_{kh}, S_j{}^i{}_{kh}$ *and d-tensor of torsion* $R^i{}_{jk}, P^i{}_{jk}$ *having the known expression (see Ch.1, §5).*

We notice that starting from the canonical metrical connection $C\Gamma(N) = (L^i{}_{jk}, C^i{}_{jk})$, other remarkable connections like the connection of Berwald $B\Gamma(N)$, Chern–Rund $R\Gamma(N)$ and Hashiguchi $H\Gamma(N)$ have the coefficients

$$B\Gamma(N) = \left(\frac{\partial N^i{}_j}{\partial y^k}, 0\right), \quad R\Gamma(N) = (L^i{}_{jk}, 0), \quad H\Gamma(N) = \left(\frac{\partial N^i{}_j}{\partial y^k}, C^i{}_{jk}\right),$$

respectively. The following commutative diagram holds.

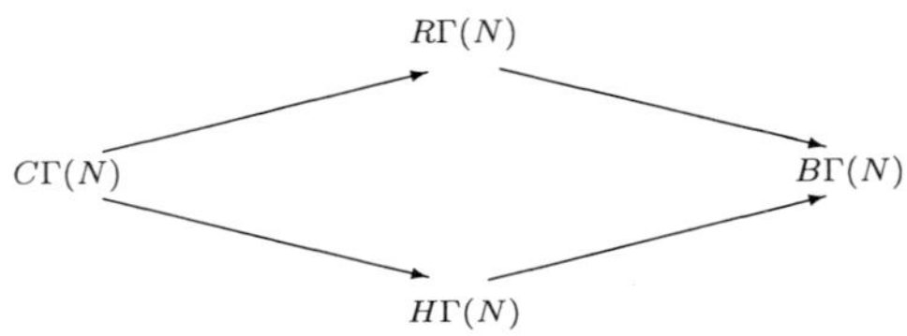

The corresponding transformations of connections from this diagram may be easily deduced from the Finslerian case.

Some properties of the canonical metrical connection $C\Gamma$ are given by

Proposition 3.7.1. *We have:*

1° $\sum_{(ijk)} R_{ijk} = 0$, $P_{ijk} = g_{is} P^s{}_{jk}$ *is totally symmetric.*

2° $C_{ijk} = \dfrac{1}{4}\, \dfrac{\partial^3 L}{\partial y^i \partial y^j \partial y^k} = g_{js} C^s{}_{ik}$.

3° *The covariant curvature d-tensors* $R_{ijkh}, P_{ijkh}, S_{ijkh}$ *(with* $R_{ijkh} = g_{js} R_i{}^s{}_{hk}$*, etc.) are skew-symmetric in the first two indices.*

4° $S_{ijkh} = C_{iks} C^s{}_{jh} - C_{ihs} C^s{}_{jk}$.

These properties can be proved using the property $d\theta = 0$, where θ is the symplectic structure (6.2), the Ricci identities applied to the fundamental tensor g_{ij} of the space and the equations $g_{ij|k} = 0$, $g_{ij}|_k = 0$.

By the same methods we can study the metrical N–linear connections $\overline{D}\Gamma(N) = (\overline{L}^i{}_{jk}, \overline{C}^i{}_{jk})$ which satisfy the axioms A_1, A_2, A_3 and have a priori given d-tensors of torsion $\overline{T}^i{}_{jk}$ and $\overline{S}^i{}_{jk}$.

Theorem 3.7.4.

1° *There exists only one N–linear* $\overline{D}\Gamma(N) = (\overline{L}^i{}_{jk}, \overline{C}^i{}_{jk})$ *connection which satisfies the following axioms:*

A'_1 *N is canonical nonlinear connection of the space* L^n.
A'_2 $g_{ij|k} = 0$ *($\overline{D}$ is h-metrical).*
A'_3 $g_{ij}|_k = 0$ *($\overline{D}$ is v-metrical).*
A'_4 *The h-tensor of torsion* $\overline{T}^i{}_{jk}$ *is apriori given.*
A'_5 *The v-tensor of torsion* $\overline{S}^i{}_{jk}$ *is apriori given.*

2° *The coefficients* $\overline{L}^i{}_{jk}$ *and* $\overline{C}^i{}_{jk}$ *of the previous connection are as follows*

$$(7.11) \qquad \begin{cases} \overline{L}^i{}_{jk} = L^i{}_{jk} + \dfrac{1}{2} g^{ih}\left(g_{jr}\overline{T}^r{}_{kh} + g_{kr}\overline{T}^r{}_{jh} - g_{hr}\overline{T}^r{}_{kj}\right) \\ \overline{C}^i{}_{jk} = C^i{}_{jk} + \dfrac{1}{2} g^{ih}\left(g_{jr}\overline{S}^r{}_{kh} + g_{kr}\overline{S}^r{}_{jh} - g_{hr}\overline{S}^r{}_{kj}\right) \end{cases}$$

$(L^i{}_{jk}, C^i{}_{jk})$ *being the coefficients of canonical metrical N–connection* $C\Gamma(N)$.

The proof is similar with that of Theorem 3.7.2.

From now on $\overline{T}^i{}_{jk}, \overline{S}^i{}_{jk}$ will be denoted simply by $T^i{}_{jk}, S^i{}_{jk}$.

Proposition 3.7.2. *The Ricci identities of the metrical N–connection* $D\Gamma(N)$ *are given by:*

$$(7.12) \qquad \begin{cases} X^i{}_{|j|k} - X^i{}_{|k|j} = X^m R_m{}^i{}_{jk} - X^i{}_{|m} T^m{}_{jk} - X^i|_m R^m{}_{jk}, \\ X^i{}_{|j}|_k - X^i|_{k|j} = X^m P_m{}^i{}_{jk} - X^i{}_{|m} C^m{}_{jk} - X^i|_m P^m{}_{jk}, \\ X^i|_j|_k - X^i|_k|_j = X^m S_m{}^i{}_{jk} - X^i|_m S^m{}_{jk}. \end{cases}$$

Indeed, we can apply Proposition 1.8.1.

Of course, the previous identities can be extended to a d-tensor field of type (r, s) on the ordinary way.

Denoting

$$(7.13) \qquad D^i{}_j = y^i{}_{|j}, \; d^i{}_j = y^i|_j$$

we have the h-deflection tensor field $D^i{}_j$ and v-deflection tensor field $d^i{}_j$.

The tensors $D^i{}_j$ and $d^i{}_j$ have the known expressions:

$$(7.13)' \qquad D^i{}_j = y^s L^i{}_{sj} - N^i{}_j; \; d^i{}_j = \delta^i{}_j + y^s C_s{}^i{}_j.$$

Applying the Ricci identities (7.12) to the Liouville vector field y^i we obtain:

Theorem 3.7.5. *If $D\Gamma(N)$ is a metrical N–linear connection then the following identities hold*

$$
\begin{aligned}
&D^i{}_{j|k} - D^i{}_{k|j} = y^m R_m{}^i{}_{jk} - D^i{}_m T^m{}_{jk} - d^i{}_m R^m{}_{jk},\\
(7.14)\qquad &D^i{}_{j}|_k - d^i{}_{k|j} = y^m P_m{}^i{}_{jk} - D^i{}_m C^m_{jk} - d^i{}_m P^m{}_{jk},\\
&d^i{}_{j}|_k - d^i{}_{k}|_j = y^m S_m{}^i{}_{jk} - d^i{}_m S^m{}_{jk}.
\end{aligned}
$$

We will apply the previous theory in next section, taking into account the canonical metrical N–connection $C\Gamma(N)$ and taking $T^i{}_{jk} = 0$, $S^i{}_{jk} = 0$. Of course, a theory of parallelism of vector field with respect to the connection $D\Gamma(N)$ can be done taking into account the considerations from §9, Ch.1.

3.8 Gravitational and electromagnetic fields

Let us consider a Lagrange space $L^n = (M, L(x,y))$ endowed with the canonical metrical N–connection $C\Gamma(N) = (L^i{}_{jk}, C^i{}_{jk})$.

The covariant deflection tensors D_{ij} and d_{ij} can be introduced by

$$
(8.1)\qquad D_{ij} = g_{is} D^s{}_j,\ \ d_{ij} = g_{is} d^s{}_j.
$$

Obviously we have

$$
D_{ij|k} = g_{is} D^s{}_{j|k},\ \ d_{ij|k} = g_{is} d^s{}_{j|k}
$$

and analogous for the v-covariant derivation. Then, Theorem 3.7.5 implies:

Theorem 3.8.1. *The covariant deflection tensor fields D_{ij} and d_{ij} of the canonical metrical N–connection $D\Gamma(N)$ satisfy the identities:*

$$
(8.2)\qquad
\begin{cases}
D_{ij|k} - D_{ik|j} = y^s R_{sijk} - d_{is} R^s{}_{jk},\\
D_{ij}|_k - d_{ik|j} = y^s P_{sijk} - D_{is} C^s{}_{jk} - d_{is} P^s{}_{jk},\\
d_{ij}|_k - d_{ik}|_j = y^s S_{sijk}.
\end{cases}
$$

Some considerations from the Lagrange theory of electrodynamics, lead us to introduce:

Definition 3.8.1. The tensor fields

$$
(8.3)\qquad F_{ij} = \frac{1}{2}(D_{ij} - D_{ji}),\ \ f_{ij} = \frac{1}{2}(d_{ij} - d_{ji})
$$

are called h- and v-*electromagnetic tensor field* of the Lagrange spaces L^n, respectively.

Proposition 3.8.1. *With respect to the canonical metrical N–connection $C\Gamma(N)$, the v-electromagnetic tensor f_{ij} vanishes.*

The Bianchi identities of $C\Gamma(N)$ and the identities (8.2) lead the following important result.

Theorem 3.8.2. *The following generalized Maxwell equations hold:*

$$
\begin{aligned}
&F_{ij|k} + F_{jk|i} + F_{ki|j} = -\sum_{(ijk)} C_{ios} R^s{}_{jk}, \\
&F_{ij}\big|_k + F_{jk}\big|_i + F_{ki}\big|_j = 0,
\end{aligned}
\tag{8.4}
$$

where $C_{ioj} = y^s C_{isj}$.

Remarks.

1° If Lagrange space L^n is a Finsler space F^n then $C_{ios} = 0$ and equations (8.4) simplifies.

2° If the canonical nonlinear connection N is flat, i.e. the distribution N is integrable, $R^h{}_{ij} = 0$ and the previous equations have a simple form.

If we put

$$F^{ij} = g^{is} g^{jr} F_{sr} \tag{8.5}$$

and

$$hJ^i = F^{ij}{}_{|j}, \; vJ^i = F^{ij}\big|_j, \tag{8.6}$$

then we can prove:

Theorem 3.8.3. *The following laws of conservation hold*

$$
\begin{cases}
hJ^i{}_{|i} = \dfrac{1}{2}\{F^{ij}(R_{ij} - R_{ji}) + F^{ij}\big|_r R^r{}_{ij}\}, \\
vJ^i\big|_i = 0,
\end{cases}
\tag{8.7}
$$

where R_{ij} is the Ricci tensor of the curvature tensor $R_i{}^h{}_{jk}$.

Remark. The electromagnetic tensor field F_{ij}, f_{ij} and the Maxwell equations were introduced by R. Miron and M. Radivoiovici, [113]. Important contributions have M. Anastasiei, K. Buchner, R. Roşca (see Ref. from the book [113]).

The curvature d-tensors of the connection $C\Gamma(N)$, $R_i{}^h{}_{jk}, P_i{}^h{}_{jk}, S_i{}^h{}_{jk}$ have the following Ricci and scalar curvatures

$$
\begin{cases}
R_{ij} = R_i{}^h{}_{jh}, \; S_{ij} = S_i{}^h{}_{jh}, \; {}'P_{ij} = P_i{}^h{}_{jh}, \; {}''P_{ij} = P_i{}^h{}_{hj}, \\
R = g^{ij} R_{ij}, \; S = g^{ij} S_{ij}.
\end{cases}
\tag{8.8}
$$

Let us denote by $\overset{H}{T}_{ij}, \overset{V}{T}_{ij}, \overset{1}{T}_{ij}, \overset{2}{T}_{ij}$ the components in the adapted basis $\left(\dfrac{\delta}{\delta x^i}, \dfrac{\partial}{\partial y^i}\right)$ of the energy momentum tensor.

Using the almost Kählerian model $K^{2n} = (\widetilde{TM}, \mathbb{G}, \mathbb{F})$ of the Lagrange space and taking into account the canonical metrical connection $C\Gamma(N)$ (see Theorem 3.7.2) we obtain[113]:

Theorem 3.8.4.

1° *The Einstein equations of the almost Kählerian space K^{2n}, $(n > 2)$, endowed with the canonical metrical connection $C\Gamma(N)$ are the following*

$$(8.9)\qquad \begin{aligned} R_{ij} - \frac{1}{2}\, R g_{ij} &= \kappa \overset{H}{T}_{ij}, \quad {}'P_{ij} = \kappa \overset{1}{T}_{ij}, \\ S_{ij} - \frac{1}{2}\, S g_{ij} &= \kappa \overset{V}{T}_{ij}, \quad {}''P_{ij} = -\kappa \overset{2}{T}_{ij}, \end{aligned}$$

where k is a real constant.

2° *The energy momentum tensors $\overset{H}{T}_{ij}$ and $\overset{V}{T}_{ij}$ satisfy the following laws of conservation*

$$(8.9)' \qquad \kappa \overset{H}{T}{}^i{}_{j|i} = -\frac{1}{2}\left(P^{ih}{}_{js} R^s{}_{hi} + 2R^s{}_{ij} P^i{}_s\right), \kappa \overset{V}{T}{}^i{}_{j}\big|_i = 0$$

The physical background of the previous theory was discussed by S. Ikeda in the last chapter of the book [112]. All this theory is very simple if the Lagrange spaces L^n have the property $P_j{}^i{}_{kh} = 0$.

Corollary 3.8.1. *If the canonical metrical connection $C\Gamma(N)$ has the property $P_i{}^h{}_{jk} = 0$ then we have*

1° *For $n > 2$, the Einstein equations of the Lagrange space L^n have the form:*

$$(8.10)\qquad \begin{cases} R_{ij} - \dfrac{1}{2}\, R g_{ij} = \kappa \overset{H}{T}_{ij} \\ S_{ij} - \dfrac{1}{2}\, S g_{ij} = \kappa \overset{V}{T}_{ij}. \end{cases}$$

2° *The following laws of conservation hold:* $\overset{H}{T}{}^i{}_{j|i} = 0$, $\overset{V}{T}{}^i{}_{j}\big|_i = 0$.

In the next section, we apply this theory for the Lagrangian of Electrodynamics.

3.9 The Lagrange space of electrodynamics

The Lagrange spaces $L_0^n = (M, L_0(x,y))$, with the Lagrangian from Electrodynamics

$$L_0(x,y) = mc\gamma_{ij}(x)y^iy^j + \frac{2e}{m} A_i(x)y^i \tag{9.1}$$

which is obtained from the Lagrangian $L(x,y)$, (1.3) for $U = 0$ is a very good example in our theory. It was studied in the book [113]. We emphasized here only the main results. The space L_0^n with the fundamental function (9.1) is called the Lagrange space of electrodynamics.

The fundamental tensor of the space L_0^n is

$$g_{ij}(x,y) = mc\gamma_{ij}(x) \tag{9.2}$$

and its contravariant is $g^{ij}(x,y) = \dfrac{1}{mc}\gamma^{ij}(x)$.

The canonical spray of the space L_0^n is given by the differential equations

$$\frac{d^2x^i}{dt^2} + 2G^i\left(x, \frac{dx}{dt}\right) = 0, \tag{9.3}$$

where

$$G^i(x,y) = \frac{1}{2}\gamma^i{}_{rs}(x)y^ry^s + \frac{e}{mc}\gamma^{ij} \overset{0}{F}_{jk}(x)y^k \tag{9.3$'$}$$

and

$$\overset{0}{F}_{jk}(x) = \frac{1}{2m}(\partial_j A_k - \partial_k A_j). \tag{9.3$''$}$$

Let us denote

$$\overset{0}{F}{}^i{}_j(x) = g^{is}(x) \overset{0}{F}_{sj}(x). \tag{9.3$'''$}$$

The canonical nonlinear connection N has the coefficients

$$N^i{}_j = \gamma^i{}_{jk}(x)y^k - \overset{0}{F}{}^i{}_j. \tag{9.4}$$

As we remarked already we have:

Theorem 3.9.1. *The autoparallel curve of the canonical nonlinear connection* (9.4) *are the solutions of the Lorentz equations*

$$\frac{d^2x^i}{dt^2} + \gamma^i{}_{jk}(x)\frac{dx^j}{dt}\frac{dx^k}{dt} = \overset{0}{F}{}^i{}_j(x)\frac{dx^j}{dt}.$$

The canonical metrical connection $C\Gamma(N)$ has the coefficients

$$L^i{}_{jk}(x,y) = \gamma^i{}_{jk}(x), \;\; C^i{}_{jk} = 0.$$

It follows that the curvature d-tensors are

$$R_j{}^i{}_{kh}(x,y) = r_j{}^i{}_{kh}(x), \;\; P_j{}^i{}_{kh} = 0, \;\; S_j{}^i{}_{kh} = 0,$$

where $r_j{}^i{}_{kh}(x)$ is the curvature tensor of the Levi–Civita connection $\gamma^i{}_{jk}(x)$.

The covariant deflection tensors are

$$D_{ij}(x,y) = \overset{0}{F}_{ij}(x), \;\; d_{ij}(x,y) = g_{ij}(x).$$

Proposition 3.9.1. *The h- and v-electromagnetic tensor field of the Lagrange space of electrodynamics are*

$$F_{ij} = \overset{0}{F}_{ij}, \;\; f_{ij} = 0.$$

Consequently, the v-covariant derivative of the tensor F_{ij} vanishes. We obtain

Theorem 3.9.2. *The generalized Maxwell equations* (8.4) *of the Lagrange space of electrodynamics reduce to the classical ones.*

Since $P_j{}^i{}_{kh} = 0$ implies $'P_{ij} ='' P_{ij} = 0$ and $C^i{}_{jk} = 0$ implies $S_{ij} = 0$, $S = 0$, we get from Corollary 3.8.1.

Theorem 3.9.3. *The Einstein equations* (8.10) *in L^n_0 reduce to the classical Einstein equations associated to the Levi–Civita connection.*

3.10 Generalized Lagrange spaces

The generalized Lagrange spaces were introduced by the first author, [95], and then was studied by many collaborators (see [113]). The applications in the general Relativity or in the relativistic optics was treated too [115].

A detailed study of these spaces one finds in the books [113].

In this section we give a short introduction in the geometrical theory of generalized Lagrange spaces, since they will be considered in the geometry of Hamilton spaces.

Definition 3.10.1. A generalized Lagrange space is a pair $GL^n = (M, g_{ij}(x,y))$, where $g_{ij}(x,y)$ is a d-tensor field on the manifold $\widetilde{TM}$, covariant, symmetric, of rank n and of constant signature on $\widetilde{TM}$.

g_{ij} is called fundamental tensor or metric tensor of the space GL^n.

Evidently, any Lagrange space $L^n = (M, L(x,y))$ is a generalized Lagrange space, whose fundamental tensor field is $g_{ij} = \frac{1}{2}\frac{\partial^2 L}{\partial y^i \partial y^j}$. But not any generalized Lagrange space $GL^n = (M, g_{ij}(x,y))$ is a Lagrange space.

Definition 3.10.2. We say that GL^n with fundamental tensor field $g_{ij}(x,y)$ is *reducible* to a Lagrange space if the system of partial differential equations

$$\frac{1}{2}\frac{\partial^2 L}{\partial y^i \partial y^j} = g_{ij}(x,y) \tag{10.1}$$

has solutions with respect to $L(x,y)$.

In order to a space GL^n be reducible to a Lagrange space is necessary that the d-tensor field $\frac{\partial g_{ij}}{\partial y^k}$ be totally symmetric. So we have

Proposition 3.10.1. *A generalized Lagrange space $GL^n = (M, g_{ij}(x,y))$ for which the tensor field $\frac{\partial g_{ij}}{\partial y^k}$ is not totally symmetric is not reducible to a Lagrange space.*

Example 1. The space $GL^n = (M, g_{ij}(x,y))$ with

$$g_{ij}(x,y) = e^{2\sigma(x,y)}\gamma_{ij}(x) \tag{10.2}$$

with nonvanishing d-covector field $\frac{\partial \sigma}{\partial y^i}$ and $\gamma_{ij}(x)$ a Riemannian metric is not reducible to a Lagrange space.

This example is strongly related to the axioms of Ehlers–Pirani–Schield and theory Tavakol–Miron, from General Relativity [121]. This example was also studied by Watanabe S., Ikeda, S. and Ikeda, F. [171].

Example 2. The space $GL^n = (M, g_{ij}(x,y))$ with the fundamental tensor field

$$g_{ij}(x,y) = \gamma_{ij}(x) + \left(1 - \frac{1}{n^2(x,y)}\right) y_i y_j, \tag{10.3}$$

where $\gamma_{ij}(x)$ is a Riemann or Lorentz metric tensor, $y_i = \gamma_{ij}(x)y^j$ and $n(x,y) > 1$ is a refractive index, is not reducible to a Lagrange space.

This metric was introduced in Relativistic Optics by J.L. Synge [156]. It was intensively studied by R. Miron and his collaborators [113], [115].

These two examples show that the nonlinear connection with the coefficients

$$N^i{}_j = \gamma^i{}_{jk}(x)y^k$$

$\gamma^i{}_{jk}(x)$ being Christoffel symbols of the metric tensor $\gamma_{ij}(x)$, can be associated to the fundamental tensor g_{ij} of the space GL^n. But, generally, we cannot derive a nonlinear connection N from the fundamental tensor g_{ij}.

So, we adopt the following postulate:

The generalized Lagrange space $GL^n = (M, g_{ij}(x,y))$ is endowed with an a priori given nonlinear connection N.

In this situation, we can develop the geometry of the pair (GL^n, N) by using the same methods as in the case of Lagrange space L^n.

For instance, considering the adapted basis $\left(\dfrac{\delta}{\delta x^i}, \dfrac{\partial}{\partial y^i}\right)$ to N and V, respectively, we can prove:

Theorem 3.10.1.

1° *For a generalized Lagrange space endowed with a nonlinear connection N, there exists a unique N–linear connection $D\Gamma(N) = (L^i{}_{jk}, C^i{}_{jk})$ satisfying the following axioms:*

$$A_1''\ \ g_{ij|k} = 0; \quad A_2''\ \ g_{ij}|_k = 0; \quad A_3''\ \ T^i{}_{jk} = 0; \quad A_4''\ \ S^i{}_{jk} = 0.$$

2° *$D\Gamma(N)$ has the coefficients given by the generalized Christoffel symbols:*

$$\begin{aligned} L^i{}_{jk} &= \frac{1}{2} g^{is}\left(\frac{\delta g_{sk}}{\delta x^j} + \frac{\delta g_{js}}{\delta x^k} - \frac{\delta g_{jk}}{\delta x^s}\right), \\ C^i{}_{jk} &= \frac{1}{2} g^{is}\left(\frac{\partial g_{sk}}{\partial y^j} + \frac{\partial g_{js}}{\partial y^k} - \frac{\partial g_{jk}}{\partial y^s}\right). \end{aligned} \tag{10.4}$$

For more details we send the reader to the books [113].

Let us end this chapter with the following important remark. The class of Riemannian spaces R^n is a subclass of the Finsler spaces F^n and this is a subclass of the class of the Lagrange spaces L^n. Moreover, the class of Lagrange spaces L^n is a subclass of the generalized Lagrange spaces GL^n. Hence, we have the following inclusions:

$$\{R^n\} \subset \{F^n\} \subset \{L^n\} \subset \{GL^n\}.$$

This sequence of inclusions is important in applications to the geometric models in Mechanics, Physics, Biology, etc., [18].

Chapter 4

The geometry of cotangent bundle

The geometrical theory of cotangent bundle (T^*M, π^*, M) of a real, finite dimensional manifold M is important in the differential geometry. Correlated with that of tangent bundle (TM, π, M) we get a framework for construction of geometrical models for Lagrangian and Hamiltonian Mechanics, as well as, for the duality between them – via Legendre transformation.

The total space T^*M can be studied by the same methods as the total space of tangent bundle TM. But there exist some specific geometric objects on T^*M. For instance the Liouville–Hamilton vector $\mathbb{C}^*$, Liouville 1-form ω, the canonical symplectic structure and the canonical Poisson structure. These properties are fundamental for introducing the notions of Hamilton space or Cartan space.

In this chapter we study some fundamental object fields on T^*M: nonlinear connections, N-linear connections, structure equations and their properties. We preserve the convention that all geometrical objects on T^*M or mappings defined on T^*M are of C^∞–class.

4.1 The bundle (T^*M, π^*, M)

Let M be a real n-dimensional differentiable manifold and let (T^*M, π^*, M) be its cotangent bundle [175]. If (x^i) is a local coordinate system on a domain U of a chart on M, the induced system of coordinates on $\pi^{*-1}(U)$ are (x^i, p_i), $(i, j, k, ... = 1, ..., n)$. The coordinates $p_1, ..., p_n$ are called "momentum variables".

A change of coordinate on T^*M is given by

$$
(1.1)\qquad \begin{cases} \widetilde{x}^i = \widetilde{x}^i(x^1, ..., x^n), \text{rank}\left(\dfrac{\partial \widetilde{x}^i}{\partial x^j}\right) = n, \\ \widetilde{p}_i = \dfrac{\partial x^j}{\partial \widetilde{x}^i} p_j \end{cases} .
$$

Therefore the natural frame $\left(\dfrac{\partial}{\partial x^i}, \dfrac{\partial}{\partial p_i}\right)$ are transformed by (1.1) in the form

$$
(1.2)\qquad \begin{cases} \dfrac{\partial}{\partial x^i} = \dfrac{\partial \widetilde{x}^j}{\partial x^i} \dfrac{\partial}{\partial \widetilde{x}^j} + \dfrac{\partial \widetilde{p}_j}{\partial x^i} \dfrac{\partial}{\partial \widetilde{p}_j} \\ \dfrac{\partial}{\partial p_i} = \dfrac{\partial \widetilde{x}^j}{\partial x^i} \dfrac{\partial}{\partial \widetilde{p}_j} \end{cases}
$$

Looking at the second formula (1.2) the following notation can be adopted

$$
(1.3)\qquad \dot{\partial}^i = \frac{\partial}{\partial p_i}.
$$

Indeed (1.2) gives us:

$$
(1.3)'\qquad \dot{\widetilde{\partial}}^i = \frac{\partial \widetilde{x}^i}{\partial x^j} \dot{\partial}^j.
$$

The natural coframe (dx^i, dp_i) is changed by (1.1) by the rule:

$$
(1.2)'\qquad d\widetilde{x}^i = \frac{\partial \widetilde{x}^i}{\partial x^j} dx^j,\ d\widetilde{p}_i = \frac{\partial x^j}{\partial \widetilde{x}^i} dp_j + \frac{\partial^2 x^j}{\partial \widetilde{x}^i \partial \widetilde{x}^r} p_j d\widetilde{x}^r
$$

The Jacobian matrix of change of coordinate (1.1) is

$$
J(u) = \begin{pmatrix} \dfrac{\partial \widetilde{x}^j}{\partial x^i} & 0 \\ \dfrac{\partial \widetilde{p}_j}{\partial x^i} & \dfrac{\partial x^j}{\partial \widetilde{x}^i} \end{pmatrix}_u
$$

It follows

$$
\det J(u) = 1
$$

We get:

Theorem 4.1.1. *The manifold T^*M is orientable.*

Like in the case of tangent bundle, we can prove:

Theorem 4.1.2. *If the base manifold M is paracompact, then the manifold T^*M is paracompact, too.*

The kernel of the differential $d\pi^* : TT^*M \longrightarrow TM$ of the natural projection $\pi^* : T^*M \longrightarrow M$ is the vertical subbundle VT^*M of the tangent bundle TT^*M. Associating to each point $u \in T^*M$ the fibre V_u of VT^*M we obtain the *vertical distribution*:

$$V : u \in T^*M \longrightarrow V_u \subset T_uT^*M$$

This distribution is locally generated by the tangent vector field $(\dot{\partial}^1, ..., \dot{\partial}^n)$. So it is an integrable distribution, of local dimension n.

Noticing the formulae (1.2) and (1.2)′ we can introduce the following geometrical object fields:

$$\mathbb{C}^* = p_i\dot{\partial}^i \tag{1.4}$$

$$\omega = p_i dx^i \tag{1.5}$$

$$\theta = d\omega = dp_i \wedge dx^i \tag{1.6}$$

Theorem 4.1.3. *The following properties hold:*

1° $\mathbb{C}^*$ *is a vertical vector field globally defined on T^*M.*

2° *The forms ω and θ are globally defined on T^*M.*

3° *θ is a symplectic structure on T^*M.*

Proof. 1°. By means of (1.1) and (1.3)′ it follows that $\mathbb{C}^*$ belongs to the distribution V and it has the property $\mathbb{C}^* = p_i\dot{\partial}^i = \widetilde{p}_i\dot{\widetilde{\partial}}^i$.

2°. ω and θ do not depend on the changes of coordinates on T^*M.

3°. θ is a closed 2-form on T^*M and $\text{rank}||\theta|| = 2n = \dim T^*M$.

$\mathbb{C}^*$ will be called the *Liouville–Hamilton vector field* on T^*M, ω is called the *Liouville 1-form* and θ is the *canonical symplectic structure* on T^*M.

The pair (T^*M, θ) is a symplectic manifold.

4.2 The Poisson brackets. The Hamiltonian systems

Let us consider the Poisson bracket $\{ , \}$ on T^*M, defined by

$$\{f, g\} = \frac{\partial f}{\partial p_i}\frac{\partial g}{\partial x^i} - \frac{\partial g}{\partial p_i}\frac{\partial f}{\partial x^i}, \forall f, g \in \mathcal{F}(T^*M). \tag{2.1}$$

Proposition 4.2.1. *We have*

$$\{f, g\} \in \mathcal{F}(T^*M), \ \forall\, f, g \in \mathcal{F}(T^*M).$$

Proof. By means of (1.1), (1.2), $\widetilde{f}(\widetilde{x}, \widetilde{p}) = f(x, p)$ and

$$\frac{\partial f}{\partial x^i} = \frac{\partial \widetilde{x}^j}{\partial x^i}\frac{\partial \widetilde{f}}{\partial \widetilde{x}^j} + \frac{\partial \widetilde{p}_j}{\partial x^i}\frac{\partial \widetilde{f}}{\partial \widetilde{p}_j};\ \frac{\partial f}{\partial p_j} = \frac{\partial x^j}{\partial \widetilde{x}^i}\frac{\partial \widetilde{f}}{\partial \widetilde{p}_i}.$$

Therefore we deduce

$$\frac{\partial f}{\partial p_i}\frac{\partial g}{\partial x^i} = \frac{\partial \widetilde{f}}{\partial \widetilde{p}_j}\frac{\partial x^i}{\partial \widetilde{x}^j}\left(\frac{\partial \widetilde{x}^m}{\partial x^i}\frac{\partial \widetilde{g}}{\partial \widetilde{x}^m} + \frac{\partial \widetilde{p}_m}{\partial x^i}\frac{\partial \widetilde{g}}{\partial \widetilde{p}_m}\right) =$$

$$= \frac{\partial \widetilde{f}}{\partial \widetilde{p}_j}\frac{\partial \widetilde{g}}{\partial \widetilde{x}^j} + \frac{\partial \widetilde{f}}{\partial \widetilde{p}_j}\frac{\partial \widetilde{g}}{\partial \widetilde{p}_m}\frac{\partial \widetilde{p}_m}{\partial \widetilde{x}^j} = \frac{\partial \widetilde{f}}{\partial \widetilde{p}_j}\frac{\partial \widetilde{g}}{\partial \widetilde{x}^j},$$

because $\dfrac{\partial \widetilde{p}_j}{\partial \widetilde{x}^i} = 0$. Consequently, $\{f, g\} = \widetilde{\{\widetilde{f}, \widetilde{g}\}}$. **q.e.d.**

Theorem 4.2.1. *The Poisson bracket* $\{\,,\,\}$ *has the properties*

1° $\{f, g\} = -\{g, f\}$

2° $\{f, g\}$ *is R-linear in every argument*

3° $\{\{f, g\}, h\} + \{\{g, h\}, f\} + \{\{h, f\}, g\} = 0$

4° $\{\cdot, gh\} = \{\cdot, g\}h + \{\cdot, h\}g.$

By a straightforward calculus, using (2.1), 1°–4° can be proved. Therefore $\{\,,\,\}$ is called the *canonical Poisson structure* on T^*M. The pair $(\mathcal{F}(T^*M), \{\ ,\ \})$ is a Lie algebra, called *Poisson–Lie algebra.*

The relation between the structures θ and $\{\,,\,\}$ can be given by means of the notion of Hamiltonian system.

Definition 4.2.1. A differentiable Hamiltonian is a function $H{:}T^*M \to R$ which is of class C^∞ on $\widetilde{T^*M} = T^*M\backslash\{0\}$ and continuous on the zero section of the projection $\pi^* : T^*M \to M$.

Definition 4.2.2. A Hamiltonian system is a triple (T^*M, θ, H) formed by the manifold T^*M, canonical symplectic structure θ and a differentiable Hamiltonian H.

Let us consider the $\mathcal{F}(T^*M)$-modules: $\mathcal{X}(T^*M)$ and $\mathcal{X}^*(T^*M)$ – of tangent vectors and covectors fields on T^*M, respectively.

Thus, the following $\mathcal{F}(T^*M)$-linear mapping:

$$S_\theta : \mathcal{X}(T^*M) \longrightarrow \mathcal{X}^*(T^*M)$$

can be defined by:

$$S_\theta(X) = i_X\theta, \quad \forall X \in \mathcal{X}(T^*M). \tag{2.2}$$

Proposition 4.2.2. *S_θ is an isomorphism.*

Indeed, S_θ is a $\mathcal{F}(T^*M)$-linear mapping and bijective, because

$$\text{rank}||\theta|| = 2n.$$

But we can remark that the local base of $\mathcal{X}(T^*M)$, $\left(\frac{\partial}{\partial x^i}, \frac{\partial}{\partial p_i}\right)$, is sent by S_θ in the local base (dx^i, dp_i) of $\mathcal{X}^*(T^*M)$ by the rule

$$S_\theta\left(\frac{\partial}{\partial x^i}\right) = -dp_i, \quad S_\theta\left(\frac{\partial}{\partial p_i}\right) = dx^i. \tag{2.3}$$

We get also

$$S_\theta(\mathbb{C}^*) = \omega. \tag{2.3$'$}$$

We can apply this property to prove:

Theorem 4.2.2. *The following properties of the Hamiltonian system (T^*M, θ, H) hold:*

1° *There exists a unique vector field $X_H \in \mathcal{X}(T^*M)$ having the property*

$$i_{X_H}\theta = -dH \tag{2.4}$$

2° *The integral curves of the vector field X_H are given by the Hamilton–Jacobi equations:*

$$\frac{dx^i}{dt} = \frac{\partial H}{\partial p_i}, \; \frac{dp_i}{dt} = -\frac{\partial H}{\partial x^i}. \tag{2.5}$$

Proof. 1°. The existence and uniqueness of the vector field X_H is assured by Proposition 4.2.2. It is given by

$$X_H = S_\theta^{-1}(-dH) \tag{2.6}$$

Using (2.3), X_H is expressed in the natural basis by

$$X_H = \frac{\partial H}{\partial p_i}\frac{\partial}{\partial x^i} - \frac{\partial H}{\partial x^i}\frac{\partial}{\partial p_i} \tag{2.6$'$}$$

2°. The integral curves of X_H from (2.6)′ are given by the equations (2.5).**q.e.d.**

X_H is called the Hamilton vector field.

Corollary 4.2.1. *The function $H(x,p)$ is constant along the integral curves of the Hamilton vector field X_H.*

Indeed, $\dfrac{dH}{dt} = \{H, H\} = 0.$

The structures θ and $\{\ ,\ \}$ have a fundamental property given by the theorem:

Theorem 4.2.3. *The following formula holds:*

$$\{f,g\} = \theta(X_f, X_g),\ \forall\,(f,g) \in \mathcal{F}(T^*M),\ \forall\,X \in \mathcal{X}(T^*M). \tag{2.7}$$

Proof. From (2.6)′ we deduce

$$\{f,g\} = X_f g = -X_g f = -df(X_g) = (i_{X_f}\theta)(X_g) = \theta(X_f, X_g).$$

q.e.d.

Corollary 4.2.2. *The Hamilton–Jacobi equations can be written in the form*

$$\frac{dx^i}{dt} = \{H, x^i\},\ \ \frac{dp_i}{dt} = \{H, p_i\}. \tag{2.8}$$

One knows, [167], the Jacobi method of integration of Hamilton–Jacobi equations (2.5). Namely, we look for a solution curve $\gamma(t)$ in T^*M, of the form

$$x^i = x^i(t),\ p_i = \frac{\partial S}{\partial x^i}(x(t)) \tag{2.9}$$

where $S \in \mathcal{F}(M)$.

Substituting in (2.5), we have

$$\frac{dx^i}{dt} = \frac{\partial H}{\partial p_i}(x(t))\frac{\partial S}{\partial x}(x(t));\ \frac{dp_i}{dt} = \frac{\partial^2 S}{\partial x^i \partial x^j}\frac{\partial H}{\partial p_j} = -\frac{\partial H}{\partial x^i}. \tag{$*$}$$

It follows
$$dH\left(x,\frac{\partial S}{\partial x}\right)=\left(\frac{\partial H}{\partial x^i}\frac{\partial H}{\partial p_i}-\frac{\partial H}{\partial p_i}\frac{\partial H}{\partial x^i}\right)dt=0.$$

Consequently, $H\left(x,\dfrac{\partial S}{\partial x}\right)=const.$, which is called the Hamilton–Jacobi equation (of Mechanics). If integrated, it defines S and $\gamma(t)$ is obtained by the integration of the first equation $(*)$.

Remarks.

1° In the next chapter we will define the notion of Hamilton space $H^n=(M,H(x,p))$, H being a differentiable regular Hamiltonian and we shall see that the Legendre mapping between the Hamilton space H^n and a Lagrange spce $L^n=(M,L(x,y))$ sent the Euler–Lagrange equations into the Hamilton–Jacobi equations. This idea is basic for considering the notion of $\mathcal{L}$-duality [66], [67], [105].

2° The Poisson bracket $\{\,,\,\}$ is also basic for quantization. The quantization of a Mechanical system is a process which associates operators on some Hilbert space with the real functions on the manifold T^*M (phase space), such as the commutator of two such operators is associated with Poisson bracket of functions (Abraham and Marsden [3], [167]).

4.3 Homogeneity

The notion of homogeneity, with respect to the momentum variables p_i, of a function $f\in\mathcal{F}(T^*M)$ can be studied by the same way as the homogeneity of functions defined on the manifold TM (see Ch.1, §2).

Let H_p be the group of homotheties on the fibres of T^*M:

$$H_p=\{h_a:(x,p)\in T^*M\longrightarrow(x,ap)\in T^*M\mid a\in R^+\}$$

The orbit of a point $u_0=(x_0,p_0)$ by H_p is given by

$$x^i=x^i_0,\ p_i=ap^0_i,\ \forall\, a\in R^+$$

The tangent vector at the point $u_0=h_1(u)$ is the Liouville–Hamilton vector field $\mathbb{C}^*(u_0)$.

A function $f\in\mathcal{F}(T^*(M)$, differentiable on $\widetilde{T^*M}$ and continuous on the zero section is called homogeneous of degree r, $(r\in\mathbb{Z})$, with respect to the variables p_i (or on the fibres of T^*M), if

$$f\circ h_a=a^rf,\ \forall\, a\in R^+ \tag{3.1}$$

This is

$$(3.1)' \qquad f(x, ap) = a^r f(x,p),\ \forall\, a \in R^+$$

Exactly as in the case of the homogeneity of the function $f : TM \to R$, one can prove:

Theorem 4.3.1. *A function $f : T^*M \to R$, differentiable on $\widetilde{T^*M}$ and continuous on the zero section of $\pi^* : T^*M \to M$ is r-homogeneous with respect to p_i if and only if we have*

$$(3.2) \qquad \mathcal{L}_{\mathbb{C}^*} f = rf,$$

where $\mathcal{L}_{\mathbb{C}^*}$ is the Lie derivation with respect to the Liouville–Hamilton vector field $\mathbb{C}^*$.

But (3.2) can be written in the form

$$(3.2)' \qquad p_i \frac{\partial f}{\partial p_i} = rf.$$

A vector field $X \in \mathcal{X}(T^*M)$ is r-homogeneous with respect to p_i if

$$X \circ h_a = a^{r-1} h_a^* \circ X,\ \forall\, a \in R^+$$

It follows :

Theorem 4.3.2. *A vector field $X \in \mathcal{X}(T^*M)$ is r-homogeneous if and only if we have:*

$$(3.3) \qquad \mathcal{L}_{\mathbb{C}^*} X = (r-1)X.$$

Evidently, $\mathcal{L}_{\mathbb{C}^*} X = [\mathbb{C}^*, X]$.

Consequently:

1° $\dfrac{\partial}{\partial x^i}, \dfrac{\partial}{\partial p_i}$ are 1– and 0–homogeneous with respect to p_i.

2° If $f \in \mathcal{F}(T^*M)$ is s-homogeneous and $X \in \mathcal{X}(T^*M)$ is r-homogeneous then fX is $s+r$ homogeneous.

3° A vector field X given in the natural frame by

$$(3.4) \qquad X = \overset{(0)}{X}{}^i \frac{\partial}{\partial x^i} + \overset{(1)}{X}_i \frac{\partial}{\partial p_i}$$

is r-homogeneous with respect to p_i, if and only if $\overset{(0)}{X}{}^i$ are $r-1$ homogeneous and $\overset{(1)}{X}_i$ are r-homogeneous with respect to p_i.

4° $\mathbb{C}^* = p_i \dfrac{\partial}{\partial p_i}$ is 1-homogeneous.

5° If X is r-homogeneous vector field on T^*M and f is a function, s-homogeneous, then Xf is $r+s-1$-homogeneous function.

6° $\dfrac{\partial f}{\partial p_i}$ are $s-1$ homogeneous and $\dfrac{\partial^2 f}{\partial p_i \partial p_j}$ are $s-2$ homogeneous, if f is s-homogeneous.

A q-form $\omega \in \Lambda^q(T^*M)$ is called s-homogeneous with respect to p_i (or on the fibres of T^*M) if:

$$\omega \circ h_a^* = a^r \omega, \quad \forall a \in R^+. \tag{3.5}$$

All properties given in Ch. 1, §2, for q-forms defined on the tangent manifold TM, concerning their homogeneity are valid in the case of q-forms from $\Lambda^q(T^*M)$.
We get:

Theorem 4.3.3. *A q-form ω on T^*M is s-homogeneous on the fibres of T^*M if and only if*

$$\mathcal{L}_{\mathbb{C}^*}\omega = s\omega. \tag{3.6}$$

Consequences.

1° $\omega \in \Lambda^q(T^*M)$, $\omega' \in \Lambda^{q'}(T^*M)$, s- respectively s'-homogeneous, imply $\omega \wedge \omega'$, $s+s'$-homogeneous.

2° $\omega \in \Lambda^q(T^*M)$, s-homogeneous and $\underset{(1)}{X}, ..., \underset{(q)}{X}$ r-homogeneous vector fields determine the function $\omega(\underset{(1)}{X}, ..., \underset{(q)}{X})$ $r+s-1$ homogeneous.

3° dx^i, dp_i are 0- respectively 1-homogeneous.

4° The Liouville 1-form ω is 1-homogeneous.

5° The canonical symplectic structure θ is 1-homogeneous.

The previous consideratiosn will be applied especially, in the study of Cartan spaces.

4.4 Nonlinear connections

On the manifold T^*M there exists a remarkable distribution $V : u \in T^*M \longrightarrow V_u \subset T_uT^*M$. As we know V is integrable, having the adapted basis $(\dot{\partial}^1, ..., \dot{\partial}^n)$ and it is of local dimension n.

Definition 4.4.1. A nonlinear connection on T^*M is a differentiable distribution $N : u \in T^*M \longrightarrow N_u \subset T_uT^*M$ which is supplementary to the vertical distribution V, i.e.:

$$T_uT^*M = N_u \oplus V_u,\ \forall\, u \in T^*M \tag{4.1}$$

Consequently, the local dimension of the distribution N is $n = \dim M$.

N will be called also the horizontal distribution.

If N is given, there are uniquely determined a system of functions $N_{ij}(x,p)$ in every domain of a local chart $\pi^{*-1}(U)$, such that the adapted basis to the distribution N has the form

$$\frac{\delta}{\delta x^i} := \frac{\partial}{\partial x^i} + N_{ij}\frac{\partial}{\partial p_j} \quad (i = 1, 2, ..., n). \tag{4.2}$$

The functions $N_{ij}(x,p)$ are called the coefficients of the nonlinear connection N.

Theorem 4.4.1. *A change of coordinate* (1.2) *on T^*M transforms the coefficients $N_{ij}(x,y)$ of a nonlinear connection N by the rule:*

$$\widetilde{N}_{ij}(\widetilde{x},\widetilde{p}) = \frac{\partial x^s}{\partial \widetilde{x}^i}\frac{\partial x^r}{\partial \widetilde{x}^j}N_{sr} + p_r\frac{\partial^2 x^r}{\partial \widetilde{x}^i \partial \widetilde{x}^j} \tag{4.3}$$

Conversely, a system of functions $N_{ij}(x,y)$ defined on each domain of local chart from T^*M, which verifies (4.3) *with respect to* (1.1)*, determines a nonlinear connection.*

Proof. By means of (1.1), (1.2), it follows from (4.2) the formula (4.3). Therefore, $\left\{\frac{\delta}{\delta x^1}, \dots, \frac{\delta}{\delta x^n}\right\}$ generate a distribution N, which is supplementary to the vertical distribution V, V being generated by $\frac{\partial}{\partial p_i}\cdot$ **q.e.d.**

The set of vector fields $\left(\frac{\delta}{\delta x^i}, \frac{\partial}{\partial p_i}\right)$ give us an adapted basis to the distributions N and V. The changes of coordinates (1.1) has the effect:

$$\frac{\delta}{\delta x^i} = \frac{\partial \widetilde{x}^j}{\partial x^i}\frac{\delta}{\delta \widetilde{x}^j},\ \frac{\partial}{\partial p_i} = \frac{\partial x^i}{\partial \widetilde{x}^j}\frac{\partial}{\partial \widetilde{p}_j} \tag{4.4}$$

The dual basis of $\left(\dfrac{\delta}{\delta x^i}, \dfrac{\partial}{\partial p_i}\right)$ is given by $(dx^i, \delta p_i)$, $(i = 1, ..., n)$ with

$$\delta p_i = dp_i - N_{ji}\, dx^j \tag{4.5}$$

and the transformations of coordinates (1.1) transform the "adapted" dual basis, $(dx^i, \delta p_i)$ in the form

$$d\widetilde{x}^i = \frac{\partial \widetilde{x}^i}{\partial x^j} dx^j, \quad \delta \widetilde{p}_i = \frac{\partial x^j}{\partial \widetilde{x}^i} \delta p_j. \tag{4.4$'$}$$

Theorem 4.4.2. *If the base manifold M is paracompact, then on the manifold T^*M there exist the nonlinear connections.*
Proof. M being paracompact, Theorem 4.1.2 affirms that T^*M is paracompact, too. Let G be a Riemannian structure on T^*M and N the orthogonal distribution, to the vetical distribution V with respect to G. Thus, the equality (4.1) holds. **q.e.d.**

Consider a nonlinear connection N with the coefficients $N_{ij}(x, p)$ and define the set of functions

$$\tau_{ij} = \frac{1}{2}(N_{ij} - N_{ji}) \tag{4.6}$$

Proposition 4.4.1. *With respect to* (1.1), τ_{ij} *is transformed by the rule*

$$\widetilde{\tau}_{ij} = \frac{\partial x^r}{\partial \widetilde{x}^i} \frac{\partial x^s}{\partial \widetilde{x}^j} \tau_{rs}. \tag{4.7}$$

Indeed, the formulae (4.3) and (4.6) lead to (4.7).

Consequently, τ_{ij} is a distinguished tensor field, covariant of order two, skew–symmetric. τ_{ij} is called the *tensor of torsion* of the nonlinear connection N. The equation $\tau_{ij} = 0$ has a geometrical meaning. In this case the nonlinear connection N is called *symmetric.*

Theorem 4.4.3. *With respect to a symmetric nonlinear connection N, the canonical symplectic structure θ and the canonical Poisson structure $\{,\}$, can be written in the following invariant form*

$$\theta = \delta p_i \wedge dx^i \tag{4.8}$$

$$\{f, g\} = \frac{\partial f}{\partial p_i} \frac{\delta g}{\delta x^i} - \frac{\partial g}{\partial p_i} \frac{\delta f}{\delta x^i} \tag{4.9}$$

Proof. Using (4.2), (4.5) and (4.6) we have

$$(4.10)\qquad \begin{cases} \theta = \delta p_i \wedge dx^i + \tau_{ij}\, dx^i \wedge dx^j \\ \{f,g\} = \dfrac{\partial f}{\partial p_i}\dfrac{\delta g}{\delta x^i} - \dfrac{\partial g}{\partial p_i}\dfrac{\delta f}{\delta x^i} - 2\tau_{ij}\dfrac{\partial f}{\partial p_i}\dfrac{\partial g}{\partial x^j} \end{cases}$$

But these formulae, for $\tau_{ij} = 0$, give us the announced formulae (4.8) and (4.9). Now, by means of (4.4) and (4.4)′ it follows the invariant form of θ and $\{\,,\,\}$.**q.e.d.**

Of course, we can define the curvature tensor field of a nonlinear connection N. Indeed, $d\theta = 0$ imply $d(\delta p_i)\wedge dx^i = 0$. But $d(\delta p_i)$ is given by

Proposition 4.4.2. *The exterior differential of δp_i are given by*

$$(4.11)\qquad d(\delta p_i) = -\left(\frac{1}{2}R_{ijm}dx^m + \dot\partial^m N_{ij}\delta p_m\right)\wedge dx^j$$

where

$$(4.12)\qquad R_{ijh} = \frac{\delta N_{ji}}{\delta x^h} - \frac{\delta N_{hi}}{\delta x^j}$$

Indeed, $d(\delta p_i) = -dN_{ji}\wedge dx^j$ and $dN_{ji} = \dfrac{\delta N_{ji}}{\delta x^m}dx^m + \dfrac{\partial N_{ji}}{\partial p_m}\delta p_m$ which determine together (4.11) and (4.12).

It follows, without difficulties:

Proposition 4.4.3. *By means of* (1.1) *we obtain:*

$$(4.13)\qquad \begin{aligned} \widetilde{R}_{ijh} &= \frac{\partial x^r}{\partial \widetilde{x}^i}\frac{\partial x^s}{\partial \widetilde{x}^j}\frac{\partial x^p}{\partial \widetilde{x}^h}R_{rsp} \\ \widetilde{\dot\partial^i N_{jh}} &= \frac{\partial \widetilde{x}^i}{\partial x^r}\frac{\partial x^s}{\partial \widetilde{x}^j}\frac{\partial x^h}{\partial \widetilde{x}^h}\dot\partial^r N_{sh} + \frac{\partial \widetilde{x}^i}{\partial x^r}\frac{\partial^2 x^r}{\partial \widetilde{x}^j \partial \widetilde{x}^h}. \end{aligned}$$

So, we can say that R_{ijh} is a d-tensor field called the *d-tensor of curvature* of N and $\dot\partial^i N_{jh}$ is a *d-connection* determined by N called the *Berwald connection.*

Analogously, we can study the Lie brackets of the vector fields from adapted basis $\left(\dfrac{\delta}{\delta x^i}, \dfrac{\partial}{\partial p_i}\right)$.

Proposition 4.4.4. *The Lie brackets of the vector fields*

$$\left(\delta_i = \frac{\delta}{\delta x^i},\ \dot\partial^i = \frac{\partial}{\partial p_i}\right)$$

are as follows

$$(4.14)\qquad [\delta_j, \delta_h] = R_{ijh}\dot\partial^i \quad [\delta_j, \dot\partial^h] = -\dot\partial^h N_{jr}\dot\partial^r.$$

Now we can give a necessary and sufficient condition for the integrability of the distribution N called the *integrability* of the nonlinear connection N.

Theorem 4.4.4. *The horizontal distribution N is integrable if and only if the d-tensor of curvature R_{ijh} vanishes.*

Indeed the Lie brackets of the vector fields from adapted basis δ_i of N belongs to N if and only if $R_{ijh} = 0$ (cf. (4.14)).

4.5 Distinguished vector and covector fields

Let N be a nonlinear connection on T^*M. It gives rise to the direct decomposition (4.1). Let h and v be the projectors defined by supplementary distributions N and V. They have the following properties

$$h + v = I,\ h^2 = h,\ v^2 = v,\ h \circ v = v \circ h = 0. \tag{5.1}$$

If $X \in \mathcal{X}(T^*M)$ we denote

$$X^H = hX,\ X^V = vX. \tag{5.2}$$

Therefore we have the unique decomposition

$$X = X^H + X^V \tag{5.3}$$

Every component X^H and X^V is called a distinguished vector field. Shortly a *d*-vector field.

In the adapted basis $\left(\delta_i, \dot{\partial}^i\right)$ we get

$$X^H = X^i(x,p)\delta_i,\ X^V = X_i(x,p)\dot{\partial}^i. \tag{5.3)'}$$

With respect to (1.1), we have, using (4.4),

$$\widetilde{X}^i = \frac{\partial \widetilde{x}^i}{\partial x^j} X^j,\ \ \widetilde{X}_i = \frac{\partial x^j}{\partial \widetilde{x}^i} X_j. \tag{5.3)''}$$

But, these are the classical rules of transformations of the local coordinates of vector and covector fields on the base manifold M. Therefore $X^i(x,p)$ is called a *d*-vector field, too and $X_i(x,p)$ is called a *d*-covector field on T^*M.

For instance $\mathbb{C}^* = p_i\dot{\partial}^i$ is a *d*-vector field and p_i is a *d*-covector field on T^*M. We have $C^{*H} = 0,\ C^{*V} = C^*$.

A similar theory can be done for distinguished 1-forms. With respect to the direct decomposition (4.1) an 1–form $\omega \in \mathcal{X}^*(T^*M)$ can be uniquely written in the form

$$\omega = \omega^H + \omega^V \tag{5.4}$$

where

$$\omega^H = \omega \circ h, \ \omega^V = \omega \circ v. \tag{5.4)'$$

In the adapted cobasis $(dx^i, \delta p_i)$, we get

$$\omega^H = \omega_i(x,p)dx^i, \ \omega^V = \omega^i(x,p)\delta p_i \tag{5.5}$$

ω^H and ω^V are called, every one, d-1 form. The coefficients $\omega_i(x,p)$ and $\omega^i(x,p)$ are d-covector and d-vector fields, respectively.

Now, let us consider a function $f \in \mathcal{F}(T^*M)$. The 1-form df can be written in the form (5.4)

$$df = (df)^H + (df)^V, \text{ where} \tag{5.6}$$

$$(df)^H = \frac{\delta f}{\delta x^i}dx^i, \ (df)^V = \frac{\partial f}{\partial p_i}\delta p_i \tag{5.6)'$$

A curve $\gamma : I \subset R \longrightarrow T^*M$, having $\operatorname{Im}\gamma \subset \pi^*(U)$ has the analytical representation:

$$x^i = x^i(t), \ p_i = p_i(t), \ t \in I. \tag{5.7}$$

The tangent vector $\dfrac{d\gamma}{dt}$, in a point of curve γ can be set in the form

$$\frac{d\gamma}{dt} = \left(\frac{d\gamma}{dt}\right)^H + \left(\frac{d\gamma}{dt}\right)^V = \frac{dx^i}{dt}\frac{\delta}{\delta x^i} + \frac{\delta p_i}{dt}\frac{\partial}{\partial p_i} \tag{5.8}$$

where

$$\frac{\delta p_i}{dt} = \frac{dp_i}{dt} - N_{ji}(x(t),p(t))\frac{dx^j}{dt}. \tag{5.8)'$$

The curve γ is called horizontal if $\dfrac{d\gamma}{dt} = \left(\dfrac{d\gamma}{dt}\right)^H, \forall\, t \in I.$

Theorem 4.5.1. *An horizontal curve γ is characterized by the system of differential equations*

$$x^i = x^i(t), \ \frac{\delta p_i}{dt} = 0. \tag{5.9}$$

Clearly, if the functions $x^i(t)$ are given the previous system of differential equations has local solutions, when a point $x^i_0 = x^i(t_0)$, p^0_i, $t_0 \in I$ is given.

4.6 The almost product structure $\mathbb{P}$. The metrical structure $\mathbb{G}$. The almost complex structure $\check{\mathbb{F}}$

A nonlinear connection N on T^*M can be characterized by some almost product structure $\mathbb{P}$.

Let us consider the $\mathcal{F}(T^*M)$-linear mapping

$$\mathbb{P} : \mathcal{X}(T^*M) \longrightarrow \mathcal{X}(T^*M)$$

defined by

$$\mathbb{P}(X^H) = X^H, \ \mathbb{P}(X^V) = -X^V, \ \forall X \in \mathcal{X}(T^*M). \tag{6.1}$$

Thus $\mathbb{P}$ has the properties:

$$\begin{aligned} &\mathbb{P} \circ \mathbb{P} = I \\ &\mathbb{P} = I - 2v = 2h - I \\ &\operatorname{rank} \mathbb{P} = 2n. \end{aligned} \tag{6.2}$$

Theorem 4.6.1. *A nonlinear connection N on T^*M is characterized by the existence of an almost product structure $\mathbb{P}$ on the manifold T^*M whose eigenspaces corresponding to the eigenvalue -1 coincide with the linear spaces of the vertical distribution V on T^*M.*

Proof. If N is a nonlinear connection, then we have the direct sum (4.1). Denoting h and v the projectors determined by (4.1) we get $\mathbb{P} = I - 2v$. Then $\mathbb{P}$ has the property (6.1). $\mathbb{P}$ is an almost product structure, for which $\mathbb{P}(X^V) = -X^V$. Conversely, if $\mathbb{P}^2 = I$ and $\mathbb{P}(X^V) = -X^V$, then $v = \dfrac{1}{2}(I - \mathbb{P})$ and $h = \dfrac{1}{2}(I + \mathbb{P})$ satisfy (5.1). Therefore, $N = \operatorname{Ker} v$ and it follows $N \oplus V = TT^*M$. **q.e.d.**

Proposition 4.6.1. *The almost product structure $\mathbb{P}$ is integrable if and only if the nonlinear connection N is integrable.*

Indeed, the Nijenhuis tensor of the structure $\mathbb{P}$, given by

$$\mathcal{N}_{\mathbb{P}}(X, Y) = \mathbb{P}^2[X, Y] + [\mathbb{P}X, \mathbb{P}Y] - \mathbb{P}[\mathbb{P}X, Y] - \mathbb{P}[X, \mathbb{P}Y] \tag{6.3}$$

vanishes if and only if $[X^H, Y^H]^V = 0$, $\forall X, Y \in \mathcal{X}(T^*M)$, because

$$\mathcal{N}_{\mathbb{P}}(X^H, Y^H) = 4v[X^H, Y^H],\ \mathcal{N}_{\mathbb{P}}(X^H, X^V) = \mathcal{N}_{\mathbb{P}}(X^V, Y^V) = 0.$$

Remark. Of course we can consider a general almost product structure on T^*M. It will determine a general concept of connection on T^*M. This idea was developed by P. Antonelli and D. Hrimiuc in the paper [14] and will be studied in a next chapter of this book.

Let us consider a Riemannian (or metrical) structure $\mathbb{G}$ on the manifold T^*M.

The following property is evident: $\mathbb{G}$ uniquely determines an orthogonal distribution N to the vertical distribution V on T^*M. Therefore N is a nonlinear connection. Let $N_{ij}(x,p)$ be the coefficients of N and h, v the supplementary projectors defined by N and V. Then $\mathbb{G}$ can be written in the form

$$\mathbb{G} = \mathbb{G}^H + \mathbb{G}^V,\ \mathbb{G}^H(X,Y) = \mathbb{G}(X^H, Y^H),\ \mathbb{G}^V(X,Y) = \mathbb{G}(X^V.X^V). \tag{6.4}$$

Or, in the adapted basis,

$$\mathbb{G} = g_{ij}(x,p)dx^i \otimes dx^j + h^{ij}(x,p)\delta p_i \otimes \delta p_j \tag{6.5}$$

where g_{ij} is a covariant nonsingular, symmetric tensor field and h^{ij} is a contravariant nonsingular, symmetric tensor field. Of course, the matrix $\|g_{ij}(x,p)\|$, $\|h^{ij}(x,p)\|$ are positively defined.

The Riemannian manifold $(T^*M, \mathbb{G})$ can be studied by means of the methods given by the geometry of the manifold T^*M.

If a tensor $g_{ij}(x,p)$ covariant, symmetric and positively defined on T^*M and a nonlinear connection N with coefficients $N_{ij}(x,p)$ are given, then we can consider the following Riemannian structure on T^*M:

$$\mathbb{G}(x,p) = g_{ij}(x,p)dx^i \otimes dx^j + g^{ij}(x,p)\delta p_i \otimes \delta p_j. \tag{6.6}$$

The tensor $\mathbb{G}$ is called the N-lift to T^*M of the d-tensor metric $g_{ij}(x,p)$.

Assuming that the d-tensor metric $g_{ij}(x,p)$ and the symmetric nonlinear connection $N(N_{ij})$ are given let us consider the following $\mathcal{F}(T^*M)$-linear mapping $\check{\mathbb{F}} : \mathcal{X}(T^*M) \longrightarrow \mathcal{X}(T^*M)$

$$\check{\mathbb{F}}(\delta_i) = -\check{\partial}_i,\quad \check{\mathbb{F}}(\check{\partial}_i) = \delta_i \tag{6.7}$$

where

$$\delta_i = \frac{\delta}{\delta x^i},\ \check{\partial}_i = g_{ij}\dot{\partial}^j. \tag{6.7'}$$

It is not difficult to prove, by a straightforward calculus:

Theorem 4.6.2.

1° $\check{\mathbb{F}}$ *is globally defined on* T^*M.

2° $\check{\mathbb{F}}$ *is the tensor field*

$$\check{\mathbb{F}} = -g_{ij}\dot{\partial}^i \otimes dx^j + g^{ij}\delta_i \otimes \delta p_j \tag{6.8}$$

3° $\check{\mathbb{F}}$ *is an almost complex structure:*

$$\check{\mathbb{F}} \circ \check{\mathbb{F}} = -I. \tag{6.9}$$

We have, too

Theorem 4.6.3. *In order to the almost complex structure* $\check{\mathbb{F}}$ *be integrable is necessary that the nonlinear connection* N *be integrable.*

Proof. The Nijenhuis tensor $\mathcal{N}_{\check{\mathbb{F}}}\ (\delta_i, \delta_j) = 0$, imply, firstly $R_{ijk} = 0$. It implies also some conditions for g_{ij} and N. **q.e.d.**

The relations between the structures $\mathbb{G}$ and $\check{\mathbb{F}}$ are as follows

Theorem 4.6.4.

1° *The pair* $(\mathbb{G}, \check{\mathbb{F}})$, $\mathbb{G}$ *given by* (6.6) *and* $\check{\mathbb{F}}$ *given by* (6.7), *is an almost hermitian structure.*

2° *The associated almost symplectic structure to* $(\mathbb{G}, \check{\mathbb{F}})$ *is the canonical symplectic structure* $\theta = \delta p_i \wedge dx^i$.

Proof. 1°. We get from (6.6), (6.7), in the adapted basis $(\delta_i, \dot{\partial}^i)$, $\mathbb{G}(\check{\mathbb{F}}X, \check{\mathbb{F}}Y) = \mathbb{G}(X, Y)$.

2°. $\theta(X, Y) = \mathbb{G}(\check{\mathbb{F}}X, Y)$. **q.e.d.**

4.7 d-tensor algebra. N-linear connections

As we know, a nonlinear connection N determines a direct decomposition (4.1) with respect to which any vector field X on T^*M can be written in the form $X = X^H + X^V$. And any 1-form ω on T^*M is uniquely decomposed: $\omega = \omega^H + \omega^V$. The components X^H, X^V are distinguished vector fields and ω^H, ω^V are distinguished 1-form field. Briefly, d-vector field, d-1-form field.

Definition 4.7.1. A distinguished field on T^*M (briefly d-tensor field) is a tensor field T of type (r,s) on T^*M with the property:

$$T(\overset{1}{\omega},...,\overset{r}{\omega},\underset{1}{X},...,\underset{s}{X}) = T(\overset{1}{\omega}^H,...,\overset{r}{\omega}^V,\underset{1}{X}^H,...,\underset{s}{X}^V) \tag{7.1}$$

for any $\overset{1}{\omega},...,\overset{r}{\omega} \in \mathcal{X}^*(T^*M)$ and for any $\underset{1}{X},...,\underset{s}{X} \in \mathcal{X}(T^*M)$.

$X = X^H$, $X = X^V$, or $\omega = \omega^H$, $\omega = \omega^V$ are examples of d-vector fields or d-1-form fields.

In the adapted basis $(\delta_i, \dot{\partial}^i)$ and adapted cobasis $(dx^i, \delta p_i)$ T can be written, uniquely, in the form

$$T = T^{i_1...i_r}_{j_1...j_s}(x,p)\delta_{i_1} \otimes \cdots \otimes \dot{\partial}^{j_s} \otimes dx^{j_1} \otimes \cdots \otimes \delta p_{i_r}. \tag{7.2}$$

It follows that the addition of d-tensors and tensor product of them lead to the d-tensor fields. So, the set $(1, \delta_i, \dot{\partial}^i)$ generates the algebra of d-tensors over the ring of functions $\mathcal{F}(T^*M)$.

Other examples $\delta_i f, \dot{\partial}^i f$ are d-covector and d-vector fields, respectively.

Clearly, with respect to (1.1) the coefficients $T^{i_1...i_r}_{j_1...j_s}(x,p)$ of a d-tensor fields are transformed by the classical rule:

$$\widetilde{T}^{i_1...i_r}_{j_1...j_s} = \frac{\partial \widetilde{x}^{i_1}}{\partial x^{h_1}} \cdots \frac{\partial \widetilde{x}^{i_r}}{\partial x^{h_r}} \frac{\partial x^{k_1}}{\partial \widetilde{x}^{j_1}} \cdots \frac{\partial x^{k_s}}{\partial \widetilde{x}^{j_s}} T^{h_1...h_r}_{k_1...k_s}. \tag{7.3}$$

the notion of N-linear connection can be defined in the known manner (see Ch.1). In the following we assume N is a symmetric nonlinear connection.

Definition 4.7.2. A linear connection D on T^*M is called an *N-linear connection* if:

1° D preserves by parallelism every distribution N and V.

2° The canonical symplectic structure $\theta = \delta p_i \wedge dx^i$ has the associate tensor $\widetilde{\theta}$ absolute parallel with respect to D:

$$D\widetilde{\theta} = 0. \tag{7.4}$$

The following properties of an N-linear connection D are immediate:

$$D_X h = D_X v = 0; \; D_X \mathbb{P} = 0 \tag{7.5}$$

Indeed, $(D_X h)(Y) = D_X(Y^H) - (D_X Y)^H$. For $Y = Y^H$ we get $(D_X h)(Y^H) = 0$ and for $Y = Y^V$ we get $(D_X h)Y^V = 0$. Similarly, $D_X v = 0$ and $D_X \mathbb{P} = D_X(I - 2v) = 0$.

For an almost Hermitian structure $(\mathbb{G}, \check{\mathbb{F}})$ given in the previous section, any N-linear connection D, with the property $D\mathbb{G} = 0$, has the property $D\check{\mathbb{F}} = 0$, too.

If $X = X^H + X^V$, then

$$D_X Y = D_{X^H} Y + D_{X^V} Y \tag{7.6}$$

We find new operators of derivation in the algebra of d-tensors defined by:

$$D^H_X = D_{X^H}, \; D^V_X = D_{X^V} \tag{7.7}$$

These operators *are not the covariant* derivatives in the d-tensor algebra, since $D^H_X f = X^H f \neq Xf$, $D^V_X f = X^V f \neq Xf$.

Theorem 4.7.1. *The operators D^H_X, D^V_X have the following properties:*

1° $D^H_X f = X^H f, \; D^V_X f = X^V f.$

2° $\begin{cases} D^H_X(fY) = X^H f \cdot Y + f D^H_X Y \\ D^V_X(fY) = X^V f \cdot Y + f D^V_X Y \end{cases}$

3° $\begin{cases} D^H_X(Y+Z) = D^H_X Y + D^H_X Z \\ D^V_X(Y+Z) = D^V_X Y + D^V_X Z \end{cases}$

4° $D^H_{X+Y} = D^H_X + D^H_Y, \; D^V_{X+Y} = D^V_X + D^V_Y$

5° $D^H_{fX} = f D^H_X, \; D^V_{fX} = f D^V_X$

6° $D^H_X \theta = 0, \; D^V_X \theta = 0$

7° *The operators D^H_X, D^V_X have the property of localization.*

The proof of the previous theorem can be done by the classical methods [113].

The operator D^H_X will be called the *operator of h-covariant derivation* and D^V_X will be called the *operator of v-covariant derivation.*

D^H_X, D^V_X act on the 1-form ω on T^*M by the rules

$$\begin{aligned} (D^H_X \omega)(Y) &= X^H \omega(Y) - \omega(D^H_X Y) \\ (D^V_X \omega)(Y) &= X^V \omega(Y) - \omega(D^V_X Y) \end{aligned} \tag{7.8}$$

Consequently, the actions of D^H_X and D^V_X on a tensor field T of type (r, s) on T^*M are well determined.

4.8 Torsion and curvature

The torsion $\mathbb{T}$ of an N-linear connection D is expressed as usually by

$$\mathbb{T}(X,Y) = D_XY - D_YX - [X,Y]. \tag{8.1}$$

It can be characterized by the following vector fields

$$\mathbb{T}(X^H,Y^H), \mathbb{T}(X^H,Y^V),\ \mathbb{T}(X^V,Y^V)$$

Taking the h- and v- components of these vectors we obtain the *d-tensors of torsion*:

$$\begin{aligned} \mathbb{T}(X^H,Y^H) &= h\mathbb{T}(X^H,Y^H) + v\mathbb{T}(X^H,Y^H),\\ \mathbb{T}(X^H,Y^V) &= h\mathbb{T}(X^H,Y^V) + v\mathbb{T}(X^H,Y^V),\\ \mathbb{T}(X^V,Y^V) &= h\mathbb{T}(X^V,Y^V) + v\mathbb{T}(X^V,Y^V). \end{aligned} \tag{8.2}$$

Since V is an integrable distribution (8.1) implies

$$h\mathbb{T}(X^V,Y^V) = 0 \tag{8.3}$$

Taking into account the definition (4.8.1) of the torsion $\mathbb{T}$ we obtain:

Proposition 4.8.1. *The d-tensors of torsion of an N-linear connection D are:*

$$\begin{aligned} h\mathbb{T}(X^H,Y^H) &= D^H_XY^H - D^H_YX^H - [X^H,Y^H]^H\\ h\mathbb{T}(X^H,Y^V) &= -(D^V_YX^H + [X^H,Y^V]^H)\\ v\mathbb{T}(X^H,Y^H) &= -[X^H,Y^H]^V\\ v\mathbb{T}(X^H,Y^V) &= D^H_XY^V - [X^H,Y^V]^V\\ v\mathbb{T}(X^V.Y^V) &= D^V_XY^V - D^V_YX^V - [X^V,Y^V]^V. \end{aligned} \tag{8.4}$$

The curvature $\mathbb{R}$ of an N-linear connection D is given by

$$\mathbb{R}(X,Y)Z = (D_XD_Y - D_YD_X)Z - D_{[X,Y]}Z \tag{8.5}$$

Remarking that the vector field $\mathbb{R}(X,Y)Z^H$ is horizontal one and $\mathbb{R}(X,Y)Z^V$ is vertical one, we have

$$v(\mathbb{R}(X,Y)Z^H) = 0,\ h(\mathbb{R}(X,Y)Z^V) = 0 \tag{8.6}$$

We will see that the *d-tensors of curvature* of the v-linear connection D are:

$$\mathbb{R}(X^H,Y^H)Z^H,\ \mathbb{R}(X^H,Y^V)Z^H,\ \mathbb{R}(X^V,Y^V)Z^H \tag{8.7}$$

Therefore, by means of (8.5)

Proposition 4.8.2.

1° *The Ricci identities of an N-linear connection $D$are*

$$[D_X, D_Y]Z = \mathbb{R}(X,Y)Z + D_{[X,Y]}Z \tag{8.8}$$

2° *The Bianchi identities are given by*

$$\begin{aligned} &\sum_{(XYZ)} \{(D_X\mathbb{T})(Y,Z) - \mathbb{R}(X,Y)Z + \mathbb{T}(\mathbb{T}(X,Y),Z)\} = 0 \\ &\sum_{(XYZ)} \{(D_X\mathbb{R})(U,Y,Z) - \mathbb{R}(\mathbb{T}(X,Y),Z)U\} = 0 \end{aligned} \tag{8.9}$$

where $\sum\limits_{(XYZ)}$ *means the cyclic sum.*

4.9 The coefficients of an N-linear connection

Let D be an N-linear connection and $\delta_i = \dfrac{\delta}{\delta x^i}$, $\dot{\partial}^i = \dfrac{\partial}{\partial p_i}$ the adapted basis to N and V. Then $D_X = D^H_X + D^V_X$. In order to determine the coefficients of D in the adapted basis, we take into account the properties:

$$D_{\delta_j} = D^H_{\delta_j},\ D_{\dot{\partial}^j} = D^V_{\dot{\partial}^j}. \tag{9.1}$$

Theorem 4.9.1.

1° *An N-linear connection D on T^*M can be uniquely represented in the adapted basis $(\delta_i, \dot{\partial}^i)$ in the following form*

$$\begin{aligned} &D_{\delta_j}\delta_i = H^h_{ij}\delta_h,\ D_{\delta_j}\dot{\partial}^i = -H^i_{hj}\dot{\partial}^h \\ &D_{\dot{\partial}^j}\delta_i = C_i{}^{hj}\delta_h,\ D_{\dot{\partial}^j}\dot{\partial}^i = -C_h{}^{ij}\dot{\partial}^h \end{aligned} \tag{9.2}$$

2° *With respect to a change of coordinate (1.1) on T^*M, the coefficients $H^i_{jh}(x,p)$ transform by the rule:*

$$\widetilde{H}^i{}_{rs}\frac{\partial x^r}{\partial \widetilde{x}^j}\frac{\partial x^s}{\partial \widetilde{x}^h} = \frac{\partial \widetilde{x}^i}{\partial x^r}H^r_{jh} - \frac{\partial^2 \widetilde{x}^i}{\partial x^j \partial x^h}, \tag{9.3}$$

while $C_i{}^{jh}(x,p)$ is a d-tensor field of type $(2,1)$.

3° *Conversely, if N is an apriori given nonlinear connection and a set of function $(H^i_{jk}, C_i{}^{jh})$, verifying 2° is given, then there exist a unique N-linear connection D with the properties (9.2).*

Proof. 1°. Setting

$$D_{\delta_j}\delta_i = H^h_{ij}\delta_h,\ D_{\delta_j}\dot{\partial}^i = -\check{H}^i_{hj}\dot{\partial}^h$$

and taking into account that $D\widetilde{\theta} = 0$ where tensor $\widetilde{\theta}$ of the 2-form $\theta = \delta p_i \wedge dx^i$ have the components

$$\widetilde{\theta}(\delta_i, \delta_j) = 0,\ \widetilde{\theta}(\delta_i, \dot{\partial}^j) = -\delta^j_i,\ \widetilde{\theta}(\dot{\partial}^j, \delta_i) = \delta^j_i,\ \widetilde{\theta}(\dot{\partial}^i, \dot{\partial}^j) = 0$$

it follows that $D_X\widetilde{\theta} = 0$, imply $H^h_{ij} = \check{H}^h_{ij}$. Analogously, setting

$$D_{\dot{\partial}^j}\delta_i = C^{hj}_i\delta_h,\ D_{\dot{\partial}^j}\dot{\partial}^i = -\check{C}^{ij}_h\dot{\partial}^h,$$

from $D_X\widetilde{\theta} = 0$, it follows$C^{ij}_h = \check{C}^{ij}_h$.

2°. By a straightforward calculus, taking into account the formulae (4.4), it follows (9.3) and $\widetilde{C}^{ij}_r \dfrac{\partial x^r}{\partial \widetilde{x}^h} = \dfrac{\partial \widetilde{x}^i}{\partial x^s}\dfrac{\partial \widetilde{x}^j}{\partial x^r} C^{sr}_h$.

3°. One demonstrates by the usually methods.

The pair $D\Gamma(N)$=(H^i_{jh}, C^{jh}_i) is called the system of coefficients of D.

Proposition 4.9.1. *The following formula holds:*

$$\begin{aligned} D_{\delta_j}dx^i &= -H^i_{hj}dx^h, \quad D_{\delta_j}\delta p_i = H^h_{ij}\delta p_h \\ D_{\dot{\partial}^j}dx^i &= -C^{ij}_h dx^h, \quad D_{\dot{\partial}^j}\delta p_i = C^{hj}_i\delta p_h \end{aligned} \tag{9.4}$$

Let us consider a d-tensor T, of type (r, s) expressed in the adapted basis in the form (7.2), and a horizontal vector field $X = X^H = X^i\delta_i$. Applying Theorems 4.7.1 and 4.9.1 we obtain the h-covariant derivation D^H_X of T in the form:

$$D^H_X T = X^m T^{i_1\ldots i_r}_{j_1\ldots j_s|m}\delta_{i_1}\otimes\cdots\otimes\dot{\partial}^{j_s}\otimes dx^{j_1}\otimes\cdots\otimes\delta p_{i_r} \tag{9.5}$$

where

$$\begin{aligned} T^{i_1\ldots i_r}_{j_1\ldots j_s|m} &= \delta_m T^{i_1\ldots i_r}_{j_1\ldots j_s} + T^{hi_2\ldots i_r}_{j_1\ldots j_s}H^{i_1}_{hm} + \cdots + T^{i_1\ldots h}_{j_1\ldots h}H^{i_r}_{hm} - \\ &-T^{i_1\ldots i_r}_{hj_2\ldots j_s}H^h_{j_1 m} - \cdots - T^{i_1\ldots i_r}_{j_1\ldots h}H^h_{j_r m}. \end{aligned} \tag{9.5$'$}$$

The operator "|" is called *h-covariant derivative* with respect to $D\Gamma(N) = (H^i_{jh}, C^{jh}_i)$.

Now, taking $X = X^V = X_i\dot{\partial}^i$, $D^V_X T$ has the following form

$$D^V_X T = X_m T^{i_1\ldots i_r}_{j_1\ldots j_s}|^m \delta_i\otimes\cdots\otimes\dot{\partial}^{j_s}\otimes dx^{j_1}\otimes\cdots\otimes\delta p_{i_r}, \tag{9.6}$$

where

$$(9.6)' \qquad \begin{aligned} T^{i_1\dots i_r}_{j_1\dots j_s}|^m &= \dot{\partial}^m T^{i_1\dots i_r}_{j_1\dots j_s} + T^{hi_2\dots i_r}_{j_1\dots j_s} C^{i_1 m}_h + \dots + T^{i_1\dots i_h}_{j_1\dots j_s} C^{i_r m}_h - \\ &- T^{i_1\dots i_r}_{hj_2\dots j_s} C^{hm}_{j_1} - \dots - T^{i_1\dots i_r}_{j_1\dots h} C^{hm}_{j_s}. \end{aligned}$$

The operator "$|$" will be called the *v-covariant derivative* with respect to $D\Gamma(N) = (H^i_{jh}, C^{jh}_i)$.

Proposition 4.9.2. *The following properties hold:*

1° $T^{i_1\dots i_r}_{j_1\dots j_s|m}$ *is a d-tensor of type* $(r, s+1)$.

1° $T^{i_1\dots i_r}_{j_1\dots j_s}|^m$ *is a d-tensor of type* $(r+1, s)$.

Proposition 4.9.3. *The operators "$_|$" and "$|$" satisfy the properties:*

1° $f_{|m} = \dfrac{\delta f}{\delta x^m}, \; f|^m = \dot{\partial}^m f$

$$(9.7) \qquad \begin{cases} X^i{}_{|m} = \delta_m X^i + X^h H_{hm}{}^i; & X^i|^m = \dot{\partial}^m + X^h C_h{}^{im} \\ \omega_{i|m} = \delta_m \omega_i - \omega_h H^h{}_{im}, & \omega_i|^m = \dot{\partial}^m \omega_i - \omega_h C_i{}^{hm} \end{cases}$$

2° *"$_|$" and "$|$" are distributive with respect to the addition of the d-tensors of the same type.*

3° *"$_|$" and "$|$" commute with the operation of contraction.*

4° *"$_|$" and "$|$" verify the Leibniz rule with respect to the tensor product of d-tensors.*

As an application let us consider the *deflection* tensors

$$(9.8) \qquad \Delta_{ij} = p_{i|j}, \; \eth^j_i = p_i|^j$$

Using (9.7) we get

$$(9.9) \qquad \Delta_{ij} = N_{ij} - p_m H^m{}_{ij}, \; \eth^j_i = \delta^j_i - p_m C^{mj}_i$$

In the particular case, $\Delta_{ij} = 0$, $\eth^j_i = \delta^j_i$, $D\Gamma(N)$ is called an N-linear connection of Cartan type. It is characterized by

$$(9.10) \qquad N_{ij} = p_m H^m{}_{ij}, \; p_m C^{mj}_i = 0.$$

We have:

Proposition 4.9.4. *If the pair* $(H^i{}_{jk}(x,p), C_i^{jk}(x,p))$ *is given and* $H^i{}_{jk}(x,p)$ *satisfy* (9.3), *then* $N_{ij} = p_m H^m_{ij}$ *are the coefficients of a nonlinear connection* N, *and* $(H^i{}_{jk}(x,p), C_i^{jk}(x,p))$ *are the coefficients of an* N*-linear connection* D.

The proof is not difficult if we apply the point 3° from Theorem 4.9.1.

Based on the previous property we can prove:

Theorem 4.9.2. *If the base manifold* M *is paracompact then on* T^*M *there exist the* N*-linear connections.*

Proof. Let $g(x)$ be a Riemannian metric on M and $\gamma^i_{jh}(x)$ its coefficients of Levi-Civita connection. Then $N_{jh} = p_m\gamma^m_{jh}$ are the coefficients of a nonlinear connection N on T^*M. So, the pair $(\gamma^i_{jh}(x), 0)$ gives us the coefficients of an N-linear connection $D\Gamma(N)$ on T^*M.

4.10 The local expressions of d-tensors of torsion and curvature

In the adapted basis $(\delta_i, \dot\partial^i)$ the Ricci identities (8.8), using the operators "$_|$" and "$|$", lead to the local expressions of the d-tensors of torsion and curvature.

Theorem 4.10.1. *For any* N*-linear connection* $D\Gamma(N) = (H^i{}_{jk}, C_i^{jh})$ *the following Ricci identities hold:*

$$
\begin{aligned}
&X^i{}_{|j|h} - X^i{}_{|h|j} = X^m R_m{}^i{}_{jh} - X^i{}_{|m}T^m{}_{jh} - X^i|^m R_{mjh}\\
&X^i{}_{|j}|^h - X^i|^h{}_{|j} = X^m P_m{}^i{}_j{}^h - X^i{}_{|m}C_j^{mh} - X^i|^m P^h{}_{mj}\\
&X^i|^j|^h - X^i|^h|^j = X^m S_m{}^{ijh} - X^i|^m S_m{}^{jh}.
\end{aligned}
\tag{10.1}
$$

where the coefficients $T^i{}_{jh}, R_{mjh}, C_j{}^{mh}, P^h{}_{mj}$ *and* $S_m{}^{jh}$ *are the d-tensors of torsion:*

$$
T^i{}_{jh} = H^i{}_{jh} - H^i{}_{hj},\ S_i^{jh} = C_i^{jh} - C_i^{hj},\ P^i{}_{jh} = H^i{}_{jh} - \dot\partial^i N_{hj}
\tag{10.2}
$$

and $R_k{}^i{}_{jh}, P_k{}^i{}_j{}^h, S_k{}^{ijh}$ *are the d-tensors of curvature:*

$$
\begin{aligned}
R_k{}^i{}_{jh} &= \delta_h H^i{}_{kj} - \delta_j H^i{}_{kh} + H^m{}_{kj}H^i{}_{mh} -\\
&\qquad - H^m{}_{kh}H^i{}_{mj} + C_k{}^{im}R_{mjh},\\
P_k{}^i{}_j{}^h &= \dot\partial^h H^i{}_{kj} - C_k{}^{ih}{}_{|j} + C_k{}^{im}P^h{}_{mj}\\
S_k{}^{ijh} &= \dot\partial^h C_k{}^{ij} - \dot\partial^j C_k{}^{ih} + C_k{}^{mj}C_m{}^{ih} - C_k{}^{mh}C_m{}^{ij}.
\end{aligned}
\tag{10.3}
$$

Proof. By a direct calculus, using (9.7) we get

$$X^i{}_{|j|h} - X^i{}_{|h|j} =$$
$$= [\delta_h, \delta_j]X^i + X^m(\delta_h H^i_{mj} - \delta_j H^i_{mh} + H^s_{mj}H^i_{sh} - H^s_{mh}H^i_{sj}) - X^i{}_{|m}T^m{}_{jh}.$$

Taking into account the formula $[\delta_h, \delta_j]X^i = R_{mhj}\dot{\partial}^m X^i$, the previous equality leads to the first Ricci formula, with the coefficients $T^i{}_{jh}, R_k{}^i{}_{jh}$ from (10.2) and (10.3). Etc.

Proposition 4.10.1. *The following formule hold:*

$$\text{(10.4)}\qquad \begin{cases} \mathbb{R}(\delta_j, \delta_h)\delta_k = R_k{}^i{}_{hj}\delta_i, \ \mathbb{R}(\delta_j, \dot{\partial}^h)\delta_k = P_k{}^h{}_j{}^i\delta_i \\ \mathbb{R}(\dot{\partial}^j, \dot{\partial}^h)\delta_k = S_k{}^{ihj}\delta_i \end{cases}$$

and

$$\begin{cases} \mathbb{R}(\delta_j, \delta_h)\dot{\partial}^k = -R_i{}^k{}_{hj}\dot{\partial}^i, \ \mathbb{R}(\delta_j, \dot{\partial}^h)\dot{\partial}^k = -P_i{}^k{}_j{}^h\dot{\partial}^i \\ \mathbb{R}(\dot{\partial}^j, \dot{\partial}^h)\dot{\partial}^k = -S_i{}^{khj}\dot{\partial}^i \end{cases}$$

The proof can be given by a direct calculus, using the formula (8.5) and the equations (8.6).

As usually, we extend the Ricci identities for any d-tensor field, given by (7.2).

For instance, if $g^{ij}(x,p)$ is a d-tensor field, the Ricci identities for g^{ij}, with respect to N-linear connection $D\Gamma(N)$ are

$$\text{(10.5)}\qquad \begin{aligned} g^{ij}{}_{|k|h} - g^{ij}{}_{|h|k} &= g^{mj}R_m{}^i{}_{kh} + g^{im}R_m{}^j{}_{kh} - \\ &\quad - g^{ij}{}_{|m}T^m{}_{kh} - g^{ij}|^m R_{mkh}, \\ g^{ij}{}_{|k}|^h - g^{ij}|^h{}_{|k} &= \\ &= g^{mj}P_m{}^i{}_k{}^h + g^{im}P_m{}^j{}_k{}^h - \\ &\quad - g^{ij}{}_{|m}C_k{}^{mh} - g^{ij}|^m P^h{}_{mk} \\ g^{ij}|^k|^h - g^{ij}|^h|^k &= g^{mj}S_m{}^{ikh} + g^{im}S_m{}^{jkh} - g^{ij}|^m S_m{}^{kh}. \end{aligned}$$

In particular, if the N-linear connection $D\Gamma(N)$ satisfy the supplementary conditions:

$$\text{(10.6)}\qquad g^{ij}{}_{|k} = 0, \ g^{ij}|^k = 0$$

then (10.5) lead to the equations

$$\text{(10.5)}'\qquad \begin{aligned} & g^{mj}R_m{}^i{}_{kh} + g^{im}R_m{}^j{}_{kh} = 0 \\ & g^{mj}P_m{}^i{}_k{}^h + g^{im}P_m{}^j{}_k{}^h = 0 \\ & g^{mj}S_m{}^{ikh} + g^{im}S_m{}^{jkh} = 0 \end{aligned}$$

Such kind of equations will be used for the N-linear connections compatible with a metric structure $\mathbb{G}$ of the form (6.6).

The Ricci identities applied to the Liouville-Hamilton vector field $C^* = p_i\dot{\partial}^i$ give us some important identities. To this aim, we take into account the deflection tensors $\Delta_{ij} = p_{i|j}$, $\delta_i^{\ j} = p_i|^j$.

Theorem 4.10.2. *Any N-linear connection D satisfies the following identities*

$$\begin{aligned}\Delta_{ij|k} - \Delta_{ik|j} &= -p_m R_i{}^m{}_{jk} - \Delta_{im}T^m{}_{jk} - \delta_i{}^m R_{mjk}\\ \Delta_{ij}|^k - \delta_i{}^k{}_{|j} &= -p_m P_i{}^m{}_j{}^k - \Delta_{im}C_j{}^{mk} - \delta_i{}^m P^k{}_{mj}\\ \delta_i{}^j|^k - \delta_i{}^k|^j &= -p_m S_i{}^{mjk} - \delta_i{}^m S_m{}^{jk}.\end{aligned} \tag{10.7}$$

In particular, if the N-linear connection D is of Cartan type, i.e. $\Delta_{ij} = 0$, $\delta_j^i = \delta_j^i$, then we have

Proposition 4.10.2. *Any N-linear connection of Cartan type satisfies the following identities*

$$p_m R_i{}^m{}_{jk} + R_{ijk} = 0,\ p_m P_i{}^m{}_j{}^k + P^k{}_{ij} = 0,\ p_m S_i{}^{mjk} + S_i{}^{jk} = 0. \tag{10.8}$$

Finally, we remark that we can explicitly write the Bianchi identities, of an N-linear connection $D\Gamma(N) = (H^i{}_{jk}, C_i^{jk})$ if we express in the adapted basis $(\delta_i, \dot{\partial}^i)$ the Bianchi identities (8.9).

4.11 Parallelism. Horizontal and vertical paths

Let D be an N-linear connection, having the coefficients $D\Gamma(N) = (H^i{}_{jk}, C_i^{jk})$ in adapted basis $(\delta_i, \dot{\partial}^i)$.

Consider a smooth parametrized curve $\gamma : I \to T^*M$, having the image in a domain of a chart of T^*M. Thus γ has the analytical expression of the form:

$$x^i = x^i(t),\ p_i = p_i(t),\ t \in I. \tag{11.1}$$

The tangent vector field $\dot{\gamma} = \dfrac{d\gamma}{dt}$ can be written in the frame $(\delta_i, \dot{\partial}^i)$ (see §5, (5.8)), as follows

$$\dot{\gamma} = \frac{dx^i}{dt}\delta_i + \frac{\delta p_i}{dt}\dot{\partial}^i \tag{11.2}$$

where

$$\frac{\delta p_i}{dt} = \frac{dp_i}{dt} - N_{ji}(x(t), p(t))\frac{dx^j}{dt} \tag{11.3}$$

We denote

$$D_{\dot\gamma}X = \frac{DX}{dt},\ DX = \frac{DX}{dt}dt,\ \forall X \in \mathcal{X}(T^*M) \tag{11.4}$$

DX is called the covariant differential of the vector field X and $\dfrac{DX}{dt}$ is the covariant differential along the curve g.

If the vector field X is written in the form

$$X = X^H + X^V = X^i\delta_i + X_i\dot\partial^i, \text{ and } \dot\gamma = \dot\gamma^H + \dot\gamma^V,$$

then we can write

$$D_{\dot\gamma} = D^H_{\dot\gamma} + D^V_{\dot\gamma} = \frac{dx^i}{dt}D_{\delta_i} + \frac{\delta p_i}{dt}D_{\dot\partial^i}$$

A straightforward calculus leads to

$$DX = (dX^i + X^m\omega^i_m)\delta_i + (dX_i - X_m\omega^m_i)\dot\partial^i \tag{11.5}$$

where

$$\omega^i{}_j = H^i{}_{jk}dx^k + C^{ik}_j\delta p_k. \tag{11.6}$$

$\omega^i{}_j$ are called 1-forms of connection of D.

Setting

$$\frac{\omega^i{}_j}{dt} = H^i{}_{jk}\frac{dx^k}{dt} + C_j{}^{ik}\frac{\delta p_k}{dt} \tag{11.6$'$}$$

the covariant differential $\dfrac{DX}{dt}$ can be written:

$$\frac{DX}{dt} = \left(\frac{dX^i}{dt} + X^m\frac{\omega^i{}_m}{dt}\right)\delta_i + \left(\frac{dX_i}{dt} - X_m\frac{\omega^m{}_i}{dt}\right)\dot\partial^i \tag{11.7}$$

The vector field X is called parallel along the curve $\gamma : I \longrightarrow T^*M$, if $\dfrac{DX}{dt} = 0$, $\forall\, t \in I$. We obtain

Theorem 4.11.1. *The vector field $X = X^i\delta_i + X_i\dot\partial^i$ is parallel along the parametrized curve γ, with respect to the N-linear connection D, if and only if its coordinates X^i, X_i are solutions of the differential equations*

$$\frac{dX^i}{dt} + X^m\frac{\omega^i{}_m}{dt} = 0,\ \frac{dX_i}{dt} - X_m\frac{\omega^m{}_i}{dt} = 0 \tag{11.8}$$

The proof is immediate means of (11.7).

A theorem of existence and uniqueness for the parallel vector fields along a given parametrized curve γ in the manifold T^*M can be formulate in the classical manner.

Let us consider the case of vector field X and N-connection D, for which $DX = 0$, for any curve γ. Remark that $DX = 0$ is equivalent to

$$(*) \qquad dX^i + X^m \omega^i{}_m = 0,\ dX_i - X_m \omega^m{}_i = 0.$$

But $dX^i = \delta_j X^i dx^j + \dot{\partial}^j X^i \delta p_j$, together with (11.6) lead to the system of differential equations, equivalent to $(*)$

$$X^i{}_{|j} dx^j + X^i|^j \delta p_j = 0,\ X_{i|j} dx^j + X_i|^j \delta p_j = 0$$

Since $dx^i, \delta p_i$ are arbitrary, it follows

$$(11.9) \qquad \begin{cases} X^i{}_{|j} = 0, & X^i|^j = 0 \\ X_{i|j} = 0, & X_i|^j = 0 \end{cases}$$

Using the Ricci identities (10.1), and taking into account (11.9) we obtain the necessary conditions for a vector field $X = X^i \delta_i + X_i \dot{\partial}^i$ be absolute parallel:

$$(11.10) \qquad \begin{array}{lll} X^h R_h{}^i{}_{jk} = 0, & X^h P_h{}^i{}_j{}^k = 0, & X^h S_h{}^{ijk} = 0 \\ X_h R_i{}^h{}_{jk} = 0, & X_h P_i{}^h{}_j{}^k = 0, & X_h S_i{}^{hjk} = 0 \end{array}$$

The h-connection D is called with "absolute parallelism of vectors if, (11.10) is verified for any vector X. It follows:

Theorem 4.11.2. *The N-linear connection D is with the absolute parallelism of vectors, if and only if the curvature of D vanishes, i.e., we have*

$$(11.11) \qquad R_h{}^i{}_{jk} = 0,\ P_h{}^i{}_j{}^k = 0,\ S_h{}^{ijk} = 0.$$

Definition 4.11.1. The curve $\gamma : I \to T^*M$ is called autoparallel curve with respect to the N-linear connection D if $D_{\dot{\gamma}} \dot{\gamma} = 0$.

Using (11.2) it follows

$$(11.12) \qquad \frac{D\dot{\gamma}}{dt} = \left(\frac{d^2 x^i}{dt^2} + \frac{dx^s}{dt} \frac{\omega^i{}_s}{dt} \right) \delta_i + \left(\frac{d}{dt} \frac{\delta p_i}{dt} - \frac{\delta p_s}{dt} \frac{\omega^s{}_i}{dt} \right) \dot{\partial}^i.$$

The previous formula leads to the property:

Theorem 4.11.3. *The curve* γ, (11.1) *is autoparallel with respect to the* N*-connection* D *if and only if the functions* $x^i(t), p_i(t)$, $t \in I$, *are the solutions of the following system of differential equations:*

$$\frac{d^2x^i}{dt^2} + \frac{dx^s}{dt}\frac{\omega^i_s}{dt} = 0, \quad \frac{d}{dt}\frac{\delta p_i}{dt} - \frac{\delta p_s}{dt}\frac{\omega^s_i}{dt} = 0 \tag{11.13}$$

Starting from (11.13) we can ennounce a theorem of existence and uniqueness for the autoparallel curve can be formulated as in the classical manner.

In Section 5, we introduced the notion of horizontal curve, the condition $\frac{d\gamma}{dt} = \left(\frac{d\gamma}{dt}\right)^H$. Theorem 4.5.1 gives us a characterization of the horizontal curves by means of the system of differential equation (5.9), i.e. $x^i = x^i(t)$, $\frac{\delta p_i}{dt} = 0$.

Definition 4.11.2. An horizontal path of an N-linear connection D is a horizontal autoparallel curve with respect to D.

Theorem 4.11.4. *The horizontal paths of an* N*-linear connection* D *are characterized by the system of differential equations:*

$$\frac{d^2x^i}{dt^2} + H^i{}_{jk}(x,p)\frac{dx^j}{dt}\frac{dx^k}{dt} = 0, \quad \frac{\delta p_i}{dt} = 0 \tag{11.14}$$

Indeed, the equations (11.13) and $\frac{\delta p_i}{dt} = 0$ lead to (11.14).

A parametrized curve $\gamma : I \to T^*M$ is called vertical in the point $x_0 \in M$, if its tangent vector field $\dot{\gamma}$ belongs to the vertical distribution V. That means γ belongs to the fibre of T^*M in the point $x_0 \in M$.

Evidently, γ is a vertical curve in the point $x_0 \in M$, has the equations (11.1) of the form

$$x^i = x^i_0, \; p_i = p_i(t), \; t \in I. \tag{11.15}$$

Definition 4.11.2. A vertical path in the point $x_0 \in M$ is a vertical curve γ in the point x_0 which is autoparallel with respect to the N-linear connection D.

Theorem 4.11.3 implies:

Theorem 4.11.5. *The vertical paths in the point* $x_0 \in M$ *with respect to the* N*-linear connection* D *are characterized by the system of differential equations:*

$$x^i = x^i_0, \; \frac{d^2p_i}{dt^2} - C_i^{jk}(x_0,p)\frac{dp_j}{dt}\frac{dp_k}{dt} = 0 \tag{11.16}$$

Of course, it is not difficult to formulate a theorem of existence and uniqueness for the vertical paths in T^*M at the point $x_0 \in M$.

4.12 Structure equations of an N-linear connection. Bianchi identities

Let us consider an N-linear connection D with the coefficients $D\Gamma(N) = = (H^i{}_{jk}, C_i{}^{jk})$. Its 1-form of connection $\omega^i{}_j$ are given by (11.6).

Proposition 4.12.1. *With respect to a change of coordinate* (1.1) *on the manifold* T^*M *we have*

$$
\begin{aligned}
d\tilde{x}^i &= \frac{\partial \tilde{x}^i}{dx^j} dx^j, \ \delta\tilde{p}_i = \frac{\partial x^j}{\partial \tilde{x}^i}\delta p_j \\
\tilde{\omega}^i{}_j &= \frac{\partial \tilde{x}^i}{\partial x^r}\frac{\partial x^s}{\partial \tilde{x}^j}\omega^r{}_s + \frac{\partial x^s}{\partial \tilde{x}^j}\frac{\partial^2 \tilde{x}^i}{\partial x^s \partial x^r}dx^r
\end{aligned} \tag{12.1}
$$

Indeed, the expression of $\omega^i{}_j$ from (11.6) and the rules of transformations of $dx^i, \delta p_i$ and $H^i{}_{jk}$, $C_i{}^{jk}$ leads to (12.1).

Now, it is not difficult to prove:

Lemma 4.12.1. *The following geometrical object fields*

$$d(dx^i) - dx^m \wedge \omega^i{}_m, \ d\delta p_i + \delta p_m \wedge \omega^m{}_i \ \text{and}\ d\omega^i{}_j - \omega^m{}_j \wedge \omega^i{}_m$$

are a d-vector field, a d-covector field and a d-tensor field of type $(1,1)$, *respectively.*

Using this lemma, we can prove by a direct calculus a fundamental result.

Theorem 4.12.1. *For any* N*-linear connection* D *with the coefficients* $D\Gamma(N) = (H^i{}_{jk}, C_i{}^{jk})$ *the following structure equations hold:*

$$
\begin{aligned}
d(dx^i) - dx^m \wedge \omega^i{}_m &= -\Omega^i \\
d(\delta p_i) + \delta p_m \wedge \omega^m{}_i &= -\Omega_i \\
d\omega^i{}_j - \omega^m{}_j \wedge \omega^i{}_m &= -\Omega^i{}_j
\end{aligned} \tag{12.2}
$$

where Ω^i, Ω_i *are the 2-forms of torsion:*

$$
\begin{aligned}
\Omega^i &= \frac{1}{2}T^i{}_{jk}dx^j \wedge dx^k + C_j{}^{ik}dx^j \wedge \delta p_k \\
\Omega_i &= \frac{1}{2}R_{ijk}dx^j \wedge dx^k + P^k{}_{ij}dx^j \wedge \delta p_k + \frac{1}{2}S_i{}^{jk}\delta p_j \wedge \delta p_k
\end{aligned} \tag{12.3}
$$

and where $\Omega^i{}_j$ *is the 2-form of curvature:*

$$\Omega^i{}_j = \frac{1}{2}R_j{}^i{}_{km}dx^k \wedge dx^m + P_j{}^i{}_k{}^m dx^k \wedge \delta p_m + \frac{1}{2}S_j{}^{ikm}\delta p_k \wedge \delta p_m. \tag{12.3$'$}$$

Proof. $\Omega^i = dx^m \wedge \omega^i_m$ and (11.6) imply the first equation (12.2); $\Omega_i = -(d(\delta p_i) + \delta p_m \wedge \omega^m{}_i)$ and (4.11) lead to the second equation (12.2). Finally, the formulae (11.6), (4.11)

$$d\omega^i{}_j = dH^i_{jm} \wedge dx^m + dC_j{}^{im} \wedge \delta p_m + C^{im}_j d(\delta p_m), \quad \text{and}$$
$$dH^i{}_{jm} = \delta_s H^i_{jm} dx^s + \dot{\partial}^s H^i_{jm} \delta p_s, \; dC^{im}_j = \delta_s C^{im}_j dx^s + \dot{\partial}^s C^{im}_j \delta p_s$$

give us the last equation (12.2), with $\Omega^i{}_j$ from (12.3)'. **q.e.d.**

Remark. The previous theorem is extremely important in the geometry of the manifold T^*M and, especially in a theory of submanifolds embedded in T^*M.

Now we remark that the exterior differential of the system (12.2), modulo the same system determines the Bianchi identities of an N-linear connection $D\Gamma(N) = (H^i{}_{jk}, C^{jk}_i)$.

Theorem 4.12.2. *The Bianchi identities of an N-linear connection $D\Gamma(N)$ are as follows:*

$$\underset{(h,r,s)}{S} \{T^i{}_{hr|s} + T^i{}_{hm} T^m{}_{rs} - R_h{}^i{}_{rs} + R_{mrh} C_s{}^{im}\} = 0$$

(12.4) $$\underset{(h,r,s)}{S} \{S_k{}^{hr}|^s - S_k{}^{hm} S_m{}^{rs} + S_k{}^{hrs}\} = 0$$

$$\underset{(h,r,s)}{S} \{R_{jrh|s} + T^m{}_{hr} R_{jms} + P^m{}_{js} R_{mrh}\} = 0$$

$$\underset{(h,r,s)}{S} \{R_j{}^i{}_{hr|s} + R_j{}^i{}_{hm} T^m{}_{rs} + P_j{}^i{}_h{}^m R_{msr}\} = 0$$

(12.5)

$$\underset{(h,r,s)}{S} \{S_j{}^{ihr}|^s + S_j{}^{ihm} S_m{}^{sr}\} = 0$$

$$\underset{(h,s)}{\mathcal{A}} \{P_h{}^i{}_s{}^r + C_h{}^{mr} T^i{}_{sm} + C_h{}^{ir}{}_{|s} - C_h{}^{im} P^r{}_{ms}\} - T^i{}_{hs}|^r + T^m{}_{hs} C_m{}^{ir} = 0$$

(12.6) $$S_h{}^{irs} - \underset{(r,s)}{\mathcal{A}} \{\dot{\partial}^s C_h{}^{ir} + C_h{}^{mr} C_m{}^{is}\} = 0$$

$$R_k{}^h{}_{rs} + R_{ksr}|^h + S_k{}^{mh} R_{msr} - P^h{}_{km} T^m{}_{rs} + \underset{(r,s)}{\mathcal{A}} \{R_{ksm} C_r{}^{mh} - P^h{}_{kr|s} - P^h{}_{mr} P^m{}_{ks}\} = 0$$

$$\underset{(h,r)}{\mathcal{A}}\{P_k{}^h{}_s{}^r - P^h{}_{ks}|^r - P^h{}_{km}C_s{}^{mr} + S_k{}^{hm}P^r{}_{ms}\}+$$

$$+P^m{}_{ks}S_m{}^{rh} - S_k{}^{rh}{}_{|s} = 0, \tag{12.7}$$

$$R_j{}^i{}_{rs}|^k - P_j{}^i{}_m{}^k T^m{}_{rs} + S_j{}^{ikm}R_{msr}+$$
$$+\underset{(r,s)}{\mathcal{A}}\{R_j{}^i{}_{ms}C_r{}^{mk} + P_j{}^i{}_s{}^k{}_{|r} + P_j{}^i{}_r{}^m P^k{}_{ms}\} = 0,$$

$$S_j{}^{isk}{}_{|r} + P_j{}^i{}_r{}^m S_m{}^{ks} + \underset{(s,k)}{\mathcal{A}}\{P_j{}^i{}_r{}^s|^k + P_j{}^i{}_m{}^s C_r{}^{mk} + S_j{}^{ism}P^k{}_{mr}\} = 0. \tag{12.8}$$

In the applications we will consider the cases:

a. $T^i{}_{jk} = 0,\ S_i{}^{jk} = 0$

b. $T^i{}_{jk} = 0,\ \overset{(r)}{C}{}_i{}^{jk} = 0.$

etc.

Of course, $\underset{(h,r,s)}{S}$ is the symbol of cyclic sum and $\underset{(r,s)}{\mathcal{A}}$ is the symbol of alternate sum.

Chapter 5

Hamilton spaces

Based of the conception of classical mechanics, the notion of Hamilton space was defined and investigated by R. Miron in the papers [97], [101], [105]. It was studied by D. Hrimiuc and H.Shimada [63], [66] et al.

The geometry of Hamilton spaces can be studied using the geometrical method of that of cotangent bundle. On the other hand, it can be derived from the geometry of Lagrange spaces via Legendre transformation, using the notion of $\mathcal{L}$–duality.

In this chapter, we study the geometry of Hamilton spaces, combining these two methods and systematically using the geometrical theory of cotangent bundle.

We start with the notion of generalized Hamilton space. And then we detect from its geometry the theory of Hamilton spaces.

5.1 The spaces GH^n

Definition 5.1.1. A generalized Hamilton space is a pair $GH^n = (M, g^{ij}(x,p))$, where M is a real n–dimensional manifold and $g^{ij}(x,p)$ is a d-tensor field of type $(2,0)$ symmetric, nondegenerate and of constant signature on $\widetilde{T^*M}$.

The tensor g^{ij} is called as usual the fundamental (or metric) tensor of the space GH^n.

In the case when the manifold M is paracompact, then there exist metric tensors $g^{ij}(x,p)$ positively defined such that (M, g^{ij}) is a generalized Hamilton space.

Definition 5.1.2. A generalized Hamilton space $GH^n = (M, g^{ij}(x,p))$ is called reducible to a Hamilton one if there exists a Hamilton function $H(x,p)$ on T^*M such that

$$g^{ij} = \frac{1}{2}\dot{\partial}^i\dot{\partial}^j H \tag{1.1}$$

Let us consider the d-tensor field

$$C^{ijk} = -\frac{1}{2}\dot{\partial}^k g^{ij}. \tag{1.2}$$

In a similar manner in the case of generalized Lagrange spaces (cf. Ch.3) we can prove:

Proposition 5.1.1. *A necessary condition that a generalized Hamilton space be reducible to a Hamilton one is that the d-tensor C^{ijk} be totally symmetric.*

Theorem 5.1.1. *Let $g^{ij}(x,p)$ be the fundamental tensor of a space GH^n, 0-homogeneous. Thus a necessary and sufficient condition that GH^n be reducible to a Hamilton space is that the d-tensor field C^{ijk} be totally symmetric.*

Indeed, in this case the Hamilton function $H(x,p) = g^{ij}(x,p)p_i p_j$ satisfies the conditions imposed by Definition 5.1.2.

Remarks.

1. The definition of Hamilton spaces is given in a next section of this chapter.
2. Let $\gamma_{ij}(x)$ be a Riemann metric tensor. It is not difficult to prove that the space GH^n with the fundamental tensor

$$g^{ij}(x,p) = e^{-2\sigma(x,p)}\gamma^{ij}(x),\ \ \sigma \in \mathcal{F}(T^*M)$$

is not reducible to a Hamilton space.

The covariant tensor g_{ij} determined by g^{ij} is obtained from the equations

$$g_{ik}g^{kj} = \delta_i^{\ j}. \tag{1.3}$$

Let us consider the following tensor field:

$$C_i^{\ jk} = -\frac{1}{2}g_{is}(\dot{\partial}^j g^{sk} + \dot{\partial}^k g^{js} - \dot{\partial}^s g^{jk}). \tag{1.4}$$

Proposition 5.1.2. *The d-tensor $C_i^{\ jk}$ gives the v-coefficients of a v-covariant derivation with the property:*

$$g^{ij}|^k = \dot{\partial}^k g^{ij} + C_s^{\ ik}g^{sj} + C_s^{\ jk}g^{is} = 0 \tag{1.5}$$

The proof can be obtained easily.

We use the coefficients (1.4) in the theory of metrical connections with respect to the fundamental tensor g^{ij}.

5.2 N–metrical connections in GH^n

In general, we cannot determine a nonlinear connection from the fundamental tensor g^{ij} of the space GH^n. Therefore we study the N-linear connections compatible with g^{ij}, N being a priori given (cf. Ch.4).

If the nonlinear connection N has the coefficients $N_{ij}(x,p)$ then an adapted basis to the horizontal distribution N and vertical distribution V on T^*M is of the form

(2.1) $$\{\delta_k = \partial_k + N_{kj}\dot{\partial}^j, \dot{\partial}^i\}.$$

And an N-linear connection D has the coefficients $D\Gamma(N) = (H^i{}_{jk}, C_i^{jk})$.

Definition 5.2.1. An N-linear connection $D\Gamma(N)$ is called metrical with respect to the fundamental tensor g^{ij} of the space GH^n if:

(2.2) $$g^{ij}{}_{|h} = 0, g^{ij}|^h = 0.$$

In the case when g^{ij} is positively defined we obtain the geometrical meaning of the conditions (2.2). In this respect we can consider "the length" of a d-covector $\omega_i(x,p)$, given by $||\omega||$ as follows

$$||\omega||^2 = g^{ij}(x,p)\omega_i\omega_j, \ \forall\,(x,p) \in \widetilde{T^*M}.$$

We can prove without difficulties the followings:

Theorem 5.2.1. *An N-linear connection $D\Gamma(N)$ is metrical with respect to the fundamental tensor g^{ij} of the space GH^n if, and only if, along any smooth curve $\gamma : I \to \widetilde{T^*M}$ and for any parallel d-covector field ω_i, i.e., $\dfrac{D\omega_i}{dt} = 0$, we have $\dfrac{d||\omega||}{dt} = 0$.*

Using the condition (1.3), the tensorial equations (2.2) are equivalent to:

(2.2)′ $$g_{ij|k} = 0, g_{ij}|^k = 0.$$

Now, by the same methods as in Ch.3 we can prove a very important theorem. We have:

Theorem 5.2.2. 1) *There exist an unique N-linear connection $D\Gamma(N) = (H^i{}_{jk}, C_i^{jk})$ having the properties:*

1° *The nonlinear connection N is a priori given.*

2° *$D\Gamma(N)$ is metrical with respect to the fundamental tensor g^{ij} of the space GH^n.*

3° *The torisons $T^i{}_{jk}$ and $S_i{}^{jk}$ vanish.*

2) *The previous connection has the v-coefficients $C_i{}^{jk}$ from (1.4) and the h-coefficients $H^i{}_{jk}$ expressed by*

$$H^i{}_{jk} = \frac{1}{2}g^{is}(\delta_j g_{sk} + \delta_k g_{js} - \delta_s g_{jk}). \tag{2.3}$$

For a generalized Hamilton space $GH^n = (M, g^{ij})$, Obata's operators [113] can be defined:

$$\Omega^{ij}_{hk} = \frac{1}{2}(\delta^i_h \delta^j_k - g_{hk}g^{ij}),\ \Omega^{*ij}_{hk} = \frac{1}{2}(\delta^i_h \delta^j_k + g_{hk}g^{ij}). \tag{2.4}$$

We can prove:

Theorem 5.2.3. *The set of all metrical N–linear connections $D\overline{\Gamma}(N) = (\overline{H}^i{}_{jk}, \overline{C}_i{}^{jk})$ with respect to the fundamental tensor g^{ij} is given by*

$$\begin{aligned} \overline{H}^i{}_{jk} &= H^i{}_{jk} + \Omega^{is}_{rj} X^r_{sk}, \\ \overline{C}_i{}^{jk} &= C_i{}^{jk} + \Omega^{is}_{rj} Z^{rk}_s, \end{aligned} \tag{2.5}$$

where $D\Gamma(N) = (H^i{}_{jk}, C_i{}^{jk})$ is from (1.4), (2.3) and $X^i{}_{jk}$, $Z_i{}^{jk}$ are arbitrary d-tensor fields.

It is important to remark that the mappings $D\Gamma(N) \longrightarrow D\overline{\Gamma}(N)$ determined by (2.5) and the compsotion of these mappings is an Abelian group.

From the previous theorem, we deduce:

Theorem 5.2.4. *There exist an unique metrical N-linear connection $D\overline{\Gamma}(N) = (\overline{H}^i{}_{jk}, \overline{C}_i{}^{jk})$ with respect to the fundamental tensor g^{ij} having a priori given torsion d-tensor fields $T^i{}_{jk}$ $(= -T^i{}_{kj})$, $S_i{}^{jk}$ $(= -S_i{}^{kj})$. The coefficients of $D\overline{\Gamma}(N)$ have the following expressions:*

$$\begin{aligned} \overline{H}^i{}_{jk} &= \frac{1}{2}g^{is}(\delta_j g_{sk} + \delta_k g_{js} - \delta_s g_{jk}) + \frac{1}{2}g^{is}(g_{sh}T^h{}_{jk} - g_{jh}T^h{}_{sk} + g_{kh}T^h{}_{js}) \\ \overline{C}_i{}^{jk} &= -\frac{1}{2}g_{is}(\dot{\partial}^j g^{sk} + \dot{\partial}^k g^{js} - \dot{\partial}^s g^{jk}) - \frac{1}{2}g_{is}(g^{sh}S_h{}^{jk} - g^{jh}S_h{}^{sk} + g^{kh}S_h{}^{js}) \end{aligned} \tag{2.6}$$

Moreover, if we denote $R^{hi}{}_{jk} = g^{hs}R_s{}^i{}_{jk}$, etc. and apply to fundamental tensor g^{ij} the Ricci identities taking into account the equations (2.2) we obtain:

Proposition 5.2.1. *For any metrical N–linear connections $D\Gamma(N)$ with respect to the fundamental tensor g^{ij} of the space GH^n the following identities hold:*

$$\begin{aligned} &R^{ij}{}_{hk} + R^{ji}{}_{hk} = 0,\ P^{ij}{}_h{}^k + P^{ji}{}_h{}^k = 0,\ S^{ijhk} + S^{jihk} = 0 \\ &R_j{}^i{}_{hk} + R_j{}^i{}_{kh} = 0,\ S_i{}^{ihk} + S_i{}^{jkh} = 0. \end{aligned} \tag{2.7}$$

The expressions of the curvature tensors are given in the formulae (10.3), Ch.4.

The notions of parallelism, horizontal or vertical paths, as well as the structure equations for an metrical N-linear connection with respect to the fundamental tensor g^{ij} can be studied, using the results obtained in the last section of Chapter 4.

5.3 The N–lift of GH^n

Let $GH^n = (M, g^{ij})$ be a generalized Hamilton space and N (N_{ij}) an apriori given non–linear connection on the manifold T^*M. Thus $(\delta_i, \dot{\partial}^i)$ from (2.1) is an adapted basis to the distributions N and V. Its dual $(dx^i, \delta p_i)$ basis is expressed by the 1-forms dx^i and by

$$\delta p_i = dp_i - N_{ji} dx^j \tag{3.1}$$

Definition 5.3.1. The N-lift of the fundamental tensor g^{ij} is:

$$\mathbb{G} = g_{ij}(x,p) dx^i \otimes dx^j + g^{ij}(x,p)\delta p_i \otimes \delta p_j. \tag{3.2}$$

Theorem 5.3.1. *We have:*

1° *The N-lift $\mathbb{G}$ is a tensor field of type (0,2) on $\widetilde{T^*M}$, symmetric, nonsingular depending only on g^{ij} and on the nonlinear connection N.*

2° *The pair $(\widetilde{T^*M}, \mathbb{G})$ is a pseudo–Riemannian space.*

3° *The distributions N and V are orthogonal with respect to $\mathbb{G}$.*

Indeed, every term from (3.2) is defined on $\widetilde{T^*M}$ and is a tensor field. In the adapted basis (2.1) the tensor field $\mathbb{G}$ has the components

$$\mathbb{G}(\delta_i, \delta_j) = g_{ij}, \ \ \mathbb{G}(\delta_i, \dot{\partial}^j) = 0, \ \ \mathbb{G}(\dot{\partial}^i, \dot{\partial}^j) = g^{ij}. \tag{3.3}$$

It follows that $\mathbb{G}$ is symmetric, nondegenerate of type (0,2) tensor field. Obviously we have $\mathbb{G}(X^H, Y^V) = 0$, $\forall X^H, Y^V$. **q.e.d.**

Now, assuming that the nonlinear connection N (N_{ij}) is symmetric, i.e. $N_{ij} = N_{ji}$, we consider the $\mathcal{F}(\widetilde{T^*M})$–linear mapping $\check{\mathbb{F}} : \mathcal{X}(\widetilde{T^*M}) \longrightarrow \mathcal{X}(\widetilde{T^*M})$ defined in (6.7), Ch.4

$$\check{\mathbb{F}}(\delta_i) = -\check{\partial}_i, \ \ \mathbb{F}(\check{\partial}_i) = \delta_i, \ \ \check{\partial}_i = g_{ij}\dot{\partial}^j. \tag{3.4}$$

Theorem 5.3.2.

1° $\check{\mathbb{F}}$ *is globally defined on* T^*M.

2° $\check{\mathbb{F}}$ *is the tensor field*

$$\check{\mathbb{F}} = -g_{ij}\dot{\partial}^i \otimes dx^j + g^{ij}\delta_i \otimes \delta p_i.$$

3° $\check{\mathbb{F}}$ *is an almost complex structure determined by the fundamental tensor* g^{ij} *and by the nonlinear connection* N.

4° *The pair* $(\mathbb{G}, \check{\mathbb{F}})$ *is an almost Hermitian structure determined only by* g^{ij} *and* N.

5° *The associated almost symplectic structure to* $(\mathbb{G}, \check{\mathbb{F}})$ *is the canonical symplectic structure* $\theta = \delta p_i \wedge dx^i = dp_i \wedge dx^i$.

It follows that the space $(\widetilde{T^*M}, (\mathbb{G}, \check{\mathbb{F}}))$ is almost Kählerian. It is called the almost Kählerian model of the generalized Hamilton space GH^n.

5.4 Hamilton spaces

A Hamilton space $H^n = (M, H(x,p))$ is a particular case of a generalized Hamilton space $GH^n = (M, g^{ij})$ in the sense that the fundamental tensor derived from a regular Hamilton function $H : T^*M \longrightarrow R$.

Since the triple (T^*M, θ, H) forms a Hamiltonian system, we can apply the theory from Chapter 4.

Definition 5.4.1. A Hamilton space is a pair $H^n = (M, H(x,p))$ where M is a real n-dimensional manifold and H is a function on T^*M having the properties:

1° $H : (x,p) \in T^*M \longrightarrow H(x,p) \in R$ is differentiable on the manifold $\widetilde{T^*M}$ and it is continuous on the null section of $\pi^* : T^*M \longrightarrow M$.

2° The Hessian of H (with respect to the momenta p_i), given by the matrix $\|g^{ij}(x,y)\|$ is nondegenerate:

$$g^{ij}(x,y) = \frac{1}{2}\dot{\partial}^i\dot{\partial}^j H, \ \operatorname{rank}\left\|g^{ij}(x,p)\right\| = n \text{ on } \widetilde{T^*M}. \tag{4.1}$$

3° The d-tensor field $g^{ij}(x,p)$ has constant signature on $\widetilde{T^*M}$.

Of course, g^{ij} from (4.1) is a d-tensor field, cf. §5, Ch.4. It is called the fundamental tensor, or metric tensor of the space $H^n = (M, H)$ and the Hamilton function H is called the fundamental function for H^n.

From the previous definition we obtain:

Theorem 5.4.1. *Every Hamilton space $H^n = (M, H)$ is a generalized Hamilton space.*

Indeed, $GH^n = (M, g^{ij})$, where g^{ij} is given by (4.1), is a generalized Hamilton space.

The converse is not true (see Proposition 5.1.1).

Theorem 5.4.2. *If the base manifold is paracompact, then there exists a Hamilton function H on T^*M, such that $H^n = (M, H)$ is a Hamilton space.*

Indeed, M being a paracompact manifold, let $g^{ij}(x)$ be a Riemannian metric tensor on M. Then we can consider the Hamilton function on T^*M:

$$H(x,p) = \frac{1}{mc} g^{ij}(x) p_i p_j,$$

where $m \neq 0$, and c is speed light. The properties 1°–3° from the last definition can be proved directly. **q.e.d.**

Let us consider the canonical symplectic structure θ on T^*M:

$$\theta = dp_i \wedge dx^i \tag{4.2}$$

By means of Definition 4.2.2, we obtain that the triple (T^*M, θ, H), where H is the fundamental function of a Hamilton space H^n, form a Hamilton system. Thus the mapping $S_\theta : \mathcal{X}(T^*M) \longrightarrow \mathcal{X}^*(T^*M)$ defined by (2.2), Ch.4, $S_\theta(X) = i_X\theta$ is an isomorphism. Applying the Theorem 4.2.2 we obtain:

Theorem 5.4.3. *For any Hamilton space $H^n = (M, H(x,p))$ the following properties hold:*

1° *There exists a unique vector field $X_H \in \mathcal{X}(\widetilde{T^*M})$ with the property:*

$$i_{X_H}\theta = -dH. \tag{4.3}$$

2° *The integral curves of the vector field X_H are given by the Hamilton–Jacobi equations:*

$$\frac{dx^i}{dt} = \frac{\partial H}{\partial p_i}, \ \frac{dp_i}{dt} = -\frac{\partial H}{\partial x^i}. \tag{4.4}$$

Indeed, from (2.6)′, the Hamilton vector field X_H can be written as:

$$X_H = \frac{\partial H}{\partial p_i}\frac{\partial}{\partial x^i} - \frac{\partial H}{\partial x^i}\frac{\partial}{\partial p_i}. \tag{4.5}$$

Also, we have:

Corollary 5.4.1. *The fundamental function $H(x,p)$ of the space H^n is constant along the integral curve of the Hamilton vector field X_H.*

Now it is easy to observe that the following formula holds:

$$\{f,g\} = \theta(X_f, X_g), \ \forall X \in \mathcal{X}(T^*M). \tag{4.6}$$

Therefore, by means of Poisson brackets, the Hamilton–Jacobi equations can be written in the form:

$$\frac{dx^i}{dt} = \{H, x^i\}, \ \frac{dp_i}{dt} = \{H, p_i\}. \tag{4.4$'$}$$

We remark that the Jacobi method of integration of Hamilton–Jacobi equations, mentioned in §2, Ch.4, works in the present case as well.

One of the important d-tensor field derived from the fundamental function H of the Hamilton space H^n is:

$$C^{ijh} = -\frac{1}{4}\dot{\partial}^k\dot{\partial}^j\dot{\partial}^i H = -\frac{1}{2}\dot{\partial}^k g^{ij}. \tag{4.7}$$

Proposition 5.4.1. *We have:*

1° *The d-tensor field C^{ijk} is totally symmetric.*

2° *C^{ijk} vanishes, if and only if the fundamental tensor field $g^{ij}(x,p)$ does not depend on the momenta p_i.*

Let us consider the coefficients $C_i^{\ jk}$ from (1.4) of the v-covariant derivation. From (4.7) it follows

$$C_i^{\ jk} = -\frac{1}{2}g_{is}\dot{\partial}^s g^{jk}; \ C^{ijk} = g^{is}C_s^{\ jk}. \tag{4.8}$$

Of course, these coefficients have the properties

$$g^{ij}|^h = 0, \ S_i^{\ jh} = 0. \tag{4.9}$$

In the next section we will use the functions $C_i^{\ jk}$ as the v-coefficients of the canonical metrical connection of the space H^n.

5.5 Canonical nonlinear connection of the space H^n

An important problem is to determine a nonlinear connection N (N_{ij}) for a Hamilton space $H^n = (M, H(x,p))$ depending on the fundamental function H, only.

A method for finding a nonlinear connection with the mentioned property was given by R. Miron in the paper [97]. This consists in the transformation of the canonical nonlinear connection N of a Lagrange space $L^n = (M, L(x,y))$, via Legendre transformation: Leg : $L^n \longrightarrow H^n$, into the canonical nonlinear connection of the Hamilton space H^n. This method will be developed in the Chapter 7, using the notion of $\mathcal{L}$–duality between the spaces L^n and H^n.

Here, we give the following result of R. Miron, without demonstration, which can be found in §2 of Chapter 7.

We have:

Theorem 5.5.1.

1° *The following set of functions*

$$N_{ij} = \frac{1}{4}\{g_{ij}, H\} - \frac{1}{4}\left(g_{ik}\frac{\partial^2 H}{\partial p_k \partial x^j} + g_{jk}\frac{\partial^2 H}{\partial p_k \partial x^i}\right) \tag{5.1}$$

determines the coefficients of a nonlinear connection N of the Hamilton space H^n.

2° *The nonlinear connection with the coefficients N_{ij}, depends only on the fundamental function H of the Hamilton space H^n.*

The brackets { } from (5.1) are the Poisson brackets (2.1), Ch.4.

Indeed, by a straightforward computation, it follows that, under a coordinate change on the total space of cotangent bundle T^*M, N_{ij} from (5.1) obeys the rule of transformation (4.3), Ch.4. Hence, the point 1° of Theorem 5.5.1 is verified. Taking into account the expressions of the coefficients N_{ij}, given in (5.1), the property 2° is also evident.

The previous nonlinear connection will be called *canonical.*

Remark. If the fundamental function $H(x,p)$ of the space H^n is globally defined on T^*M, then the canonical nonlinear connection has the same property.

Proposition 5.5.1. *The canonical nonlinear connection N of the Hamilton space $H^n = (M, H(x,p))$ has the following properties*

$$\tau_{ij} = \frac{1}{2}(N_{ij} - N_{ji}) = 0 \tag{5.2}$$

$$R_{ijk} + R_{jki} + R_{kij} = 0 \tag{5.3}$$

where R_{ijh} *is given by* (4.12), *Ch.4.*

Proof. 1° The proof follows directly by (5.1); 2° is a consequence of the formula (4.12), Ch.4. **q.e.d.**

Taking into account the fact that the canonical nonlinear connection N is a regular distribution on $\widetilde{T^*M}$ with the property:

$$T_u\widetilde{T^*M} = N_u \oplus V_u,\ \forall\, u \in \widetilde{T^*M}, \tag{5.4}$$

it follows that $(\delta_i, \dot{\partial}^i)$ is an adapted basis to the direct decomposition (5.4), where

$$\delta_i = \partial_i + N_{ij}\dot{\partial}^j. \tag{5.5}$$

The dual basis of $(\delta_i, \dot{\partial}^i)$ is $(dx^i, \delta p_i)$, where

$$\delta p_i = dp_i - N_{ji}dx^j. \tag{5.5'}$$

Therefore, we can apply the theory of N-metrical connection expounded in §2 of the present chapter for study the canonical case.

5.6 The canonical metrical connection of Hamilton space H^n

Let us consider the N-linear connections $D\Gamma(N) = (H^i{}_{jk}, C_i{}^{jk})$ with the property that N is the canonical nonlinear connection with the coefficients (5.1). It can be studied by means of theory presented in §5.2.

Consequently, we have:

Theorem 5.6.1. 1) *In a Hamilton space* $H^n = (M, H(x,p))$ *there exists a unique* N*-linear connection* $D\Gamma(N) = (H^i{}_{jk}, C_i{}^{jk})$ *verifying the axioms:*

1° *N is the canonical nonlinear connection.*

2° *The fundamental tensor g^{ij} is h-covariant constant*

$$g^{ij}{}_{|k} = 0. \tag{6.1}$$

3° *The tensor g^{ij} is v-covariant constant, i.e.*

$$g^{ij}|^k = 0. \tag{6.2}$$

4° *$D\Gamma(N)$ is h-torsion free, i.e.*

$$T^i{}_{jk} = H^i{}_{jk} - H^i{}_{kj} = 0.$$

5° $D\Gamma(N)$ *is v-torsion free, i.e.*

$$S_i^{\ jk} = C_i^{\ jk} - C_i^{\ kj} = 0.$$

2) *The connection* $D\Gamma(N)$ *has its coefficients*$(N_{ij}, H^i{}_{jk}, C_i^{\ jk})$ *given by* (5.1) *and by the following generalized Christoffel symbols:*

$$\begin{aligned} H^i{}_{jk} &= \frac{1}{2} g^{is}(\delta_j g_{sk} + \delta_k g_{js} - \delta_s g_{jk}), \\ C_i^{\ jk} &= -\frac{1}{2} g_{is}(\dot{\partial}^j g^{sk} + \dot{\partial}^k g^{js} - \dot{\partial}^s g^{jk}) = -\frac{1}{2} g_{is} \dot{\partial}^j g^{sk}. \end{aligned} \tag{6.3}$$

3) *This connection depends only on the fundamental function H of the Hamilton space* H^n.

The proof of this theorem follows the ordinary way.

A such kind of connection, determined only by the fundamental function H is called canonical and is denoted by $C\Gamma(N) = (H^i{}_{jk}, C_i^{\ jk})$ or by $C\Gamma = (N_{ij}, H^i{}_{jk}, C_i^{\ jk})$.

Now we can repeat Theorems 5.2.3 and 5.2.4, using the canonical metrical connection $C\Gamma$.

Proposition 5.6.1. *The Ricci identities, with respect to the canonical connection* $C\Gamma$ *are given by*

$$\begin{aligned} X^i{}_{|j|h} - X^i{}_{|h|j} &= X^m R_m{}^i{}_{jh} - X^i|^m R_{mjh}, \\ X^i{}_{|j}|^h - X^i|^h{}_{|j} &= X^m P_m{}^i{}_j{}^h - X^i{}_{|m} C_j^{\ mh} - X^i|^m P^h{}_{mj}, \\ X^i|^j|^h - X^i|^h|^j &= X^m S_m^{\ ijh}, \end{aligned} \tag{6.4}$$

where the d-tensors of torsion are $T^i{}_{jk} = 0$, $S_i^{\ jk} = 0$, $C_i^{\ jk}$ *and*

$$R_{ijh} = \delta_h N_{ji} - \delta_j N_{hi}, \; P^h{}_{ij} = H^h{}_{ij} - \dot{\partial}^h N_{ji}, \tag{6.5}$$

respectively, and the d-tensors of curvature of $C\Gamma$ *are given by* (10.3), *Ch.4.*

The Ricci identities can be extended as usual to any d-tensor field.

For instance, in the case of a d-tensor field $t^{ij}(x,p)$ we have

$$\begin{aligned} t^{ij}{}_{|k|h} - t^{ij}{}_{|h|k} &= t^{mj} R_m{}^i{}_{kh} + t^{im} R_m{}^j{}_{kh} - t^{ij}|^m R_{mkh} \\ t^{ij}{}_{|k}|^h - t^{ij}|^h{}_{|k} &= t^{mj} P_m{}^i{}_k{}^h + t^{im} P_m{}^j{}_k{}^h - t^{ij}{}_{|m} C_k^{\ mh} - t^{ij}|^m P^h{}_{mk} \\ t^{ij}|^k|^h - t^{ij}|^h|^k &= t^{mj} S_m^{\ ikh} + t^{jm} S_m^{\ jkh}. \end{aligned} \tag{6.4$'$}$$

Applying the identities $(6.4)'$ to the fundamental tensor g^{ij} of the Hamilton space H^n we obtain

Proposition 5.6.2. *The canonical connection $C\Gamma$ has the properties* (2.7), *i.e.*

$$R^{ij}{}_{hk} + R^{ji}{}_{hk} = 0, \;\; P^{ij}{}_{h}{}^{k} + P^{ji}{}_{h}{}^{k} = 0, \;\; S^{ijhk} + S^{jihk} = 0 \tag{6.6}$$

The Ricci identities applied to the Liouville–Hamilton vector field $C^* = p_i\dot{\partial}^i$ lead to some important identities.

Proposition 5.6.3. *The canonical metrical connection $C\Gamma$ of the Hamilton space $H^n = (M, H)$ satisfies the following identities*

$$\begin{aligned} &\Delta_{ij|k} - \Delta_{ik|j} = -p_m R_i{}^m{}_{jk} - \delta_i{}^m R_{mjk} \\ &\Delta_{ij}|^k - \delta_i{}^k{}_{|j} = -p_m P_i{}^m{}_j{}^k - \Delta_{im} C_j{}^{mk} - \delta_i{}^m P^k{}_{mj} \\ &\delta_i{}^j|^k - \delta_i{}^k|^j = -p_m S_i{}^{mjk} \end{aligned} \tag{6.7}$$

where Δ_{ij} and $\delta_i{}^k$ are the deflections tensors of $C\Gamma(N)$:

$$\Delta_{ij} = p_{i|j}; \;\; \delta^i_j = p_j|^i \tag{6.7$'$}$$

are given by (9.9), *Ch.*4.

The canonical metrical connection $C\Gamma$ is of Cartan type if $\Delta_{ij} = 0$, $\delta^i_j = \delta^i_j$. From (6.7) we obtain the following

Proposition 5.6.4. *If the canonical metrical connection of the space H^n is of Cartan type, then the identities* (10.8), *Ch.* 4, *hold good.*

Applying again the Ricci identities to the fundamental function H of the Hamilton space H^n, we obtain:

Proposition 5.6.5. *The following identities hold:*

$$\begin{aligned} &H_{|j|k} - H_{|k|j} = -H|^m R_{mjk} \\ &H_{i|j}|^k - H|^k{}_{i|j} = -H_{|m} C_j{}^{mk} - H|^m P^k{}_{mj} \\ &H|^j|^k - H|^k|^j = 0. \end{aligned} \tag{6.8}$$

5.7 Structure equations of $C\Gamma(N)$. Bianchi identities

For the canonical connection $C\Gamma = (H^i{}_{jk}, C_i{}^{jk})$ the 1-form of connection $\omega^i{}_j$ are given by (11.6), Ch.4:

$$\omega^i{}_j = H^i{}_{jk} dx^k + C_i{}^{jk} \delta p_k. \tag{7.1}$$

The structure equations of $C\Gamma$ are given by Theorem 4.12.1. In this case, we have:

Theorem 5.7.1. *The canonial metrical connection $C\Gamma$ of the Hamilton space $H^n = (M, H)$ has the following structure equations*

$$\begin{aligned} d(dx^i) - dx^m\wedge\omega^i{}_m &= -\Omega^i, \\ d(\delta p_i) + \delta p_m\wedge\omega^m{}_i &= -\Omega_i, \\ d\omega^i{}_j - \omega^m{}_j\wedge\omega^i{}_m &= -\Omega^i{}_j, \end{aligned} \tag{7.2}$$

where Ω^i, Ω_i are the 2-forms of torsion:

$$\Omega^i = C_j{}^{ik} dx^j\wedge\delta p_k, \quad \Omega_i = \frac{1}{2}R_{ijk}dx^j\wedge dx^k + P^k{}_{ij}dx^j\wedge\delta p_k \tag{7.3}$$

and where $\Omega^i{}_j$ are the 2-forms of curvature:

$$\Omega^i{}_j = \frac{1}{2}R_j{}^i{}_{km}dx^k\wedge dx^m + P_j{}^i{}_k{}^m dx^k\wedge\delta p_m + \frac{1}{2}S_j{}^{ikm}\delta p_k\wedge\delta p_m. \tag{7.4}$$

The previous theorem is very useful in the geometry of Hamilton spaces and especially in the theory of subspaces of Hamilton spaces.

We can now derive the Bianchi identities of the canonical metrical connection $C\Gamma$, taking the exterior differential of the system of equations (7.2), modulo the same system.

We obtain a particular case of Theorem 4.12.2:

Theorem 5.7.2. *The canonical metrical connection $C\Gamma$ of the Hamilton space H^n satisfies the Bianchi identities*(12.4)–(12.8), *Ch.4, with $T^i{}_{jk} = 0$, $S_i^{jk} = 0$.*

5.8 Parallelism. Horizontal and vertical paths

The notion of parallelism of vector fields in the Hamilton spaces $H^n = (M, H(x,p))$, endowed with the canonical metrical connection $C\Gamma = (N_{ij}, H^i{}_{jk}, C_i^{jk})$ can be studied as an application of the theory presented in §11, Chapter 4.

Let $\gamma : I \longrightarrow \widetilde{T^*M}$ be a parametrized curve with the analytical expression

$$x^i = x^i(t),\ p_i = p_i(t),\ t \in I,\ \text{rank}\|p_i(t)\| = 1. \tag{8.1}$$

Then it results

$$\dot\gamma(t) = \frac{d\gamma}{dt} = \frac{dx^i}{dt}\delta_i + \frac{\delta p_i}{dt}\dot\partial^i$$

where

$$\frac{\delta p_i}{dt} = \frac{dp_i}{dt} - N_{ji}\frac{dx^j}{dt}. \tag{8.2}$$

Here N_{ij} are the coefficients (5.1) of the canonical nonlinear connection. We say that the curve γ is horizontal if $\dfrac{\delta p_i}{dt} = 0$, $\forall\, t \in I$.

For a vector field X on $\widetilde{TM}$, given in adapted basis by

$$X = X^i\delta_i + X_i\dot{\partial}^i,$$

along of the curve γ, we have the covariant differential D of $C\Gamma$ of the form (11.7), Ch.4:

$$\frac{DX}{dt} = \left(\frac{dX^i}{dt} + X^m\frac{\omega^i_m}{dt}\right)\delta_i + \left(\frac{dX_i}{dt} - X_m\frac{\omega^m{}_i}{dt}\right)\dot{\partial}^i \tag{8.3}$$

where $\omega^i{}_j$ are 1-forms connection of $C\Gamma$.

It follows:

Theorem 5.8.1. *The vector field $X = X^i\delta_i + X_i\dot{\partial}^i$ is parallel along the parametrized curve γ, with respect to the canonical metrical connection $C\Gamma$, if and only if its coordinates X^i and X_i are solutions of the differential equations* (11.8), *Ch.*4, *where $\omega^i{}_j(x(t), p(t))$ are the* 1*-forms connection of $C\Gamma$.*

In particular, (see Theorem 4.11.2), we have:

Theorem 5.8.2. *The Hamilton space $H^n = (M, H)$, endowed with the canonical metrical connection $C\Gamma$, is with absolute parallelism of vectors if and only if the d-curvature tensors of $C\Gamma$ $R_i{}^j{}_{kh}$, $P_j{}^i{}_k{}^h$ and $S_i{}^{jkh}$ respectively vanish.*

We say that the curve γ is autoparallel with respect to $C\Gamma$ if $\dfrac{D\dot{\gamma}}{dt} = 0$, $\forall\, t \in I$.

Taking into account Theorem 4.11.3, we have

Theorem 5.8.3. *A curve $\gamma : I \longrightarrow \widetilde{T^*M}$ is autoparallel with respect to the canonical metrical connection $C\Gamma$ if and only if the functions $(x^i(t), p_i(t))$, $t \in I$ are the solutions of the system of differential equations* (11.13), *Ch.* 4, *in which $\omega^i{}_j$ are the* 1*-forms connection of $C\Gamma$.*

By means of Definition 4.11.2, a *horizontal path* of the canonical metrical connection $C\Gamma$ of Hamilton space H^n, is a horizontal autoparallel curve. Theorem 4.11.4 gives us:

Theorem 5.8.4. *The horizontal paths of the Hamilton space H^n, endowed with the canonial metrical connection $C\Gamma$ are characterized by the system of differential equations:*

(8.4) $$\frac{d^2x^i}{dt^2} + H^i{}_{jk}(x(t), p(t))\frac{dx^j}{dt}\frac{dx^k}{dt} = 0, \quad \frac{dp_i}{dt} - N_{ji}(x(t), p(t))\frac{dx^j}{dt} = 0$$

We recall that $\gamma : t \in I \longrightarrow \gamma(t) \in \widetilde{T^*M}$ is a vertical curve in the point $x_0 \in M$ if $\frac{dx^i}{dt} = 0$. Hence, its analytical representation in local coordinates is of the form

$$x^i = x^i_0, \; p_i = p_i(t), \; t \in I.$$

Thus, a vertical path in the point $x_0 \in M$ is a vertical curve γ in the point x_0 which is autoparallel with respect to $C\Gamma$.

Theorem 4.11.5 leads to:

Theorem 5.8.5. *The vertical paths in the point $x_0 \in M$ with respect to the canonical metrical connection $C\Gamma$ of the Hamilton space $H^n = (M, H)$ are characterized by the system of differential equations*

$$x^i = x^i_0, \; \frac{d^2p_i}{dt} - C_i^{\;jk}(x_0, p)\frac{dp_j}{dt}\frac{dp_k}{dt} = 0.$$

In next section we apply these results in some important particular cases.

5.9 The Hamilton spaces of electrodynamics

Let us consider some important examples of Hamilton spaces.

1) Gravitational field

The Lagrangian of gravitational field $L = mc\gamma_{ij}(x)y^iy^j$ (see Ch.3) is transformed, via Legendre transformation ,in the regular Hamiltonian

(9.1) $$H = \frac{1}{mc}\gamma^{ij}(x)p_ip_j.$$

Therefore the pair $H^n = (M, H)$ is a Hamilton space. Its fundamental tensor field is given by

(9.2) $$g^{ij} = \frac{1}{mc}\gamma^{ij}(x).$$

It follows that $(T^*M.\theta, H)$ is a Hamiltonian system. Using Theorem 4.2.2, we obtain

Proposition 5.9.1. *The Hamilton–Jacobi equations of the space H^n, with the fundamental function* (9.1), *are:*

$$\frac{dx^i}{dt} = g^{ij}p_j; \ \frac{dp_i}{dt} = -\frac{\partial H}{\partial x^i}. \tag{9.3}$$

Let us denote $\gamma^i{}_{jk}(x)$ the Christoffel symbols of the metric tensor $\gamma_{ij}(x)$ we obtain:

Proposition 5.9.2. *The canonical nonlinear connection N of the space $H^n = (M, H)$,* (9.1) *has the following coefficients*

$$N_{ij}(x, y) = \gamma^k{}_{ij}(x)p_k. \tag{9.4}$$

Indeed, the coefficients N_{ij} are given by the formula (5.1). Therefore (9.4) holds.

Proposition 5.9.3. *The canonical metrial connection of the Hamilton space H^n has the coefficients*

$$H^i{}_{jk} = \gamma^i{}_{jk}, \ C_i^{jk} = 0. \tag{9.5}$$

Now we can apply to this case the whole theory from the previous sections of this chapter.

2) The Hamilton space of electrodynamics

The Lagrangian of electrodynamics (9.1), Ch.3, is transformed via Legendre transformation in the following regular Hamiltonian

$$H(x, p) = \frac{1}{mc}\gamma^{ij}(x)p_ip_j - \frac{2e}{mc^2}A^i(x)p_i + \frac{e^2}{mc^3}A^i(x)A_i(x), \tag{9.6}$$

where $m, c, \gamma^{ij}(x)$ have the meanings from (9.1), e is the charge of test body, $A_i(x)$ is the vector–potential of an electromagnetic field, and $A^i(x) = \gamma^{ij}(x)A_j(x)$.

The first thing to remark is:

Proposition 5.9.4. *The pair $H^n = (M, H)$,* (9.6) *is an Hamilton space.*

Indeed, a direct calculus leads to the d-tensor

$$g^{ij} = \frac{1}{mc}\gamma^{ij}. \tag{9.7}$$

This is the fundamental tensor the space H^n. Its covariant is $g_{ij}(x) = mc\gamma_{ij}(x)$. The Hamilton–Jacobi equations, (2.2), Ch.4, can be easily written.

The velocity field of this space can be written:

$$\rho^i = \frac{1}{2}\dot{\partial}^i H = \frac{1}{mc}\gamma^{ij}(x)p_j - \frac{e}{mc^2}A^i(x). \tag{9.8}$$

And the canonical nonlinear connection is given by

Proposition 5.9.5. *The canonical nonlinear connection of the space H^n with the Hamiltonian* (9.6) *has the following coefficients:*

$$N_{ij} = \gamma^k{}_{ij}(x)p_k + \frac{e}{c}(A_{i|k} + A_{k|i}), \tag{9.9}$$

where $A_{i|k} = \dfrac{\partial A_i}{\partial x^k} - A_s\gamma^s{}_{ik}$.

Indeed, to prove the formula (9.9) we apply the formula (5.1) to the fundamental function (9.6).

Now, remarking that $\dfrac{\delta g_{ij}}{\delta x^k} = mc\dfrac{\partial \gamma_{ij}}{\partial x^k}$, we get:

Proposition 5.9.6. *The canonical metrical connection $C\Gamma$ of the Hamilton space H^n with the fundamental function $H(x,p)$ with the Hamiltonian* (9.6), *has the following coefficients:*

$$H^i{}_{jk} = \gamma^i{}_{jk},\ C_i^{jk} = 0.$$

These geometrical object fields: H, g_{ij}, N_{ij}, $H^i{}_{jk}$, C_i^{jk} allow to develop the geometry of the Hamilton space of electrodynamic.

3) In the case when we take into account the Lagrangian

$$L(x,y) = mc\gamma_{ij}(x)y^iy^j + \frac{2e}{m}A_i(x)y^i + U(x),$$

where $U(x)$ is a force function, applying Legendre mapping we get the following Hamiltonian

$$\begin{aligned} H(x,p) &= \frac{1}{c\left(m + \dfrac{U}{c^2}\right)}\gamma^{ij}(x)p_ip_j - \frac{2e}{c^2\left(m + \dfrac{U}{c^2}\right)}p_iA^i(x) + \\ &+ \frac{e^2}{2c^3\left(m + \dfrac{U}{c^2}\right)}A_i(x)A^i(x) - \frac{1}{c}\,\frac{U(x)}{c}. \end{aligned} \tag{9.10}$$

It follows

Proposition 5.9.7. *The pair $H^n = (M, H)$ with the Hamiltonian* (9.10) *is a Hamilton space, having the fundamental tensor field:*

$$g^{ij} = \frac{1}{c\left(m + \dfrac{U}{c^2}\right)} \gamma^{ij}(x) p_i p_j.$$

We can prove also:

Proposition 5.9.8. *The canonical nonlinear connection of the space H^n,* (9.10) *has the coefficients*

$$N_{ij} = \gamma^k{}_{ij} p_k + \frac{e}{2c^2\left(m + \dfrac{U}{c^2}\right)} \{g_{hi} A^h{}_{|j} + g_{hj} A^h{}_{|i} - (A_i U_j + A_j U_i)\}.$$

Proposition 5.9.9. *The canonical metrical connection of the space H^n,* (9.10), *has the coefficients $C\Gamma = (H^i{}_{jk}, C_i^{jk})$:*

$$H^i{}_{jk} = \gamma^i{}_{jk},\ C_i^{jk} = 0.$$

Now the previous theory of this chapter can be applied to the Hamilton space H^n, (9.10).

5.10 The almost Kählerian model of an Hamilton space

Let $H^n = (M, H(x,p))$ be a Hamilton space and $g^{ij}(x,p)$ its fundamental tensor field.

The canonical nonlinear connection N has the coefficients (5.1). The adapted basis to the distributions N and V is $\left(\dfrac{\delta}{\delta x^i} = \delta_i = \partial_i + N_{ij}\dot{\partial}^j, \dot{\partial}^i\right)$ and its dual basis $(dx^i, \delta p_i = dp_i - N_{ji} dx^j)$.

Thus the following tensor on $\widetilde{T^*M}$:

$$\mathbb{G} = g_{ij}(x,p) dx^i \otimes dx^j + g^{ij}(x,p) \delta p_i \otimes \delta p_j \tag{10.1}$$

gives a pseudo-Riemannian structure on $\widetilde{T^*M}$, which depends only on the fundamental function $H(x,p)$ of the Hamilton space H^n. These properties are the consequences of Theorem 5.3.1 and of the fact that N_{ij} is the canonical nonlinear connection.

The tensor $\mathbb{G}$ is called the N-lift of the fundamental tensor g^{ij}.

The distributions N and V are orthogonal with respect to $\mathbb{G}$, because the formulae (3.3) hold.

Taking into account the mapping $\check{\mathbb{F}} : \mathcal{X}(\widetilde{T^*M} \longrightarrow \mathcal{X}(\widetilde{T^*M})$ defined in (3.4) with respect to the canonical nonlinear connection, we obtain the following properties:

1° $\check{\mathbb{F}}$ is globally defined on $\widetilde{T^*M}$.

2° $\check{\mathbb{F}}$ is an almost complex structure: $\check{\mathbb{F}} \circ \check{\mathbb{F}} = -I$ on $\widetilde{T^*M}$.

3° $\check{\mathbb{F}}$ is determined only by the fundamental function $H(x,p)$ of the Hamilton space H^n.

Finally, we obtain a particular form of Theorem 5.3.2:

Theorem 5.10.1.

1° *The pair* $(\mathbb{G}, \check{\mathbb{F}})$ *is an almost Hermitian structure on the manifold* $\widetilde{T^*M}$.

2° *The structure* $(\mathbb{G}, \check{\mathbb{F}})$ *is determined only by the fundamental function* $H(x,p)$ *of the Hamilton space* H^n.

3° *The associated almost symplectic structure to* $(\mathbb{G}, \check{\mathbb{F}})$ *is the canonical symplectic structure* $\theta = dp_i \wedge dx^i = \delta p_i \wedge dx^i$.

4° *The space* $(\widetilde{T^*M}, \mathbb{G}, \check{\mathbb{F}})$ *is almost Kählerian.*

The proof is similar with that from Lagrange spaces (cf. Ch.3).

The equality $\delta p_i \wedge dx^i = (dp_i - N_{ji}dx^j)\wedge dx^i = dp_i \wedge dx^i = \theta$ follows from the fact that the torsion $\tau_{ij} = \dfrac{1}{2}(N_{ij} - N_{ji})$ of the canonical nonlinear connection vanishes (cf (5.1)).

The space $(\widetilde{T^*M}, \mathbb{G}, \check{\mathbb{F}})$ is called the almost Kählerian model of the Hamilton space H^n. This model is useful in applications.

Chapter 6

Cartan spaces

The modern formulation of the notion of Cartan spaces is due of the first author [97], [98], [99]. Based on the studies of E. Cartan, A. Kawaguchi [75], H. Rund [139], R. Miron [98], [99], D. Hrimiuc and H.Shimada [66], [67], P.L. Antonelli [21], etc., the geometry of Cartan spaces is today an important chapter of differential geometry.

In the previous chapter we have presented the geometrical theory of Hamilton spaces $H^n = (M, H(x,p))$. In particular, if the fundamental function $H(x,p)$ is 2-homogeneous on the fibres of the cotangent bundle (T^*M, π^*, M) the notion of Cartan space is obtained. It is remarkable that these spaces appear as dual of the Finsler spaces, via Legendre transformation. Using this duality several important results in the Cartan spaces can be obtained: the canonical nonlinear connection, the canonical metrical connection etc. Therefore, the theory of Cartan spaces has the same symmetry and beauty like Finsler geometry. Moreover, it gives a geometrical framework for the Hamiltonian theory of Mechanics or Physical fields.

6.1 The notion of Cartan space

As usually, we consider a real, n-dimensional smooth manifold M, the cotangent bundle (T^*M, π^*, M) and the manifold $\widetilde{T^*M} = T^*M \setminus \{0\}$.

Definition 6.1.1. A Cartan space is a pair $\mathcal{C}^n = (M, K(x,p))$ such that the following axioms hold:

1° K is a real function on T^*M, differentiable on $\widetilde{T^*M}$ and continuous on the null section of the projection π^*.

2° K is positive on T^*M.

3° K is positively 1-homogeneous with respect to the momenta p_i.

4° The Hessian of K^2, with elements

$$g^{ij}(x,p) = \frac{1}{2}\dot{\partial}^i\dot{\partial}^j K^2 \tag{1.1}$$

is positive-defined.

It follows that $g^{ij}(x,p)$ is a symmetric and nonsingular d-tensor field, contravariant of order 2. Hence we have

$$\operatorname{rank}\left\| g^{ij}(x,p) \right\| = n, \quad \text{on } \widetilde{T^*M}. \tag{1.2}$$

The functions $g^{ij}(x,p)$ are 0–homogeneous with respect to the momenta p_i.

For a Cartan space $\mathcal{C}^n = (M, K(x,p))$ the function K is called the fundamental function and g^{ij} the fundamental or metric tensor.

At the beginning we remark:

Theorem 6.1.1 *If the base manifold M is paracompact, then on the manifold T^*M there exist functions K such that the pair (M,K) is a Cartan space.*

Indeed, if M is paracompact, then T^*M is paracompact too. Let $g^{ij}(x,p)$ be a Riemann structure on M. Considering the function

$$K(x,p) = \{g^{ij}(x)p_ip_j\}^{1/2} \text{ on } T^*M \tag{1.3}$$

we obtain a fundamental function for a Cartan space $\mathcal{C}^n$.

Examples.

1. Let $(M, \gamma_{ij}(x))$ be a Riemannian manifold and

$$\alpha^* = \{\gamma^{ij}(x)p_ip_j\}^{1/2}, \; \beta^* = b^i(x)p_i.$$

Assuming $\beta > 0$ on an open set $\check{U} \subset T^*M$ it follows that

$$K(x,p) = \alpha^* + \beta^* \tag{1.4}$$

$$K(x,p) = \frac{(\alpha^*)^2}{\beta^*} \tag{1.4$'$}$$

are the fundamental functions of Cartan spaces.

The first one, (1.4), is called *Randers metric* and the second one, (1.4)$'$, is called the *Kropina metric.* They will be studied in Chapter 7.

Remark. More generally, we can consider the Cartan spaces with (α^*, β^*)–metric. They are given by the definition 6.1.1, with $K(x,p) = \check{K}(\alpha^*(x,p), \beta^*(x,p))$, $\check{K}$ being a function 1-homogeneous with respect to α^* and β^*.

2. The pair $\mathcal{C}^n = (M, K(p))$, where

$$K(p) = \{(p_1)^m + (p_2)^m + \cdots + (p_m)^m\}^{1/m}, \ (m = 2r, \ r > 1). \tag{1.5}$$

in the preferential charts of an atlas on T^*M, is a Cartan space.

The function

$$\check{K}(x, p) = e^{\sigma(x)} K(p) \tag{1.5'}$$

and $K(p)$ from (1.5) is the so–called Antonelli ecological metric [11].

A general remark is imporant:

Theorem 6.1.2. *Every Cartan space $\mathcal{C}^n = (M, K(x,p))$ uniquely determines a Hamilton space: $H^m = (M, K^2(x, y))$.*

Indeed, by means of the axioms 1°–4° from Definition 6.1.1 it follows that the pair $(M, K^2(x,p))$ is a Hamilton space.

We can apply the theory from previous chapter. So, considering the canonical symplectic structure θ on T^*M:

$$\theta = dp_i \wedge dx^i \tag{1.6}$$

we deduce that the triple (T^*M, θ, K^2) is a Hamiltonian system. We can apply Theorems 4.4.3 and 5.4.3. Therefore we can formulate:

Theorem 6.1.3. *For any Cartan space $\mathcal{C}^n = (M, K(x,p))$, the following properties hold:*

1° *There exists a unique vector field $X_{K^2} \in \mathcal{X}(\widetilde{T^*M})$ with the property*

$$i_{X_{K^2}}\theta = -dK^2. \tag{1.7}$$

2° *The integral curves of the Hamilton vector field X_{K^2} are given by the Hamilton–Jacobi equations*

$$\frac{dx^i}{dt} = \frac{\partial K^2}{\partial p_i}; \ \frac{dp_i}{dt} = -\frac{\partial K^2}{\partial x^i}. \tag{1.8}$$

Indeed, X_{K^2} is of the form

$$X_{K^2} = \frac{\partial K^2}{\partial p_i}\frac{\partial}{\partial x^i} - \frac{\partial K^2}{\partial x^i}\frac{\partial}{\partial p_i}$$

and (1.8) gives us its integral curves.

Corollary 6.1.1. *The function K^2 is constant along the integral curves of Hamilton vector field X_{K^2}.*

The Hamilton–Jacobi equations (1.8) of the Cartan space $\mathcal{C}^n$ are fundamental for the geometry of $\mathcal{C}^n$. They are the dual of the Euler–Lagrange equations of the Finsler space. Therefore (1.8) are called the equations of geodesics of the Cartan space $\mathcal{C}^n$. This theory will be developed in the next chapter.

6.2 Properties of the fundamental function K of Cartan space $\mathcal{C}^n$

Proposition 6.2.1. *The following properties hold:*

1° $p^i = \frac{1}{2}\dot{\partial}^i K^2$ *is 1–homogeneous d-vector field on* $\widetilde{T^*M}$.

2° $g^{ij} = \dot{\partial}^j p^i = \frac{1}{2}\dot{\partial}^i\dot{\partial}^j K^2$ *is 0-homogeneous d-tensor field.*

3° $C^{ijk} = -\frac{1}{4}\dot{\partial}^i\dot{\partial}^j\dot{\partial}^h K^2$ *is* (-1)*-homogeneous, symmetric d-tensor field.*

Indeed, if $f(x,p)$ is r-homogeneous with respect of p_i, then $\frac{\partial f}{\partial p_i} = \dot{\partial}^i f$ is $r-1$–homogeneous. Therefore 1°–3° follows. Let g_{ij} be the covariant tensor of g^{ij}, i.e.: $g_{ij}g^{jh} = \delta_i^k$.

Proposition 6.2.2. *We have the following formulae:*

$$p^i = g^{ij}p_j,\ p_i = g_{ij}p^j \tag{2.1}$$

$$K^2 = g^{ij}p_ip_j = p_ip^j \tag{2.2}$$

$$C^{ijk}p_k = C^{ikj}p_k = C^{kij}p_k = 0. \tag{2.3}$$

These formulae are consequences of 1°–3° from the previous proposition.

The fundamental tensor $g^{ij}(x,p)$ depends only on the point $x = \pi^*(x,p) \in M$ if and only if $\dot{\partial}^h g^{ij} = 0$. In this case the pair $(M, g^{ij}(x))$ is a Riemannian space. So we have

Proposition 6.2.3. *The Cartan space* $\mathcal{C}^n = (M, K(x,p))$ *is Riemannian if, and only if, the d–tensor field* $C^{ijk}(x,p)$ *vanished.*

Indeed, $\dot{\partial}^h g^{ij} = -2C^{ijh} = 0$ holds, if and only if $g^{ij}(x,p) = g^{ij}(x)$.

Let us consider the coefficients C_i^{jk} from (1.4), Ch.5. of the v-covariant derivation. They are given by (4.8) Ch.5 for a Cartan space, i.e.:

$$C_i^{jk} = -\frac{1}{2}g_{is}\dot{\partial}^s g^{jk} = g_{is}C^{sjk}. \tag{2.4}$$

Naturally, these coefficients have the following properties:

$$g^{ij}|^k = 0,\ S_i^{jk} = 0 \tag{2.5}$$

$$p_i|^k = \delta_i^k. \tag{2.6}$$

Other properties of the fundamental function K of the Cartan space $\mathcal{C}^n$ will be expressed by means of the canonical nonlinear connection and canonical metrical connection.

6.3 Canonical nonlinear connection of a Cartan space

Since the Cartan space $\mathcal{C}^n = (M, K)$ is Hamilton space $H^n = (M, K^2)$, the canonical nonlinear connection of the space $\mathcal{C}^n$ has the coefficients N_{ij}, from (5.1), Ch.5.

If we consider the Christoffel symbols of $g^{ij}(x,p)$ given by

$$\gamma^i{}_{jk} = \frac{1}{2} g^{ih}(\partial_k g_{hj} + \partial_j g_{hk} - \partial_h g_{jk}), \tag{3.1}$$

then the following contractions by p_i or p^i:

$$\gamma^0_{jk} = \gamma^i{}_{jk} p_i, \gamma^0_{j0} = \gamma^i{}_{jk} p_i p^k \tag{3.2}$$

lead us to

Theorem 6.3.1. (Miron [98], [99]) *The canonical nonlinear connection of the Cartan space $\mathcal{C}^n = (M, K)$ is given by the following coefficients*

$$N_{ij} = \gamma^0_{ij} - \frac{1}{2}\gamma^0_{h0}\dot{\partial}^h g_{ij}. \tag{3.3}$$

Proof. By means of formula (5.1), with $H = K^2$, and $K^2 = g^{ij} p_i p_j$ we obtain

$$N_{ij} = \gamma^0_{ij} + \frac{1}{4}\dot{\partial}^s g_{ij} \partial_s g^{kh} p_k p_h.$$

But $\gamma^0_{j0} = -\frac{1}{2}\partial_j g^{kh} p_k p_h$. The two last equations imply the formula (3.3). **q.e.d.**

Remark. The coefficients (3.3) can be obtained from the coefficients of the Cartan nonlinear connection of a Finsler spaces by means of the so called $\mathcal{L}$–duality (see Ch.7).

Let us consider the adapted basis $(\delta_i, \dot{\partial}^i)$ to the distributions N and V where N is determined by the canonical nonlinear connection of the Cartan space $\mathcal{C}^n$. Its dual basis is $(dx^i, \delta p_i)$.

Then the d-tensor of integrability of N is (4.12), Ch.4:

$$R_{ijh} = \delta_h N_{ji} - \delta_j N_{hi}. \tag{3.4}$$

By a direct calculus we have:

Proposition 6.3.1. *In a Cartan space $\mathcal{C}^n$ the following properties hold:*

$$\tau_{ij} = \frac{1}{2}(N_{ij} - N_{ji}) = 0 \tag{3.5}$$

$$R_{ijk} + R_{jki} + R_{kij} = 0. \tag{3.6}$$

Proposition 6.3.2. *The distribution N determined by the canonical nonlinear connection of a Cartan space $\mathcal{C}^n$ is integrable if and only if the d-tensor field R_{ijh} vanishes.*

Other consequence of the previous theorem is given by

Proposition 6.3.3. *The canonical nonlinear connection of the Cartan space $\mathcal{C}^n$ depends only on the fundamental function K.*

6.4 The canonical metrical connection

Let us consider the N-linear connection $D\Gamma(N) = (H^i{}_{jk}, C_i{}^{jk})$ of the Cartan space $\mathcal{C}^n = (M, K(x,p))$ in which N is the canonical nonlinear connection, with the coefficients N_{ij} from (3.3). The h- and v-covariant derivatives of the fundamental tensor $g^{ij} = \frac{1}{2}\dot{\partial}^i\dot{\partial}^j K^2$ of the space $\mathcal{C}^n$ are expressed by

$$\begin{cases} g^{ij}{}_{|k} = \delta_k g^{ij} + g^{sj}H^i_{sk} + g^{is}H^j_{sk}, \\ g^{ij}|^k = \dot{\partial}^k g^{ij} + g^{sj}C_s{}^{ik} + g^{is}C_s{}^{jk} \end{cases} \tag{4.1}$$

In particular, in the case of Cartan spaces, Theorem 6.6.1 implies:

Theorem 6.4.1. 1) *In a Cartan space $\mathcal{C}^n = (M, K(x,p))$ there exists a unique N-linear connection $C\Gamma(N) = (H^i{}_{jk}, C_i{}^{jk})$ verifying the axioms:*

1° *N is the canonical nonlinear connection of th espace $\mathcal{C}^n$,*

2° *The equation $g^{ij}{}_{|k} = 0$ holds with respect to $C\Gamma(N)$,*

3° *The equation $g^{ij}|^k = 0$ holds with respect to $C\Gamma(N)$,*

4° *$C\Gamma(N)$ is h-torsion free: $T^i{}_{jk} = 0$,*

5° *$C\Gamma(N)$ is v-torsion free: $S_i{}^{jk} = 0$.*

2) *The connection $C\Gamma(N)$ has the coefficients given by the generalized Christoffel symbols:*

$$(4.2)\qquad \begin{aligned} H^i{}_{jk} &= \frac{1}{2}g^{is}(\delta_j g_{sk} + \delta_k g_{js} - \delta_s g_{jk}) \\ C_i^{jk} &= -\frac{1}{2}g_{is}(\dot\partial^j g^{sk} + \dot\partial^k g^{js} - \dot\partial^s g^{jk}) = g_{is}C^{sjk}. \end{aligned}$$

3) *$C\Gamma(N)$ depends only on the fundamental function K of the Cartan space $\mathcal{C}^n$.*

The connection $C\Gamma(N)$ from the previous theorem will be called the canonical metrical connection of the Cartan space $\mathcal{C}^n$.

The connection $C\Gamma(N)$ is called metrical since the conditions $g^{ij}{}_{|k} = 0$, $g^{ij}|^k = 0$ hold good. But these two conditions have a geometrical meaning.

Let us consider "the square of the norm" of a d-covector field $\lambda_i(x,p)$ on $\widetilde{T^*M}$:

$$(4.3)\qquad \|\lambda\|^2 = g^{ij}(x,p)\lambda_i\lambda_j$$

and a parametrized curve $c : t \in I \longrightarrow \widetilde{T^*M}$, given in a local chart of $\widetilde{T^*M}$ by

$$x^i = x^i(t),\ p_i = p_i(t),\ t \in I.$$

We have the folowing

Theorem 6.4.2. *An N-linear connection $D\Gamma(N) = (H^i{}_{jk}, C_i^{jk})$ on $\widetilde{T^*M}$ has the property $\frac{d}{dt}\|\lambda\|^2 = 0$, along any curve c, and for any parallel covector field $\lambda_i(x(t),p(t))$ on c, if and only if the following equations $g^{ij}{}_{|k} = 0$, $g^{ij}|^k = 0$ hold.*

Proof.

$$(*)\qquad \frac{d}{dt}\|\lambda\|^2 = \frac{D}{dt}(g^{ij}\lambda_i\lambda_j) = \frac{Dg^{ij}}{dt}\lambda_i\lambda_j + 2g^{ij}\lambda_i\frac{D\lambda_j}{dt}.$$

But

$$\frac{d}{dt}\|\lambda\|^2 = 0,\ \frac{D\lambda^i}{dt} = 0,\ \forall\,\lambda_i \text{ imply } \frac{Dg^{ij}}{dt} = 0.$$

Or, the curve c being arbitrary we get $Dg^{ij} = 0 \Longleftrightarrow \{g^{ij}{}_{|k} = 0,\ g^{ij}|^k = 0\}$. Conversely, if $g^{ij}{}_k = 0$, $g^{ij}|^k = 0$, then $\frac{Dg^{ij}}{dt} = 0$, $\frac{D\lambda_i}{dt} = 0$ and $(*)$ imply $\frac{d}{dt}\|\lambda\|^2 = 0$.

q.e.d.

We will denote $C\Gamma(N) = (H^i{}_{jk}, C_i^{jk})$ by $C\Gamma = (N_{ij}, H^i{}_{jk}, C_i^{jk})$, pointed out all coefficients of the canonical metrical connection.

The first property of $C\Gamma$ is as follows:

Proposition 6.4.1. *The coefficients* $(N_{ij}, H^i{}_{jk}, C_i{}^{jk})$ *of the canonical metrical connection* $C\Gamma$ *are homogeneous with respect to momenta* p_i, *of degree* $1, 0, -1$, *respectively.*

Indeed, g^{ij} are 0–homogeneous imply $\gamma^i{}_{jk}$ are 0-homogeneous, γ^0_{ij} are 1-homogeneous and, using (3.3), it results N_{ij} are 1-homogeneous etc.

Proposition 6.4.2. *The canonical metrical connection* $C\Gamma$ *of the Cartan space* $\mathcal{C}^n$ *is of Cartan type. Namely its deflection tensors have the property*

$$\Delta_{ij} = p_{i|k} = 0, \ \ \delta^j_i = p_i|^j = \delta^j_i. \tag{4.4}$$

Proof. First of all we remark that the h-coefficients $H^i{}_{jk}$ of $C\Gamma$ give us

$$p_i H^i{}_{jk} = \gamma^0_{jk} - \frac{1}{2} N_{0r} \dot{\partial}^r g_{jk}. \tag{4.5}$$

But N_{0j} can be calculated from (3.3). We obtain $N_{0j} = \gamma^0_{0j}$. Therefore

$$\Delta_{ij} = N_{ij} - p_m H^m_{ij} = \gamma^0_{ij} - \frac{1}{2}\gamma^0_{h0}\dot{\partial}^h g_{ij} - \left(\gamma^0_{ij} - \frac{1}{2} N_{0h}\dot{\partial}^h g_{ij}\right) = 0$$

Similarly, $\delta^j_i = \delta^i_j - p_m C_i{}^{mj} = \delta^j_i$. **q.e.d.**

The first equality (4.4) is an important one. It can be substituted with the axiom 1° from Theorem 6.4.1. We obtain a system of axioms of Matsumoto type (cf. 2.5, Ch.2) for the canonical metrical connection $C\Gamma$ of the Cartan space $\mathcal{C}^n$.

Theorem 6.4.3. 1) *For a Cartan space* $\mathcal{C}^n = (M, K(x,p))$ *there exists a unique linear connection* $D\Gamma = (N_{ij}, H^i{}_{jk}, C_i{}^{jk})$ *which satisfies the following axioms*

A_1 $\Delta_{ij} = 0$ (*h–deflection tensor field of* $D\Gamma$ *vanishes*)
A_2 $g^{ij}{}_{|k} = 0$ ($D\Gamma$ *is h-metrical*)
A_3 $g^{ij}|^k = 0$ ($D\Gamma$ *is v-metrical*)
A_4 $T^i{}_{ij} = H^i{}_{jk} - H^i{}_{kj} = 0$
A_5 $S_i{}^{jk} = C_i{}^{jk} - C_i{}^{kj} = 0$

2) *The previous metrical connection is exactly the canonical metrical connection* $C\Gamma$.

Proof. Assume that the nonlinear connection N, with the coefficients N_{ij} satisfies the first axiom A_1 then the connection $D\Gamma(N) = (H^i{}_{jk}, C_i{}^{jk})$ satisfies the axioms A_2–A_5 given by Theorem 6.4.1.

Now, using $H^i{}_{jk}$ from (4.2) we obtain $p_s H^s_{jk}$ given by (4.5). So, for $\Delta_{ij} = 0$, we get $N_{ij} - p_s H^s_{ij} = 0$ and

$$N_{ij} = \gamma^0_{ij} - \frac{1}{2} N_{0r}\dot{\partial}^r g_{ij}. \tag{$*$}$$

Contracting by p^i, we deduce $N_{0j} = \gamma^0_{0j}$. Substituting in $(*)$ we have

$$N_{ij} = \gamma^0_{ij} - \frac{1}{2}\gamma^0_{0r}\dot{\partial}^r g_{ij}.$$

But these are the coefficients of the canonical nonlinear connection. The uniqueness of the connection $C\Gamma = (N_{ij}, H^i{}_{jk}, C_i^{jk})$, which satisfies the axioms A_1–A_5 can be obtained by usual way [88]. **q.e.d.**

One can prove that the axioms A_1–A_5 are independent [88].
Finally of this section we obtain without difficulties

Proposition 6.4.3. *The canonical metrical connection $C\Gamma$ of the Cartan space $\mathcal{C}^n = (M, K(x,p))$, has the properties:*

1° $K_{|j} = 0,\ K|^j = \dfrac{p^j}{K}$

2° $K^2{}_{|j} = 0,\ K^2|^j = 2p^i$

3° $p_{i|j} = 0,\ p_i|^j = \delta^j_i$

4° $p^i{}_{|j} = 0,\ p^i|^j = g^{ij}.$

Proposition 6.4.4. *The d-torsions of the canonical metrical connection $C\Gamma$ are given by the following:*

$$R_{ijk} = \delta_k N_{ij} - \delta_j N_{ik},\ C_i^{jk}, T^i{}_{jk} = 0,\ S_i^{jk} = 0,\ P^i{}_{jk} = H^i{}_{jk} - \dot{\partial}^i N_{jk}. \tag{4.6}$$

Of course, we have

$$R_{ijk} = -R_{ikj},\ P^i{}_{jk} = P^i{}_{kj},\ C_i^{jk} = C_i^{kj}. \tag{4.7}$$

Proposition 6.4.5. *The d-tensors of curvatures of the canonical metrical connection are given by the formula* (10.3), *Ch.4, where*

$$S_k{}^{ijh} = C_k{}^{mh}C_m{}^{ij} - C_k{}^{mj}C_m{}^{ih}. \tag{4.8}$$

Indeed, by means of $C_k{}^{ij} = g_{km}C^{mij}$ and

$$\dot{\partial}^h C_k{}^{ij} - \dot{\partial}^j C_k{}^{ih} = 2(C_k{}^{mh}C_m{}^{ij} - C_k{}^{mj}C_m{}^{ih}),$$

we obtain the expression (4.8) of the d-tensor of curvature $S_k{}^{ijh}$.

Applying the Ricci identities (10.5), Ch.4 to the fundamental tensor field g^{ij}, and denoting as usual $R^{ij}{}_{kh} = g^{is}R_s{}^j{}_{kh}$, etc. we obtain:

Theorem 6.4.4. *The d-tensors of curvature of the canonical metrical connection $C\Gamma$ have the properties:*

$$R^{ij}{}_{kh} + R^{ji}{}_{kh} = 0,\ P^{ij}{}_k{}^h + P^{ji}{}_k{}^h = 0,\ S^{ijkh} + S^{jikh} = 0. \tag{4.9}$$

The Ricci identities applied to the Liouville covector field p_i and taking into account the equations: $p_{i|j} = 0$, $p_i|^j = \delta_i^j$ we get some important identities.

Theorem 6.4.5. *The canonical metrical connection $C\Gamma$ of the Cartan space $\mathcal{C}^n$ satisfies the identities:*

$$R_i{}^0{}_{jk} + R_{ijk} = 0,\ P_i{}^0{}_j{}^k + P^k{}_{ij} = 0,\ S_i{}^{0jk} = 0. \tag{4.10}$$

We derive from (4.10):

Corollary 6.4.1. *The canonical nonlinear connection $C\Gamma$ has the property:*

$$R_{0jk} = 0,\ P^k{}_{0j} = 0. \tag{4.11}$$

Of course, the index "0" means the contraction by p_i or p^i.

6.5 Structure equations. Bianchi identities

Taking into account the general theory of structure equations and Bianchi identities of a general N–linear connection $D\Gamma(N) = (H^i{}_{jk}, C_i{}^{jk})$, in the case of Cartan spaces $\mathcal{C}^n = (M, K(x,p))$, we obtain, for the canonical metrical connection $C\Gamma(N)$ with the coefficients (4.2), the following results.

The 1-form connections of $C\Gamma$ are:

$$\omega^i{}_j = H^i{}_{jh}dx^h + C_j{}^{ih}\delta p_h. \tag{5.1}$$

Taking into account the fact that the torsion tensors $T^i{}_{jk}$ and $S_i{}^{jk}$ vanish, we obtain:

Theorem 6.5.1. *The structure equations of the canonical metrical connection $C\Gamma$ of the Cartan space $\mathcal{C}^n = (M, K(x,p))$ are:*

$$\begin{aligned} d(dx^i) - dx^m \wedge \omega^i{}_m &= -\Omega^i \\ d(\delta p_i) + \delta p_m \wedge \omega_i^m &= -\Omega_i \end{aligned} \tag{5.2}$$

$$d\omega^i{}_j - \omega^m_j \wedge \omega^i{}_m = -\Omega^i{}_j \tag{5.3}$$

Ω^i, Ω_i being the 2-forms of torsion:

$$\begin{aligned} \Omega^i &= C_j{}^{ik} dx^j \wedge \delta p_k \\ \Omega_i &= \frac{1}{2} R_{ijk} dx^j \wedge dx^k + P^k{}_{ij} dx^j \wedge \delta p_k \end{aligned} \tag{5.4}$$

and $\Omega^i{}_j$ is the 2-form of curvature:

$$\Omega^i{}_j = \frac{1}{2} R_j{}^i{}_{km} dx^k \wedge dx^m + P_j{}^i{}_k{}^m dx^k \wedge \delta p_m + \frac{1}{2} S_j{}^{ikm} \delta p_k \wedge \delta p_m. \tag{5.5}$$

Applying Theorem 4.12.2, we get:

Theorem 6.5.2. *The Bianchi identities of the canonical metrical connection $C\Gamma$ of Cartan space $\mathcal{C}^n = (M, K(x,p))$ are the following:*

$$\begin{cases} \underset{(h,r,s)}{\mathcal{S}} \{R_h{}^i{}_{rs} - R_{mrh} C_s{}^{im}\} = 0, \\ \underset{(h,r,s)}{\mathcal{S}} S_k{}^{hrs} = 0, \\ \underset{(h,r,s)}{\mathcal{S}} \{R_{jrh|s} + P^m{}_{js} R_{mrh}\} = 0, \end{cases} \tag{5.6}$$

$$\begin{cases} \underset{(h,r,s)}{\mathcal{S}} \{R_j{}^i{}_{hr|s} + P_j{}^i{}_h{}^m R_{msr}\} = 0, \\ \underset{(h,r,s)}{\mathcal{S}} S_j{}^{ihr}|^s = 0, \end{cases} \tag{5.7}$$

$$\begin{cases} \underset{(h,s)}{\mathcal{A}} \{P_h{}^i{}_s{}^r + C_h{}^{ir}{}_{|s} - C_h{}^{im} P^r{}_{ms}\} = 0, \\ S_h{}^{irs} - \underset{(r,s)}{\mathcal{A}} \{\dot\partial^s C_h{}^{ir} + C_h{}^{mr} C_m{}^{is}\} = 0, \\ R_k{}^h{}_{rs} + R_{ksr}|^h + \underset{(r,s)}{\mathcal{A}} \{R_{ksm} C_r{}^{mh} - P^h{}_{kr|s} - P^h{}_{mr} P^m{}_{ks}\} = 0, \end{cases} \tag{5.8}$$

$$\begin{cases} \underset{(h,r)}{\mathcal{A}} \{P_k{}^h{}_s{}^r - P^h{}_{km} C_s{}^{mr}\} = 0, \\ R_j{}^i{}_{rs}|^k + S_j{}^{ikm} R_{msr} + \underset{(r,s)}{\mathcal{A}} \{R_j{}^i{}_{ms} C_r{}^{mk} + P_j{}^i{}_s{}^k{}_{|r} + P_j{}^i{}_r{}^m P^k{}_{ms}\} = 0, \end{cases} \tag{5.9}$$

and

$$S_j{}^{isk}{}_{|r} + \underset{(s,k)}{\mathcal{A}} \{P_j{}^i{}_r{}^s|^k + P_j{}^i{}_m{}^s C_r{}^{mk} + S_j{}^{ism} P^k{}_{mr}\} = 0, \tag{5.10}$$

where the symbol $\underset{(h,r,s)}{\mathcal{S}}$ is of cyclic sum and $\underset{(i,j)}{\mathcal{A}}$ is for alternate sum.

6.6 Special N-linear connections of Cartan space $\mathcal{C}^n$

Let $\mathcal{C}^n = (M, K)$ be a Cartan space, N its canonical nonlinear connection and $C\Gamma(N)$ the canonical metrical connection of $\mathcal{C}^n$.

The set of all N–linear connections $D\overline{\Gamma}(N) = (\overline{H}^i{}_{jk}, \overline{C}_i{}^{kj})$ which are metrical with respect to the fundamental tensor field g^{ij} of the Cartan spce $\mathcal{C}^n$ is given by Theorem 5.2.3 with some particular references.

We have

Theorem 6.6.1. *In a Cartan space $\mathcal{C}^n = (M, K(x,p))$ the set of all N–linear connection $D\overline{\Gamma}(N) = (\overline{H}^i{}_{jk}, \overline{C}_i{}^{jk})$, metrical with respect to the fundamental tensor g^{ij} of $\mathcal{C}^n$ is given by*

$$\text{(6.1)}\qquad \begin{aligned} \overline{H}^i{}_{jk} &= H^i_{jk} + \Omega^{is}_{rj} X^r_{sk} \\ \overline{C}_i{}^{jk} &= C_i{}^{jk} + \Omega^{is}_{rj} Z_s^{rk} \end{aligned}$$

where $(H^i{}_{jk}, C_i{}^{jk}) = C\Gamma(N)$ is the canonical metrical connection and X^r_{sk} is an arbitrary 0–homogeneous d-tensor field and Z_s^{rk} is an arbitrary (-1)-homogeneous tensor field.

In this case we can remark that the mappings $D\Gamma(N) \longrightarrow D\overline{\Gamma}(N)$ determined by (6.1) form an Abelian group, isomorphic to the additive groups of the pairs of d-tensor field $(\Omega^{is}_{rj} X^r_{sk}, \Omega^{is}_{rj} Z_s{}^{rk})$. Theorem 5.2.4 has a particular case for Cartan spaces.

Theorem 6.6.2. *Let $\mathcal{C}^n = (M, K(x,p))$ be a Cartan space and N its canonical nonlinear connection. There exists an unique metrical N-linear connection $D\overline{\Gamma}(N) = (\overline{H}^i{}_{jk}, \overline{C}_i{}^{jk})$ with respect to the fundamental tensor g^{ij}, having a priori given torsions fields: $T^i{}_{jk}$ $(= -T^i{}_{kj})$, 0–homogeneous and $S_i{}^{jk}$ $(= -S_i{}^{kj})$, (-1)-homogeneous. The coefficients of $D\overline{\Gamma}(N)$ have the expressions*

$$\text{(6.2)}\qquad \begin{aligned} \overline{H}^i{}_{jk} &= H^i{}_{jk} + \frac{1}{2} g^{is}(g_{sh} T^h{}_{jk} - g_{jh} T^h{}_{sk} + g_{kh} T^h{}_{js}) \\ \overline{C}_i{}^{jk} &= C_i{}^{jk} - \frac{1}{2} g_{is}(g^{sh} S_h{}^{jk} - g^{jh} S_h{}^{sk} + g^{kh} S_h{}^{js}) \end{aligned}$$

where $(H^i{}_{jk}, C_i{}^{jk})$ are the coefficients of the canonical metrical connection $C\Gamma(N)$.

If we take in particular case from (6.2) as follows

$$\text{(6.3)}\qquad \begin{aligned} T^i{}_{jk} &= \delta^i_j \sigma_k - \delta^i_k \sigma_j, \quad \sigma_k \text{is 0-homogeneous;} \\ S_i{}^{jk} &= \delta^j_i \tau^k - \delta^k_i \tau^j, \quad \tau^k \text{is } -1\text{-homogeneous} \end{aligned}$$

we get, for σ_k, τ^k arbitrary, the set of all metrical semisymmetric connections of Cartan space $\mathcal{C}^n$.

Let us consider the following N-linear connections of $\mathcal{C}^n$:

1) $C\Gamma(N) = (H^i{}_{jk}, C_i{}^{jk})$–canonical metrical connection

2) $B\Gamma(N) = (\dot{\partial}^i N_{jk}, 0)$–Berwald connection

3) $RC\Gamma(N) = (H^i{}_{jk}, 0)$–Rund–Chern connection

4) $H\Gamma(N) = (\dot{\partial}^i N_{jk}, C_i{}^{jk})$–Hashiguchi connection

These connections are determined only by the fundamental function $\mathcal{K}$ of the Cartan space $\mathcal{C}^n$. Every connection 1)–4) can be defined by a specific system of independent axioms, [88].

We denote by $\overset{c}{|}, \overset{c}{|}$ or $\overset{b}{|}, \overset{b}{|}$ etc. the h- and v-covariant derivatives with respect to $C\Gamma, B\Gamma, ...$ etc.

Proposition 6.6.1. *The properties of metrizability of the connection $C\Gamma(N)$–$H\Gamma(N)$ are given by the following table*

$C\Gamma(N)$	$g^{ij}{}_{\overset{c}{\mid} k} = 0$	$g^{ij}\overset{c}{\mid}{}^k = 0$
$B\Gamma(N)$	$g^{ij}{}_{\overset{b}{\mid} k} = -2C_k{}^{ij}{}_{\overset{c}{\mid} 0}$	$g^{ij}\overset{b}{\mid}{}^k = -2C^{ijk}$
$RC\Gamma(N)$	$g^{ij}{}_{\overset{r}{\mid} k} = 0$	$g^{ij}\overset{r}{\mid}{}^k = -2C^{ijk}$
$H\Gamma(N)$	$g^{ij}{}_{\overset{H}{\mid} k} = -2C_k{}^{ij}{}_{\overset{c}{\mid} 0}$	$g^{ij}\overset{H}{\mid}{}^k = 0$

The calculation of $g^{ij}{}_{\overset{b}{\mid} k}$ or $g^{ij}{}_{\overset{H}{\mid} k}$ is similar with Finslerian case (cf. Ch.2).

Let us consider a transformation of N-linear connection

$$t(\sigma^i{}_{jk}, \tau_i{}^{jk}) : D\Gamma(N) = (H^i{}_{jk}, C_i{}^{jk}) \longrightarrow D\overline{\Gamma}(N) = (\overline{H}^i{}_{jk}, \overline{C}_i{}^{jk})$$

defined by

$$\overline{H}^i{}_{jk} = H^i{}_{jk} + \sigma^i{}_{jk}, \quad \overline{C}_i{}^{jk} = C_i{}^{jk} + \tau_i{}^{jk} \tag{6.4}$$

where $\sigma^i{}_{jk}, \tau_i{}^{jk}$ are d-tensors 0-homogeneous and -1–homogeneous, respectively.

The following particular transformations are remarkable:

$$t_1(0, -C_i{}^{jk}) : C\Gamma(N) \longrightarrow RC\Gamma(N)$$
$$t_2(-P^i{}_{jk}, 0) : C\Gamma(N) \longrightarrow H\Gamma(N)$$
$$t_3(-P^i{}_{jk}, -C_i{}^{jk}) : C\Gamma(N) \longrightarrow B\Gamma(N)$$

Now, it follows easily:

Proposition 6.6.2. *The following diagram holds good*

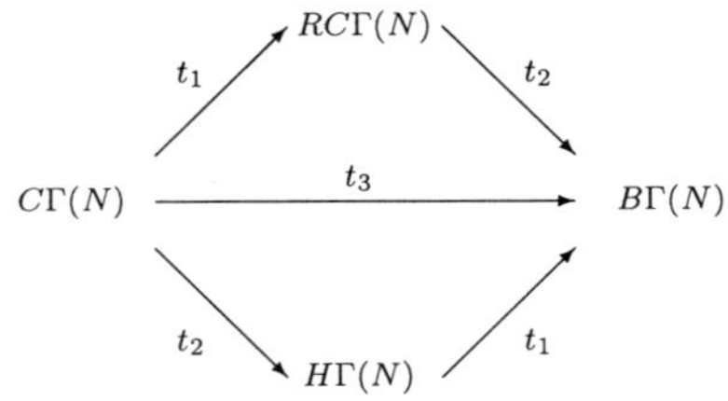

The existence of this commutative diagram shows us that the N-linear connections $C\Gamma(N), B\Gamma(N), RC\Gamma(N)$ and $H\Gamma(N)$ are important in the geometry of Cartan spaces.

6.7 Some special Cartan spaces

The Berwald connection $B\Gamma(N) = (\dot{\partial}^i N_{jk}, 0)$ of the Cartan spaces has d-tensors of the torsion:

$$\overset{b}{T}{}^i{}_{jk} = 0,\ \overset{b}{S}{}_i{}^{jk} = 0,\ \overset{b}{C}{}_i{}^{jk} = 0,\ \overset{b}{P}{}^i{}_{jk} = 0,\ \text{and } R_{ijk} = \delta_k N_{ij} - \delta_j N_{ik}$$

The d-tensors of curvature of $B\Gamma(N)$ are:

$$\begin{aligned} &\overset{b}{R}{}_j{}^i{}_{kj} = \delta_h B^i{}_{jk} - \delta_k B^i{}_{jh} + B^s{}_{jk} B^i{}_{sh} - B^s{}_{jh} B^i{}_{sk} \\ &\overset{b}{P}{}_j{}^i{}_k{}^h = \dot{\partial}^h B^i{}_{jk} \\ &\overset{b}{S}{}_j{}^{ikh} = 0 \end{aligned} \tag{7.1}$$

where

$$B^i{}_{jk} = \dot{\partial}^i N_{jk} \tag{7.2}$$

are the coefficients of the $B\Gamma(N)$–connection.

The Bianchi identities of the Berwald connection $B\Gamma(N)$ are given by Theorem 4.12.2.

Now, let us give the following definition:

Definition 6.7.1. A Cartan space $\mathcal{C}^n$ is called *Berwald–Cartan space* if the coefficients $B^i{}_{jk}$ of Berwald connection $B\Gamma$ are functions of position alone:

$$(*)\qquad B^i{}_{jk}(x,p) = B^i{}_{jk}(x).$$

The Berwald–Cartan spaces can be characterized by:

Theorem 6.7.1. *A Cartan space is a Berwald–Cartan space if and only if the following tensor equation holds:*

$$(7.3)\qquad C^{ijk}{}_{\overset{c}{|h}} = 0.$$

The proof is similar with the one of Theorem 2.3.1.
From the formula (7.1) we obtain:

Corollary 6.7.1. *A Cartan space is a Berwald–Cartan space if and only if the d-tensor of curvature $\overset{b}{P}{}_j{}^i{}_k{}^h$ vanishes.*

Definition 6.7.2. A Cartan space is called a Landsberg–Cartan space if its Berwald connection is h-metrical, i.e.

$$g^{ij}_{\overset{}{b}\,|k} = 0.$$

Theorem 6.7.2. *A Cartan space is a Landsberg–Cartan space if and only if the following tensor equation holds:*

$$(7.4)\qquad C^{ij}_{k\,\overset{c}{|0}} = 0.$$

As in the case of Landsberg–Finsler space we can prove:

Theorem 6.7.3. *A Cartan space is a Landsberg–Cartan space if and only if the d-curvature tensor $P_i{}^j{}_k{}^h$ of the canonical metrical connection $C\Gamma(N)$ vanishes identically.*

Corollary 6.7.2. *If a Cartan space is a Berwald–Cartan space, then it is a Landsberg–Cartan space.*

Remark. We will study again these spaces, in next chapter using the theory of $\mathcal{L}$–duality of Finsler spaces and Cartan spaces. Also, it will be introduced the Cartan space of scalar and constant curvature.

Definition 6.7.3. A Cartan space $\mathcal{C}^n = (M, K(x,p))$ is called locally Minkowski–Cartan space if in every point $x \in M$ there is a coordinate system (U, x^i) such that on $\pi^{*-1}(U) \subset T^*M$ the fundamental function $K(x,p)$ depends only on the momenta (p_i).

Exactly as in the case of Finsler spaces we can proof the following important result:

Theorem 6.7.4. *A Cartan space $\mathcal{C}^n = (M, K(x,p))$ is a locally Minkowski–Cartan space if and only if the d-tensor of curvature $R_j{}^i{}_{kh}$ of the canonical metrical connection $C\Gamma$ vanishes and $C_i{}^{jk}{}_{|h} = 0$.*

Examples.

1° The Cartan spaces $\mathcal{C}^n = (M, K)$, where K is given by (1.5), is a locally Minkowski–Cartan space.

2° The Cartan spaces $\mathcal{C}^n = (M, K)$, with the fundamental function (Berwald–Moór)

$$K = \{p_1 p_2 ... p_n\}^{\frac{1}{n}},$$

is a locally Minkowski–Cartan space.

6.8 Parallelism in Cartan space. Horizontal and vertical paths

In a Cartan space $\mathcal{C}^n = (M, K(x,p))$ endowed with the canonical metrical connection $C\Gamma(N) = (H^i{}_{jk}, C_i{}^{jk})$, the notion of parallelism of vector fields along a curve $\gamma : I \longrightarrow \widetilde{T^*M}$ can be studied using the associate Hamilton space $H^n = (M, K^2(x,p))$, (cf.§5.8, Ch.5).

Let $\gamma : I \longrightarrow \widetilde{T^*M}$ be a parametrized curve expressed in a local chart of $\widetilde{T^*M}$ by

$$x^i = x^i(t),\ p_i = p_i(t),\ t \in I,\ \text{rank}\|p_i(t)\| = 1. \tag{8.1}$$

The tangent vector field $\dfrac{d\gamma}{dt}$ is

$$\frac{d\gamma}{dt} = \frac{dx^i}{dt}\delta_i + \frac{\delta p_i}{dt}\dot{\partial}^i \tag{8.2}$$

where

$$\frac{\delta p_i}{dt} = \frac{dp_i}{dt} - N_{ji}\frac{dx^j}{dt}. \tag{8.3}$$

The curve γ is called horizontal if $\left(\frac{d\gamma}{dt}\right)^{v} = 0$. So, an horizontal curve γ is characterized by the equations:

$$x^i = x^i(t),\ \frac{\delta p_i}{dt} = \frac{dp_i}{dt} - N_{ji}\frac{dx^j}{dt} = 0. \tag{8.4}$$

Taking into account that $N_{ij} = p_m H^m{}_{ij}$ it follows that:

Proposition 6.8.1. *The horizontal curves γ in the Cartan space $\mathcal{C}^n$ are characterized by the equations:*

$$x^i = x^i(t),\ \frac{Dp_i}{dt} = \frac{dp_i}{dt} - p_m H^m{}_{ij}\frac{dx^j}{dt} = 0. \tag{8.5}$$

The first consequence is follows:

Theorem 6.8.1. *The geodesics of the Cartan space $\mathcal{C}^n$ are the horizontal curves.*

Proof. The geodesics of the space $\mathcal{C}^n$ are given by the Hamilton–Jacobi equations

$$\frac{dx^i}{dt} = \frac{\partial F^2}{\partial p_i};\ \frac{dp_i}{dt} = -\frac{\partial F^2}{\partial x^i}.$$

Therefore, we have $\frac{\delta p_i}{dt} = \frac{dp_i}{dt} - N_{ji}\frac{dx^j}{dt} = -\left(\frac{\partial H}{\partial x^i} + N_{ji}\frac{\partial H}{\partial p_j}\right) = -H_{|i} = 0.$ **q.e.d.**

Corollary 6.8.1. *The geodesics of the space $\mathcal{C}^n$ are characterized by the equations:*

$$\frac{dx^i}{dt} = \frac{\partial F^2}{\partial p_i},\ \frac{Dp_i}{dt} = 0. \tag{8.6}$$

For a vector field $X \in \mathcal{X}(T^*M)$, which is locally expressed by:

$$X = X^i(x,p)\delta_i + X_i(x,p)\dot{\partial}^i, \tag{8.7}$$

we have:

Theorem 6.8.2. *The vector field X, given by* (8.7) *is parallel along the curve γ, with respect to the canonical metrical connection $C\Gamma$ of the Cartan space $\mathcal{C}^n$, if and only if its coordinate X^i, X_i are solutions of the differential equations* (11.8), *Ch.4, where $\omega^i{}_j(x(t),p(t))$ is the 1-forms connection of $C\Gamma$.*

In particular, Theorem 4.11.2 can be applied:

Theorem 6.8.3. *The Cartan space $\mathcal{C}^n = (M, K)$, endowed with the canonical metrical connection $C\Gamma$ is with the absolute parallelism of vectors, if and only if the d-tensors of curvature $R_j{}^i{}_{kh}$, $P_j{}^i{}_k{}^h$ and $S_i{}^{jkh}$ vanish.*

Theorem 4.11.3 can be particularized, too:

Theorem 6.8.4. *The curve γ given by (8.1) is autoparallel with respect to the canonical metrical connection $C\Gamma$ of the Cartan space $\mathcal{C}^n$ if, and only if, the functions $x^i(t)$, $p_i(t)$, $t \in I$ are the solutions of the system of differential equations*

$$\frac{d^2x^i}{dt^2} + \frac{dx^s}{dt}\frac{\omega^i{}_s}{dt} = 0, \quad \frac{d}{dt}\frac{\delta p_i}{dt} - \frac{\delta p_s}{dt}\omega^s_i = 0. \tag{8.8}$$

As is known, the horizontal autoparallel curves of the Cartan space $\mathcal{C}^n$ are *the horizontal paths.*

Theorem 6.8.5. *The horizontal paths of the connection $C\Gamma$ of a Cartan space are characterized by the system of differential equations*

$$\frac{d^2x^i}{dt^2} + H^i{}_{jk}(x,p)\frac{dx^j}{dt}\frac{dx^h}{dt} = 0, \quad \frac{\delta p_i}{dt} = 0. \tag{8.9}$$

Now, taking into account Theorem 4.11.5, we get:

Theorem 6.8.6. *The vertical paths in the point $x_0 \in M$ with respect of a Cartan space $\mathcal{C}^n$ are characterized by the system of differential equations*

$$x^i = x^i_0, \quad \frac{d^2p_i}{dt^2} - C_i^{jk}(x_0,p)\frac{dp_j}{dt}\frac{dp_k}{dt} = 0.$$

6.9 The almost Kählerian model of a Cartan space

To a Cartan space $\mathcal{C}^n = (M, K(x,p))$ we can associate some important geometrical object fields on the manifold $\widetilde{T^*M}$. Namely, the N–lift Γ of the fundamental tensor g^{ij}, the almost complex structure $\check{\mathbb{F}}$, etc. If N is the canonical nonlinear connection of $\mathcal{C}^n$, thus $(\mathbb{G}, \check{\mathbb{F}})$ determine an almost Hermitian structure, which is derivated only from the fundamental function K of the Cartan space. This structure gives us the so–called [99] geometrical model on $\widetilde{T^*M}$ of the Cartan space $\mathcal{C}^n$.

Let $(\delta_i, \dot{\partial})$ be the adapted basis to the distribution N and V, N being the canonical nonlinear connection of the space $\mathcal{C}^n$. Its dual basis is $(dx^i, \delta p_i)$.

The N-lift of the fundamental tensor field g^{ij} of the space $\mathcal{C}^n$ is defined by

$$\mathbb{G}(x,p) = g_{ij}(x,p)dx^i \otimes dx^j + g^{ij}(x,p)\delta p_i \otimes \delta p_j. \tag{9.1}$$

We obtainas usual

Theorem 6.9.1. *The following properties hold:*

1° $\mathbb{G}$ *is a Riemannian structure globally defined on* $\widetilde{T^*M}$.

2° $\mathbb{G}$ *is determined only by the fundamental function* K *of the Cartan space* $\mathcal{C}^n$.

3° *The distributions* N *and* V *are orthogonal.*

Now, considering the covector fields $\check{\partial}_i = g_{ij}\dot{\partial}^j$ in every point $u \in \widetilde{T^*M}$, we can define the $\mathcal{F}(\widetilde{T^*M})$–linear mapping $\mathbb{F} : \mathcal{X}(\widetilde{T^*M}) \longrightarrow \mathcal{X}(T^*M)$, defined in (6.7), Ch.4, by

$$\check{\mathbb{F}}(\delta_i) = -\check{\partial}_i, \ \check{\mathbb{F}}(\check{\partial}_i) = \delta_i \tag{9.2}$$

By means of Theorem 4.6.2, we obtain

Theorem 6.9.2. *We have the followings:*
1° $\check{\mathbb{F}}$ *is globally defined on* $\widetilde{T^*M}$.
2° $\check{\mathbb{F}}$ *is the tensor field of type* $(1,1)$*:*

$$\check{\mathbb{F}} = -g_{ij}\dot{\partial}^i \otimes dx^j + g^{ij}\delta_i \otimes \delta p_j. \tag{9.3}$$

3° $\check{\mathbb{F}}$ *is an almost complex structure on* $\widetilde{T^*M}$*:*

$$\check{\mathbb{F}} \circ \check{\mathbb{F}} = -I. \tag{9.4}$$

4° $\check{\mathbb{F}}$ *depends only on the fundamental function* K *of the Cartan space* $\mathcal{C}^n$.

Theorem 4.6.4 lead to

Theorem 6.9.3.

1° *The pair* $(\mathbb{G}, \check{\mathbb{F}})$ *is an almost Hermitian structure on* $\widetilde{T^*M}$.

2° *The almost Hermitian structure* $(\mathbb{G}, \check{\mathbb{F}})$ *depends only on the fundamental function* K *of the Cartans space* $\mathcal{C}^n$.

3° *The associate almost symplectic structure to the structure* $(\mathbb{G}, \check{\mathbb{F}})$ *is the canonical symplectic structure* $\theta = \delta p_i \wedge dx^i = dp_i \wedge dx^i$.

Corollary 6.9.1. *The space* $(\widetilde{T^*M}, \mathbb{G}, \check{\mathbb{F}})$ *is almost Kählerian and it is determined only by the Cartan space* $\mathcal{C}^n$.

Finally, we remark:

Theorem 6.9.4. *The N-linear connection D, determined by the canonical metrical connection $C\Gamma$ of the Cartan space $\mathcal{C}^n$ is an almost Kählerian one, i.e.*

$$D\mathbb{G} = 0, \;\; D\check{\mathbb{F}} = 0.$$

Due to the last property, we call the space $(\widetilde{T^*M}, \mathbb{G}, \check{\mathbb{F}})$ the almost Kählerian model of the Cartan space $\mathcal{C}^n$.

It is extremely useful in the applications in mechanics, theoretical physics, etc.

Chapter 7

The duality between Lagrange and Hamilton spaces

In this chapter we develop the concept of $\mathcal{L}-$duality between Lagrange and Hamilton spaces (particularly between Finsler and Cartan spaces) investigated in [66], [67], [97] and a new technique in the study of the geometry of these spaces is elaborated. We will apply this technique to study the geometry of Kropina spaces (especially the geometric objects derived from the Cartan Connection) via the geometry of Randers spaces.

These spaces are already used in many applications.

7.1 The Lagrange-Hamilton $\mathcal{L}$-duality

Let L be a regular Lagrangian on a domain $D \subset TM$ and let H be a regular Hamiltonian on a domain $D^* \subset T^*M$.

Hence, the matrices with entries

$$g_{ab}(x,y) := \dot{\partial}_a \dot{\partial}_b L(x,y) \tag{1.1}$$

and

$$g^{*ab}(x,p) := \dot{\partial}^a \, \dot{\partial}^b \, H(x,p) \tag{1.2}$$

are everywhere nondegenerate on D and respectively D^*, $(a, b, c, ... \in \{1, .., n\})$.

Note: The metric tensors (1.1) and (1.2) used in this chapter are those of Chapter 3 and Chapter 5 multiplied by a factor 2. This notation allows us to simplify a certain number of equations and to preserve the classical Legendre duality encounterd in Mechanics (see[3]).

If $L \in \mathcal{F}(D)$ is a differentiable map, we can consider the *fiber derivative of* L, locally given by

$$\varphi(x,y) = (x^i,\, \dot{\partial}_a L(x,y)) \tag{1.3}$$

which will be called the *Legendre transformation.*

It is easily seen that L is a regular Lagrangian if and only if φ is a local diffeomorphism [3].

In the same manner if $H \in \mathcal{F}(D^*)$ the fiber derivative is given locally by

$$\psi(x,p) = (x^i,\, \dot{\partial}^a\, H(x,p)) \tag{1.4}$$

which is a local diffeomorphism if and only if H is regular.

Let us consider a regular Lagrangian L. Then φ is a diffeomorphism between the open sets $U \subset D$ and $U^* \subset T^*M$. We can define in this case the function $H : U^* \longrightarrow \mathbb{R}$:

$$H(x,p) = p_a y^a - L(x,y), \tag{1.5}$$

where $y = (y^a)$ is the solution of the equations

$$p_a = \dot{\partial}_a L(x,y). \tag{1.5'}$$

Also, if H is a regular Hamiltonian on M, ψ is a diffeomorphism between same open sets $U^* \subset D^*$ and $U \subset TM$ and we can consider the function $L : U \longrightarrow \mathbb{R}$:

$$L(x,y) = p_a y^a - H(x,p), \tag{1.6}$$

where $p = (p_a)$ is the solution of the equations

$$y^a = \dot{\partial}^a H(x,p). \tag{1.6'}$$

It is easily verified that H and L given by (1.5) and (1.6) are regular.

The Hamiltonian given by (1.5) will be called the *Legendre transformation of the Lagrangian* L (also L given by (1.6) will be called the *Legendre transformation of the Hamiltonian* H.

Examples:

1. If L is m–homogeneous, $m \neq 1$, regular Lagrangian, then locally,

$$H(x,p) = (m-1)L(x,y),\ p_a = \dot{\partial}_a L(x,y).$$

2. If $L(x,y) = \frac{1}{2}a_{ij}(x)y^iy^j + b_iy^i + c$ then its Legendre transformation is the Hamiltonian
$$H(x,p) = \frac{1}{2}a^{ij}(x)p_ip_j - b^ip_i + d$$
where $b^i := a^{ij}b_j$ and $d := b_ib^i - c$.

In the following, we will restrict our attention to the diffeomorphisms

$$\varphi : U \longrightarrow U^* \text{ and } \psi : U^* \longrightarrow U \tag{1.7}$$

(where ψ is the Legendre transformation associated to the Hamiltonian given in (1.6)).

We remark that U and U^* are open sets in TM and respectively T^*M and generally are not domains of charts.

The following relations can be checked directly

$$\varphi \circ \psi = 1_{U^*}, \quad \psi \circ \varphi = 1_U \tag{1.8}$$

$$\partial_i H(x,p) = -\partial_i L(x,y);\ \partial_i\dot{\partial}_a L(x,y) = -\partial_i\dot{\partial}^b H(x,p)g^*_{ab}(x,p) \tag{1.9}$$

$$g_{ab}(x,y)g^{*bc}(x,p) = \delta^c_a \tag{1.10}$$

where $p_a = \dot{\partial}_a L(x,y)$, $y^a = \dot{\partial}^a H(x,p)$.

Using the diffeomorphism φ (or ψ) we can *pull–back* or *push–forward* the geometric structures from U to U^* or from U^* to U.

A) if $f \in \mathcal{F}(U)$ we consider the pull-back of f by ψ (or push–forward by φ)

$$f^* := f \circ \psi = f \circ \varphi^{-1},\ f^* \in \mathcal{F}(U^*). \tag{1.11}$$

Also, if $f \in \mathcal{F}(U^*)$, we get $f^0 \in \mathcal{F}(U)$

$$f^0 := f \circ \varphi = f \circ \psi^{-1}. \tag{1.12}$$

We have the following properties:

(i) $(\lambda_f + \mu g)^* = {\lambda_f}^* + \mu g^*,\ (fg)^* = f^*g^*;\ \forall\, \lambda,\mu \in \mathbb{R}, \forall\, f,g \in \mathcal{F}(U)$,

(ii) $(\lambda_f + \mu g)^0 = {\lambda_f}^0 + \mu g^0,\ (fg)^0 = f^0g^0;\ \forall\, \lambda,\mu \in \mathbb{R}, \forall\, f,g \in \mathcal{F}(U^*)$,

(iii) $(f^*)^0 = f,\ (g^0)^* = g,\ f \in \mathcal{F}(U), g \in \mathcal{F}(U^*)$,

(iv) $(g^{ab})^* = g^{*ab},\ (g_{ab})^* = g^*_{ab}$.

B) If $X \in \mathcal{X}(U)$ the push–forward of X by φ (or pull-back by ψ) is $X^* \in \mathcal{X}(U^*)$

$$X^* := T\varphi \circ X \circ \varphi^{-1} = T\psi^{-1} \circ X \circ \psi. \tag{1.13}$$

(T_φ is the tangent map of φ.)

Also, if $X \in \mathcal{X}(U^*)$ we can consider the push–forward of X by ψ (or pull-back by φ), $X^0 \in \mathcal{X}(U)$

$$X^0 := T\psi \circ X \circ \psi^{-1} = T\varphi^{-1} \circ X \circ \varphi \tag{1.14}$$

The following relations are easily checked:

(i) $(fX + gY)^* = f^*X^* + g^*Y^*,\ \forall f, g \in \mathcal{F}(U), \forall X, Y \in \mathcal{X}(U)$,

(ii) $(fX + gY)^0 = f^0X^0 + g^0Y^0,\ \forall f, g \in \mathcal{F}(U^*), \forall X, Y \in \mathcal{X}(U^*)$,

(iii) $[X, Y]^* = [X^*, Y^*],\ \forall X, Y \in \mathcal{X}(U)$,
$[X, Y]^0 = [X^0, Y^0],\ \forall X, Y \in \mathcal{X}(U^*)$,

(iv) $(X^*)^0 = X,\ (Y^0)^* = Y,\ X \in \mathcal{X}(U), Y \in \mathcal{X}(U^*)$.

C) If $\theta \in \mathcal{X}^*(U)$ the push–forward of θ by φ (or pull-back by ψ) is $\theta^* \in \mathcal{X}^*(U^*)$

$$\theta^* = (T\varphi)^* \circ \theta \circ \varphi^{-1} = (T\psi^{-1})^* \circ \theta \circ \psi. \tag{1.15}$$

and if $\theta \in \mathcal{X}^*(U^*)$ we can consider

$$\theta^0 = (T\psi)^* \circ \theta \circ \psi^{-1} = (T\varphi^{-1})^* \circ \theta \circ \varphi \tag{1.16}$$

where $(T\varphi)^*$ denotes the cotangent map of φ.

We have similar properties as (i), (ii), (iv) above.

D) Generally if $K \in \mathcal{T}^r_s(U)$ is a tensor field on U we can define similar the push–forward of K by φ, $K^* \in \mathcal{T}^r_s(U^*)$ and, for $K \in \mathcal{T}^r_s(U^*)$ we get $K^* \in \mathcal{T}^r_s(U)$ (see also [3]) we have

$$(K \otimes T)^* = K^* \otimes T^*, (K' \otimes T')^0 = K'^{C} \otimes T'^0. \tag{1.17}$$

Let ∇ be a linear connection on U. We define a linear connection ∇^* on U^* as follows:

$$\nabla^*_X Y := (\nabla_{X^0} Y^0)^*,\ X, Y \in \mathcal{X}(U^*). \tag{1.18}$$

Also, if ∇ is a linear connection on U^* we get a linear connection ∇^0 on U

$$\nabla^0_X Y := (\nabla_{X^*} Y^*)^0,\ X, Y \in \mathcal{X}(U). \tag{1.19}$$

It is easily checked, using the above examples, that ∇^* and ∇^0 are indeed linear connections on U^* and U.

For the torsion and curvature tensors of ∇^* we have

$$T^*(X, Y) = [T(X^0, Y^0)]^*,\ \forall X, Y \in \mathcal{X}(U^*), \tag{1.20}$$

$$R^*(X, Y)Z = [R(X^0, Y^0)Z^0]^*,\ \forall X, Y, Z \in \mathcal{X}(U^*). \tag{1.21}$$

Generally, if $K \in \mathcal{T}^r_s(U)$ and $K^* \in \mathcal{T}^r_s(U^*)$ is its push–forward by φ, then

$$\nabla^* K^* = (\nabla K)^*. \tag{1.22}$$

Definition 7.1.1. We will say that f and f^* (or f and f^0) X and X^* (or X and X^0), K and K^* (or K and K^0), ∇ and ∇^* (or ∇ and ∇^0) are dual by the Legendre transformation or are $\mathcal{L}$–dual.

In the next section we will look for geometric objects on U and U^* which are $\mathcal{L}$–dual. These geometric objects will be obtained easily each one from the other.

7.2 $\mathcal{L}$–dual nonlinear connections

Definition 7.2.1. Let HTU and HTU^* be two nonlinear connections on the open sets U and U^*. We say that HTU and HTU^* are *$\mathcal{L}$–dual* if

$$T\varphi(HTU) = HTU^*. \tag{2.1}$$

Let $N = (N^a_i)$ and $\bar{N} = (\bar{N}_{ia})$ be the coefficients of two nonlinear connections on U and U^*.

Theorem 7.2.1. *The following statements are equivalent:*

(i) N *and* $\bar{N}$ *are* $\mathcal{L}$*-dual*;

(ii) $N_i^{a*} = -N_{ib}g^{*ab} - \partial_i\dot{\partial}^a H$ or $N_{ia}^0 = -N_i^b g_{ba} + \partial_i\dot{\partial}_a L$;

(iii) $\delta_i y^{*a} = -N_i^{a*}$;

(iv) $\delta_i p_a^0 = N_{ia}^0$;

(v) $\delta_i^* f^* = (\delta_i f)^*,\ \forall\, f \in \mathcal{F}(U)$.

Proof. N and $\bar{N}$ are $\mathcal{L}$–dual$\Longleftrightarrow T\varphi(HTU) = HTU^* \Longleftrightarrow T\varphi(\delta_i) \in H_{(x,p)}TU^*$, $\forall\, i \in \overline{1,n}$. We must have

$$T\varphi(\delta_i) = a_i^j\delta_j^* = a_i^j(\partial_j + N_{ja}\dot{\partial}^a).$$

On the other hand,

$$\begin{aligned} T\varphi(\delta_i) &= T\varphi(\partial_i - N_i^a\dot{\partial}_a) = T\varphi(\partial_i) - N_i^a T\varphi(\dot{\partial}_a) \\ &= \partial_i + \partial_i\dot{\partial}_b L\dot{\partial}^b - N_i^a\dot{\partial}_a\dot{\partial}_b L\dot{\partial}^b = \partial_i + (\partial_i\dot{\partial}_b L - N_i^a g_{ab})\dot{\partial}^b. \end{aligned}$$

Therefore, we get

$$a_i^j = \delta_i^j \quad \text{and} \quad N_{ia}^0 = -N_i^b g_{ab} + \partial_i\dot{\partial}_a L$$

(or, equivalent, $N_i^{a*} = -N_{ib}g^{*ba} - \partial_i\dot{\partial}^a H$) and we have proved that (i)$\Longleftrightarrow$(ii).

Now we have

$$\begin{aligned} \delta_i^* y^{a*} &= \partial_i y^{a*} + N_{ib}\dot{\partial}^b y^{a*} = \partial_i\dot{\partial}^a H + N_{ib}\dot{\partial}^b\dot{\partial}^a H = N_{ib}g^{*ba} + \partial_i\dot{\partial}^a H \\ \delta_i p_a^0 &= \partial_i p_a^0 - N_i^b\dot{\partial}_b p_a^0 = \partial_i\dot{\partial}_a L - N_i^b\dot{\partial}_b\dot{\partial}_a L = -N_i^b g_{ba} + \partial_i\dot{\partial}_a L \\ \delta_i^* f^* &= (\delta_i f)^* \Longleftrightarrow (\partial_i f)^* + (\dot{\partial}_a f)^*\partial_i\dot{\partial}^a H + N_{ia}(\dot{\partial}_b f)^*\dot{\partial}^b\dot{\partial}^a H \\ &= (\partial_i f)^* - N_i^{k*}(\dot{\partial}_k f)^*. \end{aligned}$$

Using these relations we get the proof.

The $\mathcal{L}$–dual of the nonlinear connection N will be denoted by N^*. (Similarly, the $\mathcal{L}$–dual of $\bar{N}$ will be denoted by $\bar{N}^0$.)

Corollary 7.2.1. *If* $N = (N_i^a)$ *and* $N^* = (N_{ia})$ *are two* $\mathcal{L}$*–dual nonlinear connections we have*

$$(\delta_i)^* = \delta_i^*;\ (\delta y^a)^* = g^{*ab}\delta^* p_b.$$

(These properties are characteristic for two $\mathcal{L}$–dual nonlinear connections, too.)

Proposition 7.2.1. *The following equalities hold good:*

$$\begin{aligned} &(dx^i)^* = dx^i; && (\partial_a)^* = g_{ab}^*\dot{\partial}^b \\ &\dot{\partial}^a f^* = g^{*ab}(\dot{\partial}_b f)^*; && \partial_i f^* = (\partial_i f)^* + \partial_i\dot{\partial}^a H(\dot{\partial}_a f)^*. \end{aligned}$$

Corollary 7.2.2. *Let $N = (N^a_i), N^* = (N_{ia})$ be two $\mathcal{L}$–dual nonlinear connections. The following assertions hold:*

(i) *If $X = X^i\delta_i + X^a\dot{\partial}_a$ then $X^* = X^{i*}\delta^*_i + g^*_{ab}X^{a*}\dot{\partial}^b$ and $(X^H)^* = (X^*)^H, (X^V)^* = (X^*)^V$.*

(ii) *If $\omega = \omega_{ia}dx^i \wedge \delta y^a$ then $\omega^* = (\omega_{ia})^* g^{*ab} dx^i \wedge \delta^* p_b$.*

(iii) *If $K = K^i{}_j{}^a{}_b\delta_i \otimes \dot{\partial}_a \otimes dx^j \otimes \delta y^b$ then $K^* = (K^i{}_j{}^a{}_b)^* g^*_{ac} g^{*bd}\delta_i \otimes \dot{\partial}^c \otimes dx^j \otimes \delta^* p_d$.*

Remark. We have $K^{*i}{}_j{}^a{}_b = (g^{ac}g_{bd}K^i{}_j{}^d{}_c)^* = g^{*ac}g_{*bd}(K^i{}_j{}^d{}_c)^*$. Therefore the components of the $\mathcal{L}$–dual of K in (x,p) are obtained from the components $K^i{}_j{}^a{}_b$ of K in (x,y), $p^0_a = \dot{\partial}_a L(x,y)$, unchanging the horizontal part and raising and lowering of indices for vertical part by using g_{ab}.

Examples.

1. The $\mathcal{L}$–dual of the metric tensor g_{ab} has the components g^{*ab}.

2. If $C_{abc} = \frac{1}{2}\dot{\partial}_a g_{bc}$ then $C^{*abc} = -\frac{1}{2}\dot{\partial}^a g^{*bc}$.
   If $C^a{}_{bc} = \frac{1}{2}g^{ad}\dot{\partial}_d g_{bc}$ then $C^*{}_a{}^{bc} = -\frac{1}{2}g^*_{ad}\dot{\partial}^d g^{*bc}$.

3. If δ is the Kronecker delta, with components δ^i_a then the components of its dual δ^* are as follows:
$$g^{*ia} = (g^{ab}\delta^i_b)^* = g^{*ia}.$$

Let $c(t) = (x(t), y(t))$, $t \in I \subset \mathbb{R}$ be a differentiable curve on U. The tangent vector can be written as follows:

$$\dot{c}(t) = \frac{dx^i}{dt}\delta_i + \frac{\delta y^a}{dt}\dot{\partial}_a. \tag{2.2}$$

We say that c is a *horizontal curve* if $\frac{\delta y^a}{dt} = 0$.
A similar definition holds for differentiable curves on U^*.

Proposition 7.2.2. *$(N, \bar{N})$ is a pair of $\mathcal{L}$–dual nonlinear connections if and only if the $\mathcal{L}$–dual of every horizontal curve is also a horizontal curve.*

Proof. Let $c(t) = (x(t), y(t))$, $c^*(t) = (x(t), p(t))$, $t \in I \subset \mathbb{R}$ two $\mathcal{L}$–dual curves, therefore $y^a(t) = \dot{\partial}^a H(x(t), p(t))$. We have:

$$\begin{aligned} \frac{\delta y^a}{dt} = 0 \iff & \partial_i \dot{\partial}^a H \frac{dx^i}{dt} + \dot{\partial}^b \dot{\partial}^a H \frac{dp_b}{dt} + N_i^a \frac{dx^i}{dt} = 0 \\ \iff & g^{*ba} \frac{dp_b}{dt} + (\partial_i \dot{\partial}^a H + N_i^a) \frac{dx^i}{dt} = 0 \\ \iff & \frac{dp_a}{dt} + (g^*_{ab} \partial_k \dot{\partial}^b H + g^*_{ab} N_k^b) \frac{dx^k}{dt} = 0. \end{aligned}$$

Let suppose that N and $\bar{N}$ are $\mathcal{L}$–dual nonlinear connections. Using Theorem 7.2.1, (ii), and the above relation we get $\dfrac{\delta^* p_a}{dt} = 0$. Conversely, from $\dfrac{\delta y^a}{dt} = 0$, $\dfrac{\delta^* p_a}{dt} = 0$ we obtain easily that N and $\bar{N}$ are $\mathcal{L}$–dual.

Example. For a Lagrange manifold the geodesics are extremals of the action integral of L and coincide with the integral curves of the *semispray*

$$X_L = y^i \partial_i - 2G^a \dot{\partial}_a, \tag{2.3}$$

where

$$G^a = \frac{1}{2} g^{ab} (y^k \dot{\partial}_b \partial_k L - \partial_b L). \tag{2.4}$$

This semispray generates a notable nonlinear connection, called *canonical*, whose coefficients are given by

$$N_j^a = \dot{\partial}_j G^a \tag{2.5}$$

(see Section 3.3, Ch. 3).

Using (ii) from Theorem 7.2.1, we get the coefficients N_{ia} of its $\mathcal{L}$–dual nonlinear connection:

$$N^0_{ia} = -\dot{\partial}_i G^b g_{ba} + \partial_i \dot{\partial}_b L$$

and after a straightforward computation we obtain

$$N_{ij} = \frac{1}{2}(\dot{\partial}^k g^*_{ij} \partial_k H - \partial_k g^*_{ij} \dot{\partial}^k H) - \frac{1}{2}(g^*_{ik} \dot{\partial}^k \partial_j H + g^*_{jk} \dot{\partial}^k \partial_j H). \tag{2.6}$$

We remark that N_{ij} is expressed here only using the Hamiltonian. This is the *canonical nonlinear connection of the Hamilton manifold* (M, H) obtained by R.Miron in [97].

We also remark that the canonical nonlinear connection (2.3) is symmetrical, that means

$$\tau_{ij} = N_{ij} - N_{ji} = 0. \tag{2.7}$$

Taking the $\mathcal{L}$–dual of (2.7) we get the "symmetry" condition for (N_i^a)

$$N_j^b g_{bi} - N_i^b g_{bj} = \partial_j \dot{\partial}_i L - \partial_i \dot{\partial}_j L \tag{2.8}$$

((2.8) can be also checked directly and thus (2.7) may be obtained as a consequence of (2.8)).

Now, let us fix the nonlinear connection given by (2.5) and (2.6) on U and respectively U^*.

The *canonical two form*

$$\theta = \delta p_a \wedge dx^a$$

is just the canonical symplectic form of T^*M.

The Hamilton vector field X_H can be obtained from the condition:

$$i_{X_H}\omega = -dH \Longleftrightarrow i_{X_H}(dx^i \wedge \delta^* p_i) = \delta_i^* H dx^i + \dot{\partial}^i H \delta^* p_i.$$

Consequently,

$$X_H = \dot{\partial}^i H \delta_i^* - \delta_i^* H \dot{\partial}^i. \tag{2.9}$$

The integral curves of X_H are solutions of Hamilton–Jacobi equations

$$\frac{dx^i}{dt} = \dot{\partial}^i H, \ \frac{\delta^* p_i}{dt} = -\delta_i^* H \tag{2.10}$$

(equivalently with $\dfrac{dx^i}{dt} = \dot{\partial}^i H, \ \dfrac{dp_i}{dt} = -\dot{\partial}^i H$).

The $\mathcal{L}$–dual of X_H is just X_L, the Lagrange vector field from (2.3).

In adapted frames (2.3) one reads

$$X_L = y^i \delta_i + (y^j N_j^a - 2G^a)\dot{\partial}_a$$

and we remark that X_L is horizontal iff G^a is 2–homogeneous. An integral curve of X_L verifies the Euler–Lagrange equations:

$$\frac{dx^i}{dt} = y^i(t), \frac{\delta y^a}{dt} = N_j^a(x(t), \frac{dx}{dt})\frac{dx^j}{dt} - 2G^a(x(t), \frac{dx}{dt}) \tag{2.11}$$

which are $\mathcal{L}$–dual of (2.10).

The $\mathcal{L}$–dual of the *canonical one form* $\omega = p_i dx^i$ is the *canonical* 1*–form of the Lagrange manifold*

$$\omega = \dot{\partial}_i L dx^i \tag{2.12}$$

and the $\mathcal{L}$–dual of ω is the *canonical* 2*–form of* (M, L)

$$\theta_L = g_{ia} \delta y^a \wedge dx^i. \tag{2.13}$$

Proposition 7.2.3.

(i) *If N and $\bar{N}$ are $\mathcal{L}$–dual then*

$$(\bar{R}_{ajk})^0 = (R^*_{ajk})^0 = g_{ab}R^b{}_{jk},$$

(ii) $[\dot{\partial}^*_a, \dot{\partial}^*_b] = 0,$

(iii) $(\dot{\partial}^a N_{ib})^0 = g^{ac}(\delta_i g_{cb} - g_{bd}\dot{\partial}_c N^d_i).$

Proof.

(i) $[\delta^*_j, \delta^*_k] = [\delta_j, \delta_k]^* = (R^b{}_{jk}\dot{\partial}_b)^* = \bar{R}_{ajk}\dot{\partial}^a.$

(ii) We use the symmetry of the tensor $\dot{\partial}_i g_{jk}$ in all indices.

(iii) $[\dot{\partial}^b, \delta^*_j] = \dot{\partial}^b(N_{ja})\dot{\partial}^a$. On the other hand,
$[\dot{\partial}^b, \delta^*_j]^0 = [g^{ba}\dot{\partial}_a, \delta_j] = -g^{ba}\dot{\partial}_a N^c_j \dot{\partial}_c - \delta_j(g^{ba})\dot{\partial}_a$
and then we will get (iii).

Proposition 7.2.4. *Let N and $\bar{N}$ be two $\mathcal{L}$–dual nonlinear connections. Then*

(i) $N_{ij} = N_{ji} \Longleftrightarrow N^k_j g_{ki} - N^k_i g_{kj} = \partial_j\dot{\partial}_i L - \partial_i\dot{\partial}_j L,$

(ii) $\dot{\partial}^s N_{ij} = \dot{\partial}^s N_{ji} \Longleftrightarrow g_{ih}\dot{\partial}_k N^h_j - g_{jh}\dot{\partial}_k N^h_i = \delta_j g^*_{ik} - \delta_i g^*_{jk},$

(iii) $g^*_{ih}\dot{\partial}^h N_{jk} - g^*_{jh}\dot{\partial}^h N_{ik} = \delta_j g^*_{ik} - \delta_i g^*_{jk} \Longleftrightarrow \dot{\partial}_i N^k_j = \dot{\partial}_j N^k_i.$

Proof. (i) follows from Theorem 7.2.1, (ii), and (ii), (iii) are direct consequences of (iii) of Proposition 7.2.3.

7.3 $\mathcal{L}$–dual d–connections

Let (N, N^*) be a pair of $\mathcal{L}$–dual nonlinear connections. Then $\mathcal{L}$–dual of the almost product structure $P = \delta_i \otimes dx^i - \dot{\partial}_a \otimes \delta y^a$ on U is $P^* = \delta^*_i \otimes dx^i - \dot{\partial}^a \otimes \delta p_a$.

Let ∇ be a linear connection on U and ∇^* its $\mathcal{L}$–dual on U^*, given by (1.18).

Definition 7.3.1. A linear connection ∇ (∇^*) on TM (T^*M) is called *d-connection* if $\nabla P = 0$ ($\nabla^* P^* = 0$).

Proposition 7.3.1. ∇ *is a d–connection if and only if* ∇^* *is a d–connection..*
Proof. We use $\nabla^*\bar{P} = \nabla^* P^* = (\nabla P)^*$.

Theorem 7.3.1. *Let $C\Gamma(N) = (L^i_{jk}, \widetilde{L}^a_{bc}, \widetilde{C}^i{}_{jc}, C^a{}_{bc})$ be a d–connection on U, and $\overline{C\Gamma}(N^*) = (H^i_{jk}, \widetilde{H}^a_{bc}, \widetilde{V}^i{}_j{}^a, V_a{}^{bc})$ be a d–connection on U^*. Then $C\Gamma(N)$ and $\overline{C\Gamma}(N^*)$ are $\mathcal{L}$–dual if and only if the following relations hold:*

(i) $(H^i_{jk})^0 = L^i_{jk}$

(ii) $(\widetilde{H}^b_{ak})^0 = g^{bc}(\delta_k g_{ac} - g_{ad}\widetilde{L}^d_{ck})$

(iii) $(\widetilde{V}^i{}_j{}^a)^0 = g^{ab}\widetilde{C}^i{}_{jb}$

(iv) $(V^{bc}_a)^0 = g^{be}g^{cd}[\dot{\partial}_d(g_{ea}) - g_{af}C^f{}_{ed}]$.

Remark. We see that H^i_{jk} is obtained very simple from L^i_{jk} and $\widetilde{V}^i{}_j{}^a$ as the $\mathcal{L}$–dual of $\widetilde{C}^i{}_{jb}$ that is

$$\widetilde{V}^i{}_j{}^a = \widetilde{C}^{*i}{}_j{}^a.$$

On the other hand,

$$V^{bc}_a = -C^{*bc}_a - \overset{*}{g}_{ad}\dot{\partial}^d g^{*bc}$$

and therefore $V_a{}^{bc} = C^{*bc}_a \iff C^{*bc}_a = -\frac{1}{2}\overset{*}{g}_{ad}\dot{\partial}^d g^{*bc} \iff C^a{}_{bc} = \frac{1}{2}g^{ad}\dot{\partial}_d g_{bc}$.

Corollary 7.3.1.

(i) $\nabla^{*H} = \nabla^{H*}$, $\nabla^{*V} = \nabla^{V*}$.

(ii) *Let K^* be a d^*–tensor on U^*, the $\mathcal{L}$–dual of the d–tensor K on U, $K^* \underset{|}{*} k$ and $K^* \underset{|}{*}{}^c$ its h– and v–covariant derivative with respect to ∇^*, $T := K_{|k}$, $T' := K|^c$. Then*

$$K^* \underset{|}{*}k = T^*, \quad K^* \underset{|}{*}{}^c = T'^*.$$

A consequence of (1.20) and (1.21) is the following result:

Proposition 7.3.2. *Let ∇ and ∇^* be two $\mathcal{L}$–dual d–connections. Then, the torsion and curvature tensors of ∇^* are $\mathcal{L}$–dual of torsion and curvature tensors of ∇.*

Remark. The proposition above states that the torsion and curvature tensors of ∇^* can be obtained from those of ∇ by lowering or raising vertical indices, using g_{ab}.

As we have seen (Theorem 7.3.1), the $\mathcal{L}$–dual of a N-connection generally is not a N^*-connection.

Proposition 7.3.3. *Let $C\Gamma(N) = (L^i_{jk}, C^a{}_{bc})$ be a N-connection on U and $C\Gamma^*(N^*) = (H^i_{jk}, \widetilde{H}^a_{bk}, \widetilde{V}^i{}_j{}^a, V_a{}^{bc})$ its $\mathcal{L}$–dual. Then*

$$\widetilde{H}^a_{bk} = g^{*ah}(\delta^*_k g^*_{bh} - g^*_{bi} H^i_{hk}), \tag{3.1}$$

$$\widetilde{V}^i{}_j{}^a = g^{*ic}(\dot{\partial}^a g^*_{jc} - g^*_{jb} V_c{}^{ba}). \tag{3.2}$$

Conversely, if the coefficients of ∇^* *(the* $\mathcal{L}$*–dual of* ∇*) verify* (3.1) *and* (3.2) *then* ∇ *is a N-connection.*

Proof. Let $F^* = -g^*_{ia}\dot{\partial}^a \otimes dx^i + g^{*ia}\delta_i \otimes \delta p_a$ be the $\mathcal{L}$–dual of the almost complex structures

$$F = -\dot{\partial}_i \otimes dx^i + \delta_i \otimes \delta y^i.$$

Then the d-connection ∇ is a N-connection if and only if

$$\nabla F = 0 \Longleftrightarrow \nabla^* F^* = 0 \Longleftrightarrow \text{(3.1) and (3.2)}.$$

On the tangent bundle we have the *metrical structure*

$$G = g_{ij} dx^i \otimes dx^j + g_{ab}\delta y^a \otimes \delta y^b. \tag{3.3}$$

The $\mathcal{L}$–dual of this metric tensor is

$$G^* = g^*_{ij} dx^i \otimes dx^j + g^{*ab}\delta p_a \otimes \delta p_b. \tag{3.4}$$

Therefore we are in the position to apply Theorem 3.10.1 and Theorem 7.3.1 and so we will get the canonical d–connection of the Lagrange and Hamilton manifolds (here restricted to U and U^*).

Theorem 7.3.2. *The* $\mathcal{L}$*–dual of the canonical N-connection of a Lagrange manifold is just the canonical* N^**-connection of its associated Hamilton manifold. (Only in this case* $V_a{}^{bc} = C^{*bc}_a$*.)*

Proof. Using Theorem 3.10.1 and Theorem 7.3.1, (i) we get

$$(H^i_{jk})^0 = \frac{1}{2} g^{ih}(\delta_j g_{hk} + \delta_k g_{jh} - \delta_h g_{jk}),$$

and using Theorem 7.2.1 we obtain

$$H^i_{jk} = \frac{1}{2} g^{*ih}(\delta^*_j g^*_{hk} + \delta^*_k g^*_{jh} - \delta^*_h g^*_{jk}).$$

Also, making use of Theorem 7.3.1, (iv) and Theorem 3.10.1, we have

$$V_a{}^{bc} = -\frac{1}{2} g^*_{ad}(\dot{\partial}^b g^{*dc} + \dot{\partial}^c g^{*bd} - \dot{\partial}^d g^{*bc}) = -\frac{1}{2} g^*_{ad}\dot{\partial}^b g^{*cd} = C^*{}_a{}^{bc}$$

and from (ii) and (ii) of Theorem 7.3.1 we obtain

$$\widetilde{H}^a_{bk} = H^i_{jk}, \ \widetilde{V}^i{}_j{}^a = V_j{}^{ia}.$$

Theorem 7.3.3. *Let ∇^* be the $\mathcal{L}$–dual of a N-connection $C\Gamma(N) = (L^i_{jk}, C^a{}_{bc})$ on U. Then we have:*

(i) $\widetilde{H}^i_{jk} = H^i_{jk}$ *iff* ∇^* *is h–metrical* $(g^*_{ij} \overset{*k}{|} = 0, g^*_{ab} \overset{*k}{|} = 0)$;

(ii) $\widetilde{V}^i{}_j{}^c = V_j{}^{ic}$ *iff* ∇^* *is v–metrical* $(g^*_{ij} \overset{*c}{|} = 0, g^*_{ab} \overset{*c}{|} = 0)$.

Proof. (i) The h–metrical condition for ∇^* and (3.1) can be rewritten in the following forms:

$$\begin{aligned} g^*_{hj} H^h_{ik} + g^*_{ih} H^h_{jk} &= \delta^*_k g^*_{ij} \quad (\Longleftrightarrow g^*_{ij} \overset{*k}{|} = 0) \\ g^*_{cb} \widetilde{H}^c_{ak} + g^*_{ac} \widetilde{H}^c_{bk} &= \delta^*_k g^*_{ab} \quad (\Longleftrightarrow g^*_{ab} \overset{*k}{|} = 0) \\ g^*_{ha} \widetilde{H}^a_{bk} + g^*_{bi} H^i_{hk} &= \delta^*_k g^*_{bh}. \end{aligned}$$

If $\widetilde{H}^i_{jk} = H^i_{jk}$, from the last equality we get the first two and from the last condition and the first we get $\widetilde{H}^i_{jk} = H^i_{jk}$. By a similar argument we can prove (ii).

Consequently, we can conclude finally:

Theorem 7.3.4. *The class of N-connections which is preserved by $\mathcal{L}$–duality is only the class of metrical N-connections.*

Let θ_L be the canonical symplectic form of (M, L) given by (2.13). The following result is a consequence of the above theorems:

Theorem 7.3.5.

(i) *If ∇ is a N-connection on $U \subset TM$, then*

$$\nabla \theta_L = 0 \Longleftrightarrow \nabla G = 0.$$

(ii) *If ∇^* is a N^*-connection on $U^* \subset T^*M$ then*

$$\nabla F^* = 0 \Longleftrightarrow \nabla G^* = 0.$$

Proof. (i) Let ∇^* be the $\mathcal{L}$–dual of ∇. Then $\nabla G = 0 \Longleftrightarrow \nabla^* G^* \Longleftrightarrow$
$\Longleftrightarrow \nabla^*\theta = 0 \Longleftrightarrow \nabla\theta_L = 0$.

(ii) We use a similar argument.

Let us stand out some problems connected with the *deflection* tensor field.

The *h–deflection* tensor field of a d–connection ∇, on the tangent bundle can be defined as follows:

$$D : \mathcal{X}(U) \longrightarrow \mathcal{X}(U),\ D(X) = \nabla^H_X C$$

where C is the Liouville vector field.

Locally we have:

$$\begin{aligned} C &= y^a\dot{\partial}_a,\ D = D^a{}_i\dot{\partial}_a \otimes dx^i, \\ D^a{}_i &= y^a_{|i} = \widetilde{L}^a_{bi} y^b - N^a_i. \end{aligned} \tag{3.5}$$

The deflection tensor field of a $\bar{d}$–connection $\bar{\nabla}$, on the cotangent bundle, $\bar{D}(X) = \bar{\nabla}^H_X \bar{C}$ where $\bar{C}$ is the Liouville vector field on T^*M, $\bar{C} = p_a\dot{\partial}^a$, has the local form

$$\bar{D} = \bar{D}_{ai}\dot{\partial}^a \otimes dx^i,\ \bar{D}_{ai} = p_{a||i} = N_{ia} - \widetilde{H}^b_{ai} p_b. \tag{3.6}$$

Also we can consider the *v–deflection* tensor field

$$d(X) = \nabla^V_X C,\ d = d_a{}^b\dot{\partial}_b \otimes \delta y^a,\ d_a{}^b = y^b_{|a} = \delta^a_b + C^b{}_{ca} y^c \tag{3.7}$$

and its correspondent for cotangent bundle

$$\bar{d}(X) = \bar{\nabla}^V_X \bar{C},\ \bar{d} = \bar{d}^a{}_b\dot{\partial}^b \otimes \delta p_a,\ \bar{d}^a{}_b = p_b||^a = \delta^a_b - V_b{}^{ca} p_c. \tag{3.8}$$

Using the $\mathcal{L}$–duality we see that generally, the $\mathcal{L}$–duals of D and d are different by $\bar{D}$ and respectively $\bar{d}$.

We have

$$D^*(X) = \nabla^{*H}_X C^* \tag{3.9}$$

where C^* is the *$\mathcal{L}$–dual of the Liouville* vector field,

$$C^* = y^*_a\dot{\partial}^a,\ y^*_a := g^*_{ab} y^{*b} \tag{3.10}$$

and locally

$$D^* = D^*_{ai}\dot{\partial}^a \otimes dx^i,\ D^*_{ai} = g^*_{ab}(D^b{}_i)^* \quad \text{or}$$

$$D^*_{ai} = y^*_a \underset{|}{*}{}_i = \delta^*_i y^*_a - \widetilde{H}^b_{ai} y^*_b. \tag{3.11}$$

The $\mathcal{L}$–dual of d is $d^* = d^{*a}{}_b\dot{\partial}^b \otimes \delta p_a$ where

$$d^{*a}{}_b = g_{*bc} g^{*ae} (d_e{}^c)^* = y^*_b \underset{|}{*}{}^a. \tag{3.12}$$

The following result holds:

Proposition 7.3.4. *$C^* = \bar{C}$ if and only if $L(x,y) = \frac{1}{2}F^2(x,y) + u(x)$ where F is 1–homogeneous and u is a scalar field. In this case $\bar{D} = D^*$ and $\bar{d} = d^*$.*

Proof. $C^* = C \iff g^*_{ab}y^{b*} = p_a \iff y^a\dot{\partial}_a\dot{\partial}_b L = \dot{\partial}_b L \iff \dot{\partial}_a L$ is 1–homogeneous $\iff L(x,y) = \frac{1}{2}F^2(x,y) + u(x)$, and F 1–homogeneous.

Remark. As we have seen from the last two sections there exists many geometric objects (nonlinear connections, linear connections, metrical structures and so on) which can be transfered by using $\mathcal{L}$–duality from U to U^* and also from U^* to U.

Now let suppose we have a regular Hamiltonian defined on a domain $D^* \subset T^*M$. The Legendre transformation $\psi : U^* \longrightarrow U$ is a diffeomorphism between some open subsets U^*, U of D^* and TM. Taking the Lagrangian $L(x,y) = p_a y^a - H(x,y)$, $y^{a*} = \dot{\partial}^a H(x,p)$ we can construct a Lagrange geometry restricted to U and then we pull–back by ψ the geometric objects on U, to U^*. These will depend only by Hamiltonian, therefore we will be able to extend them on the whole domain D^*.

7.4 The Finsler–Cartan $\mathcal{L}$–duality

In this section we will give an idea for the study of the geometry of a Cartan space using the $\mathcal{L}$–duality and the geometry of its associated Finsler space.

Let H be a 2–homogeneous Hamiltonian on a domain of T^*M,
$\psi : U^* \longrightarrow U$ the Legendre transformation and

$$L(x,y) = p_i y^i - H(x,p),\ y^{i*} = \dot{\partial}^i H(x,p) \tag{4.1}$$

its associate Lagrangian.

We remark, using the 2–homogeneous property of H, that

$$L(x,y) = H(x,p). \tag{4.2}$$

Proposition 7.4.1. *The Lagrangian given by* (4.1) *is a 2–homogenous Lagrangian.*

Proof. Let us put

$$f^i(x,p) := y^{i*} = \dot{\partial}^i H(x,p),\ g_i(x,y) := p^0_i = \dot{\partial}_i L(x,y).$$

We know that f^i is 1–homogeneous, then

$$g_i(x,\lambda y) = g_i(x,\lambda f^j(x,p)) = g_i(x, f^j(x,\lambda p)) = \lambda p^0_i = \lambda g_i(x,y)$$

and thus g_i is 1–homogeneous so, L is 2–homogeneous.

Therefore, using the theory made in the previous sections we may carry some geometry of Finsler spaces on 2–homogeneous Hamilton manifolds.

Remark. For 2–homogeneous Hamiltonian we have

(4.3) $p_i^0 = y_i$ or $p_i = y_i^*$

(4.4) $p^{i0} := g^{ij}p_j^0 = g^{ij}y_j = y^i$

(4.5) $p^i := g^{*ij}p_j = g^{*ij}y_j^* = y^{i*}$.

Among the nonlinear connections of a Finsler space one has the most interest. It is the *Cartan nonlinear connection*

$$N_j^i = \gamma_{j0}^i - C^i{}_{jk}\gamma^k{}_{00} \tag{4.6}$$

where

$\gamma^i{}_{jk} = \frac{1}{2}g^{ih}(\partial_j g_{hk} + \partial_k g_{jh} - \partial_h g_{jk})$, $C^i{}_{jk} = \frac{1}{2}g^{ih}\dot{\partial}_h g_{jk}$, $\gamma^i{}_{00} = \gamma^i_{jk}y^j y^k$, $\gamma^i_{j0} = \gamma^i_{jk}y^k$.

Theorem 7.4.1. *The $\mathcal{L}$–dual of the Cartan nonlinear connection* (4.6) *is*

$$N_{ij} = \gamma_{ij}^{*0} - V^k{}_{ij}\gamma_{k0}^{*0} \tag{4.7}$$

where we have put

$$\gamma_{ij}^{*k} = \frac{1}{2}g^{*kh}(\partial_i g_{hj}^* + \partial_j g_{ih}^* - \partial_h g_{ij}^*),\ \gamma_{ij}^{*0} = \gamma_{ij}^{*k}p_k,$$
$$\gamma_{i0}^{*0} = \gamma_{ij}^{*k}p_k p^j \quad \text{and} \quad V^k{}_{ij} = \frac{1}{2}\dot{\partial}^k g_{ij}.$$

Proof. We have $(N_{ij})^0 = -N_i^k g_{kj} + \partial_i\dot{\partial}_j L$ and making use of (4.2) we get

$$N_{ij} = -(\gamma_{jir}y^r)^* + \frac{1}{2}g_{ha}^*(\gamma_{rs}^h y^r y^s)^*\dot{\partial}^a g_{ij}^* - A_i^a g_{aj}^* \tag{4.8}$$

where $A_i^a := \partial_i\dot{\partial}^a H$.

On the other hand,

$$2(\gamma_{jir})^* = (\partial_i g_{jr})^* + (\partial_r g_{ij})^* - (\partial_j g_{ir})^* = (\partial_i g_{jr}^* + \partial_r g_{ij}^* - \partial_j g_{ir}^*) -$$
$$- (A_i^a(\dot{\partial}_a g_{jr})^* + A_r^a(\dot{\partial}_a g_{ij})^* - A_j^a(\dot{\partial}_a g_{ir})^*).$$

So, we will have

$$2(\gamma_{jir})^* y^{r*} = -p_k\gamma_{ij}^{*k} + y^{*r}(\partial_i g_{jr}^* - \frac{1}{2}A_r^0(\partial_a g_{ij})^*)$$

and

$$\begin{aligned}(\gamma_{asr}y^r y^s)^* &= -p_k\gamma^{*k}_{as}y^{*s} + y^{*s}y^{*r}(\partial_s g^*_{ar} - \frac{1}{2}A^b_r(\dot{\partial}_b g_{as})^*)\\ &= -p_k\gamma^{*k}_{as} + y^{*s}y^{*r}\partial_s g^*_{ar}.\end{aligned}$$

Then (4.8) becomes

$$\begin{aligned}N_{ij} = p_k\gamma^{*k}_{ij} &- \frac{1}{2}\dot{\partial}^a g^*_{ij}\gamma^{*k}_{as}p_k p^s - (A^a_i g^*_{aj} + y^{*r}\partial_i g^*_{jr})\\ &+ \frac{1}{2}\dot{\partial}^a g^*_{ij}y^s(y^{*r}\partial_s g^*_{ar} + A^b_s g^*_{ba}).\end{aligned}$$

But we can write:

$$\begin{aligned}y^{*r}\partial_i g^*_{jr} &= g^{*rk}p_k\partial_i g^*_{jr} = -p_k g^{*kr}g^*_{jh}g^*_{rs}\partial_i g^{*hs}\\ &= -g^*_{jh}\partial_i(g^{*hk}p_k) = -g^*_{jh}\partial_i\dot{\partial}^h H = -A^h_i g_{hj}\end{aligned}$$

and substituting it in the above equality we get (4.7).

Among Finsler connections, the Cartan connection is without doubt very important.

The following result is also well known [88].

Theorem 7.4.2. *On a Finsler space there exists only one Finsler connection which verifies the following axioms (Matsumoto's axioms)*

C_1) $g_{ij|k} = 0$ *(h–metrical);* C_2) $g_{ij}|_k = 0$ *(v–metrical);*

C_3) $D^i_k = -N^i_k + y^j F^i_{jk} = 0$; C_4) $T^i{}_{jk} = 0$; C_5) $S^i{}_{jk} = 0$.

The coefficients of this Finsler connection are:

$$F^i_{jk} = \frac{1}{2}g^{ih}(\delta_j g_{hk} + \delta_k g_{jh} - \delta_h g_{jk}) \tag{4.9}$$

$$C^i{}_{jk} = \frac{1}{2}g^{ih}\dot{\partial}_h g_{jk} \tag{4.10}$$

and the nonlinear connection is given by (4.6).

The Finsler connection given by (4.9), (4.6), (4.10) is the *Cartan connection* of the Finsler space $F^n = (M, F)$.

Using the results of the previous sections we can state the $\mathcal{L}$–dual of Theorem 7.4.2 for the Cartan space $C^n = (M, \bar{F})$.

Theorem 7.4.3. *On a Cartan space* $(M, \bar{F})$ *there exists only one N-connection* $C\bar{\Gamma}(\bar{N}) = (H^i_{jk}, V_i^{jk})$ *which satisfies the following axioms:*

$C_1^*)$ $g^{*ij}{}_{\|k} = 0$; $C_2^*)$ $g^{*ij}\|^k = 0$; $C_3^*)$ $\bar{D}_{ik} = -N_{ik} + p_j H^j_{ik} = 0$;
$C_4^*)$ $\bar{T}^i{}_{jk} = 0$; $C_5^*)$ $\bar{S}_i^{jk} = 0$.

It is the $\mathcal{L}$–dual of the Cartan connection above. That is:

$$H^i_{jk} = \frac{1}{2} g^{*ih}(\delta_j^* g^*_{hk} + \delta_k^* g^*_{jh} - \delta_h^* g^*_{jk}) \tag{4.11}$$

$$V_i^{jk} = -\frac{1}{2} g^*_{ih} \dot{\partial}^h g^{*jk} \tag{4.12}$$

and the nonlinear connection is given by (4.7).

Proof. The connection given by (4.11), (4.12) and (4.7), verifies $C_1^*)$–$C_5^*)$. For this connection $\bar{T}^i{}_{jk} = T^{*i}{}_{jk}$, $\bar{S}_i^{jk} = S^*{}_i^{jk}$ and $\bar{D}_{ik} = D^*{}_{ik}$ (Proposition 7.3.4).

This connection is unique. Indeed, if there exists another one, taking the $\mathcal{L}$–dual of it we will get two Finsler connections (restricted to an open set) which satisfy Matsumoto's axioms.

The linear connection of Theorem 7.4.3 is just the *Cartan connection of the Cartan space* $C^n = (M, \bar{F})$. We remark that conditions $C_1^*)$–$C_5^*)$ are all the $\mathcal{L}$–duals of $C_1)$–$C_5)$.

Consequently, all properties of the Cartan connection from Finsler spaces can be transfered on the Cartan spaces only by using the $\mathcal{L}$–duality.

Remark. When we look for a $\mathcal{L}$–dual of a d–tensor field we must pay attention to the vertical indices; in this section (and sometimes in the other sections) for the sake of simplicity we have omitted to use indices a, b, c, d, e, f to stand out the vertical part.

Let $c^*(t) = (x(t), p(t))$, $t \in I \subset \mathbb{R}$ a differentiable curve on D^*. c will be called *h–path* (with respect to Cartan connection $C\Gamma^*$) if it is horizontal and

$$\frac{d^2x^i}{dt^2} + H^i_{jk}\left(\frac{dx}{dt}, \frac{dp}{dt}\right)\frac{dx^j}{dt}\frac{dx^k}{dt} = 0 \ (\Longleftrightarrow \nabla^*_{\dot{c}(t)}\dot{c}(t) = 0).$$

Theorem 7.4.4. *The $\mathcal{L}$–duals of h–paths of $C\Gamma^*(N^*) = (H^i_{jk}, V_i^{jk})$ are h–paths of $C\Gamma(N) = (F^i_{jk}, C^i{}_{jk})$.*

Proof. It follows from Proposition 7.2.2 and Theorem 7.3.1, (i).

Corollary 7.4.1. *Let $c^*(t) = (x(t), p(t))$ be an integral curve of the Hamilton vector field* (2.9). *then $c^*(t)$ is an h–path of $C\Gamma^*$.*

Proof. $c^*(t) = (x(t), p(t))$ is an integral curve of X_H iff its dual $c(t) = (x(t), y(t))$ is an integral curve of X_L, and therefore an h–path so its $\mathcal{L}$–dual c^* will be also an h–path.

The next results will give us an interesting field where the $\mathcal{L}$–dual theory can be applied.

As we know a *Randers space* is a Finsler space where the metric has the following form

$$F(x, y) = \alpha + \beta \tag{4.13}$$

(Randers metric) and a *Kropina space* is a Finsler space with the fundamental function

$$F(x, y) = \alpha^2/\beta \tag{4.14}$$

(Kropina metric) where $\alpha = \sqrt{a_{ij}(x)y^i y^j}$ is a Riemannian metric and $\beta = b_i(x)y^i$ is a differential 1–form.

We can also consider Cartan spaces having the metric functions of the following forms

$$\bar{F}(x, p) = \sqrt{a^{ij}p_i p_j} + b^i p_i \quad \text{or} \tag{4.15}$$

$$\bar{F}(x, p) = \frac{a^{ij}p_i p_j}{b^i p_i} \tag{4.16}$$

and we will again call these spaces Randers and, respectively, Kropina spaces on the cotangent bundle T^*M.

Theorem 7.4.5. *Let (M, F) be a Randers space and $b = (a_{ij}b^i b^j)^{1/2}$ the Riemannian length of b_i. Then*

(i) *If $b^2 = 1$, the $\mathcal{L}$–dual of (M, F) is a Kropina space on T^*M with*

$$H(x, p) = \frac{1}{2}\left(\frac{a^{ij}p_i p_j}{2b^i p_i}\right)^2. \tag{4.17}$$

(ii) *If $b^2 \neq 1$, the $\mathcal{L}$–dual of (M, F) is a Kropina space on T^*M with*

$$H(x, p) = \frac{1}{2}\left(\sqrt{\bar{a}^{ij}p_i p_j} \pm \bar{b}^i p_i\right)^2, \tag{4.18}$$

where $\bar{a}^{ij} = \dfrac{1}{1 - b^2}a^{ij} + \dfrac{1}{(1 - b^2)^2}b^i b^j,\ \bar{b}^i = \dfrac{1}{1 - b^2}b^i$
(in (4.18) "$-$" corresponds to $b^2 < 1$ and "$+$" corresponds to $b^2 > 1$).

Proof. We put $\alpha^2 = y_i y^i$, $b^i = a^{ij} b_j$, $\beta = b_i y^i$, $\beta^* = b^i p_i$, $p^i = a^{ij} p_j$, $\alpha^{*2} = p_i p^i = a^{ij} p_i p_j$.

We have

$$(4.19) \qquad F = \alpha + \beta, \quad p_i = \frac{1}{2}\dot{\partial}_i F^2 = (\alpha + \beta)\left(\frac{y_i}{\alpha} + b_i\right) = F\left(\frac{y_i}{\alpha} + b_i\right) \cdot$$

Contracting in (4.19) by p^i and b^i we get:

$$(4.20) \qquad \alpha^{*2} = F\left(\frac{F^2}{\alpha} + \beta^*\right), \; \beta^* = F\left(\frac{\beta}{\alpha} + b^2\right) \cdot$$

Therefore,

$$(4.21) \qquad \beta^* = F\left(\frac{F}{\alpha} + b^2 - 1\right) \cdot$$

(i) If $b^2 = 1$, from (4.21) we obtain $\beta^* = \dfrac{F^2}{\alpha}$ and using (4.20), we get

$$\bar{F}(x,p) = \frac{\alpha^{*2}}{2\beta^*} = \frac{a^{ij} p_i p_j}{2 b^i p_i}.$$

(ii) If $b^2 \neq 1$, from (4.20) and (4.21) we have:

$$\frac{1}{F}\alpha^{*2} = \frac{F}{\alpha} + \beta^*; \; \beta^* = \frac{F}{\alpha} + F(b^2 - 1)$$

and by substitution

$$\beta^* - \frac{1}{F}\alpha^{*2} = (b^2 - 1)F - \beta^* \Longleftrightarrow \left(F + \frac{\beta^*}{1 - b^2}\right)^2 = \frac{1}{1 - b^2}\alpha^{*2} + \left(\frac{\beta^*}{1 - b^2}\right)^2.$$

From this last relation we obtain (4.18).

Theorem 7.4.6. *The $\mathcal{L}$-dual of a Kropina space is a Randers space on T^*M with the Hamiltonian*

$$(4.22) \qquad H(x,p) = \frac{1}{2}\left(\sqrt{\bar{a}^{ij} p_i p_j} \pm \bar{b}^i p_i\right)^2 \quad \text{where}$$

$$\bar{a}^{ij} = \frac{b^2}{4} a^{ij}, \; \bar{b}^i = \frac{1}{2} b^i.$$

(Here "+" corresponds to $\beta > 0$ and "−" to $\beta < 0$.)

Proof. We use the same notations as in the proof of Theorem 7.4.5. We have

$$p_i = F\dot{\partial}_i F = \frac{F}{\beta}(2y_i - Fb_i). \tag{4.23}$$

Contracting by p^i and then by b^i we get:

$$\alpha^{*2} = \frac{F^2}{\beta}(2F - \beta^*),\ \beta^* = \frac{F}{\beta}(2\beta - Fb^2). \tag{4.24}$$

Using these relations, after a simple computation, we obtain (4.22).

We must have $b^2 \neq 0$ for regularity of $\bar{a}^{ij}$. But, the regularity condition for the Kropina metric leads to $b^2 \neq 0$.

Remark. Using the Theorem 7.4.6, we can derive the geometric properties of Kropina spaces, very simply, from those of Randers spaces, by using $\mathcal{L}$–duality.

We will explain this more precisely in the next section.

7.5 Berwald connection for Cartan spaces. Landsberg and Berwald spaces. Locally Minkowski spaces.

Berwald connection $\bar{B\Gamma} = (N_{ij}, \dot{\partial}^k N_{ij}, 0)$ of Cartan space (here N_{ij} are given by (4.6)) is not the $\mathcal{L}$–dual of Berwald connection of its associated Finsler space like Cartan connection. There exist some important distinctions here, which are consequences of the nonexistence of a spray and thus, the nonlinear connection cannot be obtained as a partial derivative of a spray.

Theorem 7.5.1. *On Cartan space $(M, \bar{F})$ there exists only one Finsler connection with the following properties:*

$B_1^*)\quad \delta_i^* \bar{F} = 0,$ $\qquad\qquad B_2^*)\quad \bar{D}_{ij} = 0,$

$B_3^*)\quad g^*_{ih}\dot{\partial}^h N_{jk} - g^*_{jh}\dot{\partial}^n N_{ik} = \delta^*_j g^*_{ik} - \delta^*_i g^*_{jk},$

$B_4^*)\quad \bar{P}_{ij}{}^k = 0,$ $\qquad\qquad B_5^*)\quad V_i^{jk} = 0.$

Proof. It is easily checked that the connection $\bar{B\Gamma} = (N_{ij}, \dot{\partial}^k N_{ij}, 0)$ with N_{ij} given by (4.6) satisfies all $B_1^*) - B_5^*)$. Indeed, let us take the local diffeomorphism $\psi : U^* \to U$ and consider $N = (N^i_j)$, the $\mathcal{L}-$dual of $\bar{N} = (N_{ij})$. $N = (N^i_j)$ is Cartan nonlinear connection of the associated Finsler space (M, F). We have

$$\bar{F}_{||i} = \delta_i^*\, \bar{F} = (\delta_i F)^* = 0.$$

B_4^*) and B_5^*) are obvious and B_3^*) is equivalent to $\dot{\partial}_i N_j^k = \dot{\partial}_j N_i^k$ (see (2.8)).

Let us prove the uniqueness of the Finsler connection which satisfies $B_1^*) - B_5^*)$. If $\widetilde{B\Gamma} = (\widetilde{N}_{ij}, \widetilde{H}^i_{jk}, \widetilde{V}_i^{jk})$ is another connection, then it must have the following form:

$$\widetilde{B\Gamma} = (\widetilde{N}_{ij}, \dot{\partial}^k \widetilde{N}_{ij}, 0).$$

Taking the lift of this connection on U^*, we get a N-connection $(\dot{\partial}^i \widetilde{N}_{ij}, \dot{\partial}^k \widetilde{N}_{ij}, 0, 0)$, and $((\dot{\partial}^k \widetilde{N}_{ij})^\circ, \partial_k \widetilde{N}^i_j, 0, g^{kh} \dot{\partial}_h g_{ij})$, is the $\mathcal{L}-$dual of this connection, where $(\widetilde{N}^i_j)$ is the $\mathcal{L}-$dual of $(\widetilde{N}_{ij})$.

This $d-$connection provides a Finsler connection of U

$$B\Gamma' = (\widetilde{N}^i_j, \dot{\partial}_k \widetilde{N}^i_j, 0)$$

which has the following properties:

$$F_{|i} = 0, \quad T^i{}_{jk} = 0, \quad D^i_j = 0, \quad P^i{}_{jk} = 0, \quad C^i{}_{jk} = 0.$$

These conditions are sufficient to assure the uniqueness of Finsler connection of U. Now, we can easily prove the uniqueness.

Let us put

$$\bar{G}_j{}^i{}_k := \dot{\partial}^i N_{jk}.$$

Berwald connection for Cartan spaces has the following, generally nonvanishing curvature tensors:

$$(5.1)\qquad \begin{aligned} \widetilde{H}_h{}^i{}_{jk} &= \coprod_{(k,j)} \{\delta^*_k \bar{G}_h{}^i{}_j + \bar{G}_h{}^\ell{}_j \bar{G}_\ell{}^i{}_k\} && (h-\text{curvature}),\\ \bar{G}_j{}^i{}_k{}^h &= \dot{\partial}^h \bar{G}_j{}^i{}_k && (hv-\text{curvature}) \end{aligned}$$

and a torsion tensor $\bar{R}_{ijk} = \delta^*_j N_{ki} - \delta^*_k N_{ji}$.

Proposition 7.5.1. *The following relations hold good:*

$$\text{(i) } \bar{H}_h{}^i{}_{jk} = -\dot{\partial}^i \bar{R}_{hjk}, \qquad \text{(ii) } p_i \bar{H}_h{}^i{}_{jk} = -\bar{R}_{hjk}.$$

Proposition 7.5.2. *Let $(M, \bar{F})$ be a Cartan space and $\bar{H}_h{}^i{}_{jk}$, $H_h{}^i{}_{jk}$ the $h-$curvature tensors of $\overline{B\Gamma}$ and of Berwald connection of its (locally) associated Finsler space, respectively. Then we have in U^**

$$(5.2)\qquad \bar{H}_{hijk} = -(H_{ihjk})^* - 2V_{ih}{}^s \bar{R}_{sjk},$$

where $(H_{ihjk})^*$ means that the value of H_{ihjk} is calculated in (x,p), $y^i = \dot{\partial}^i H(x,p)$. $(\bar{H}_{hijk} = g^*_{is} \bar{H}_h{}^s{}_{jk}$, $H_{ihjk} = g_{hs} H_i{}^s{}_{jk})$.

Let us denote by "$*$", the $h-$covariant derivative with respect to Berwald connection

$$\overline{B\Gamma} = (N_{ij}, \dot{\partial}^i N_{ij}, 0).$$

Theorem 7.5.2. *Let* $\overline{C\Gamma}$ *and* $\overline{B\Gamma}$ *be the Cartan and Berwald connections of* $(M, \bar{F})$, *respectively. Then*

$$\text{(5.3)} \qquad \text{(i) } \bar{G}_j{}^i{}_k = H^i_{jk} - V_{jk}{}^i{}_{\|0}, \qquad \text{(ii) } g^{*ij}{}_{,k}^{\ *} = -2V_k{}^{ij}{}_{\|0}.$$

Proof. (i) Let us restrict our considerations to the open sets U^*, U such that Legendre transformation $\psi : U^* \to U$ is a diffeomorphism.

If we consider the $\mathcal{L}-$dual of Cartan connection $\overline{C\Gamma}$, $C\Gamma = (N^i_j, F^i_{jk}, C^i{}_{jk})$, we have in U

$$P^s{}_{jk} = C^s{}_{jk|0} \qquad ([88], \text{page}114).$$

Taking the $\mathcal{L}-$dual of this relation, we get

$$\bar{P}_{sj}{}^k = V_{sj}{}^k{}_{\|0} \Longleftrightarrow H^k_{sj} - \dot{\partial}^k N_{js} = V_{sj}{}^k{}_{\|0},$$

where $V_{sj}{}^k{}_{\|0} = V_{sj}{}^k{}_{\|h} p^h$, $p^h = g^{*hr} p_r$.

(ii) We have

$$g^{*ij}{}_{,k}^{\ *} = \delta^*_k(g^{*ij}) + \bar{G}_h{}^i{}_k g^{*hj} + \bar{G}_h{}^j{}_k g^{*ih} = g^{*ij}{}_{\|k} - V_{hk}{}^i{}_{\|0} g^{*hj} - V_{hk}{}^j{}_{\|0} g^{*ih} = -2V_k{}^{ij}{}_{\|0}.$$

Definition 7.5.1. A Cartan space is called a *Landsberg space* if $V^{ijk}{}_{\|0} = 0$. It is called a *Berwald space* if $V^{ijk}{}_{\|h} = 0$ ($V^{ijk} = g^{*i\ell} V_\ell{}^{jk} = -\frac{1}{2}\dot{\partial}^i g^{*jk}$, $V^{ijk}{}_{\|0} = V^{ijk}{}_{\|h} p^h$).

Using the $\mathcal{L}-$duality between Cartan and Finsler spaces, we can easily prove:

Proposition 7.5.3. *A Cartan space is a Landsberg (Berwald) space if and only if every associated Finsler space is locally Landsberg (Berwald) space.*

The following theorems characterize Cartan spaces which are Landsberg and Berwald spaces.

Theorem 7.5.3. *A Cartan space is a Landsberg space if and only if one of the following conditions holds:*

$$\text{(a) } H^i_{jk} = \bar{G}_j{}^i{}_k, \qquad \text{(b) } \bar{P}_h{}^i{}_j{}^k = 0, \qquad \text{(c) } \bar{P}_{ij}{}^k = 0.$$

Proof. Using the $\mathcal{L}-$duality, we get $V^{ijk}{}_{\|0} = 0 \Longleftrightarrow$ (c) $\Longleftrightarrow$ (b). Then (a) $\Longleftrightarrow$ (b) follows from (5.3).

Theorem 7.5.4. *A Cartan space is a Berwald space if and only if one of the following conditions is true:*

(a) $\bar{G}_j{}^i{}_k{}^h = 0$,

(b) $\overline{B\Gamma}$ *is a linear connection (that is* $\bar{G}_j{}^i{}_k$ *are functions of position only),*

(c) H^i_{jk} *are functions of position only.*

Proof. Obviously (a) $\Longleftrightarrow$ (b). Now let us prove

$$V^{ijk}{}_{\|h} = 0 \Longleftrightarrow \bar{G}_j{}^i{}_k{}^h = 0.$$

The associated Finsler space (U, F) is Berwald space because of $C_{ijk|h} = 0$ ($\mathcal{L}-$dual of $V^{ijk}{}_{\|h} = 0$), therefore the coefficients F^i_{jk} of Cartan connection are functions of position only. From

$$H^i_{jk}(x,p) = F^i_{jk}(x,y), \quad y^i = \dot{\partial}^i H(x,p)$$

we obtain also that H^i_{jk} are functions of position only.

Now, using (5.3) and again $V^{ijk}{}_{\|0} = 0$, we get (a).

Conversely, we obtain $\bar{G}_j{}^i{}_k{}^h = 0$, and (5.3) yields

$$\dot{\partial}^h H^i_{jk} - \dot{\partial}^h (V_{jk}{}^i{}_{\|0}) = 0.$$

Taking the $\mathcal{L}-$*dual* of this equation, we obtain

$$\dot{\partial}_h F^i_{jk} - \dot{\partial}_h (C^i{}_{jk\|0}) = 0 \quad or \quad F_j{}^i{}_{kh} - C^i{}_{jk\|0 \cdot h} = 0,$$

where $F_j{}^i{}_{kh}$ is the hv-curvature tensor of the Rund connection of (U, F). From this relation we obtain $C_{ijk|h} = 0$, following the same way as in [88], page 161, and by $\mathcal{L}-$dualization we get $V^{ijk}{}_{\|h} = 0$.

Finally, from the above considerations we can easily prove (a) $\Longleftrightarrow$ (c).

Definition 7.5.2. A Cartan space $(M, \bar{F})$ is called *locally Minkowski space* if there exists a covering of coordinate neighborhoods in which g^{*ij} depends on p_k only.

Proposition 7.5.4. *A Cartan space* $(M, \bar{F})$ *is a locally Minkowski space if and only if every locally associated Finsler space is locally Minkowski space.*

The following result characterizes the Cartan spaces which are locally Minkowski spaces.

Theorem 7.5.5. *A Cartan space is locally Minkowski space if and only if one of the following conditions holds:*

$$\text{(i) } \bar{H}_h{}^i{}_{jk} = 0, \quad \bar{G}_h{}^i{}_j{}^k = 0, \qquad \text{(ii) } \bar{R}_h{}^i{}_{jk} = 0, \quad V_i^{hj}{}_{\|k} = 0.$$

Proof. If $(M, \bar{F})$ is locally Minkowski space, then $N_{ij} = 0$ and $\delta_i^* g_{hk}^* = 0$. Therefore (i) and (ii) hold.

If (ii) is true, using the $\mathcal{L}-$duality we can easily prove that $(M, \bar{F})$ is locally Minkowski space (see also [130]).

If (i) is true, $\bar{G}_h{}^i{}_j{}^k = 0$ yields that $(M, \bar{F})$ is Berwald space and thus $V_i^{jk}{}_{\|h} = 0$.

On the other hand, $\bar{H}_h{}^i{}_{jk} = 0$ and from (ii) and Proposition 7.5.1, it follows $R_{hjk} = 0$.

In the same time $\bar{G}_j{}^i{}_k = H^i_{jk}$ holds and therefore, we get $\bar{R}_h{}^i{}_{jk} = \bar{H}_h{}^i{}_{jk} = 0$ from (5.1).

Definition 7.5.3. a) Cartan space $(M, \bar{F})$ is said to be *of scalar curvature* if there exists a scalar function $\bar{K} = \bar{K}(x,p)$ such that

$$\bar{H}_{hijk} p^i p^j X^h X^k = \bar{K}(g^*_{hj} g^*_{ik} - g^*_{hk} g^*_{ij}) p^i p^j X^h X^k \tag{5.4}$$

for every $(x,p) \in D^*$ and $X = (X^i) \in T_x M$.

b) A Cartan space $(M, \bar{F})$ of scalar curvature is said to be *of constant curvature* $\bar{K}$ if the scalar function from a) is constant

From Proposition 7.5.1, (ii) and (5.4) we easily obtain that $(M, \bar{F})$ is of constant curvature $\bar{K}$ if and only if $\bar{R}_{ijk} p^j = \bar{K}\, \bar{F}^2 h^*_{ik}$, where $h^*_{ik} = g^*_{ik} - \frac{1}{F^2} p_i p_k$ is the angular metric tensor of Cartan spaces.

Theorem 7.5.6 (i) *A Cartan space is of scalar curvature $K(x,p)$ if and only if every associated Finsler space is of scalar curvature $K(x,y)$, $y^i = \dot{\partial}^i H(x,p)$.*

(ii) *A Cartan space is of constant curvature K if and only if every associated Finsler space is of constant curvature K.*

Proof. Contracting (7.2) by p^i, p^j, X^h, X^k, we get

$$\bar{H}_{hijk} p^i p^j X^h X^k = -(X_{ihjk})^* p^i p^j X^h X^k$$

or

$$\bar{H}_{hijk} p^i p^j X^h X^k = -(H_{ihjk} y^i y^j X^h X^k)^*. \tag{5.5}$$

(Here the $\mathcal{L}-$dual of $X = X^i(x)\delta_i^*$ is $X^\circ = X^i(x)\delta_i$.) But a Finsler space is said to be of scalar curvature $K(X,y)$ if $H_{ihjk} y^i y^j X^h X^k = K(g_{ij} g_{hk} - g_{ik} g_{hj}) y^i y^j X^h X^k$ and if $K(x,y) =$ const., it is said to be of constant K (see [88], page 167).

Now, using (5.5), we obtain the proof.

Remark. We can get some similar results as in Proposition 7.5.3, Proposition 7.5.4 and Theorem 7.5.6 for a Finsler space, considering Cartan spaces, locally associated to it. Therefore, some nice results in Finsler space can be obtained as the $\mathcal{L}-$dual of those from Cartan spaces.

7.6 Applications of the $\mathcal{L}$-duality

In this section, we shall give some applications of the *$\mathcal{L}-$duality* between Finsler and Cartan spaces.

In terms of the Cartan connection a Landsberg space is a Finsler space such that the $hv-$curvature tensor $P_h{}^i{}_{jk} = 0$ [90].

A Cartan space is called Landsberg if $P^*{}_h{}^i{}_j{}^k = 0$. Using the *$\mathcal{L}-$duality* it is clear that a Finsler space is a Landsberg space iff its $\mathcal{L}-$dual is a Landsberg one.

In [151] (see also [90], [77]) was proved that a Randers space is a Landsberg space iff $b_{i;j} = 0$ (here ";" stands for covariant derivative with respect to Levi-Cività connection of the Riemannian manifold (M, a_{ij})).

For Kropina spaces we have a "dual" of the above result:

Theorem 7.6.1. *A Kropina space is a Landsberg space if and only if*

$$b_{i;k} = b_i f_k - b_k f_i + a_{ik} f^j b_j, \quad f^j := a^{ji} f_i. \tag{6.1}$$

Proof. The Randers metric

$$\bar{F}(x,p) = \sqrt{\bar{a}^{ij} p_i p_j} \pm \bar{b}^i p_i$$

is a Landsberg metric iff

$$\bar{b}^i_{\|K} = 0$$

(here "$\|$" stands for covariant derivative with respect to the Levi-Cività connection of the Riemannian manifold $(M, \bar{a}_{ij})$).

Hence, we have the following equivalent statements:

The Kropina space is Landsberg $\Longleftrightarrow$ its $\mathcal{L}-$dual is Landsberg $\Longleftrightarrow$

$$\Longleftrightarrow \bar{b}^i_{\|k} = 0 \Longleftrightarrow \bar{b}_{i\|k} = 0 \Longleftrightarrow \left(\frac{1}{b^2} b_i\right)_{\|k} = 0.$$

But $\overline{\{ {i \atop jk} \}} = \{ {i \atop jk} \} - \{\delta^i_k f_j + \delta^i_j f_k - a_{jk}a^{is}f_s\}$, where $f_k := \partial_k(\log b)$ and $\overline{\{ {i \atop jk} \}}$, $\{ {i \atop jk} \}$ are the coefficients of the Levi-Cività connection of $\bar{a}_{ij}$, respectively a_{ij} ($\bar{a}_{ij} = \frac{4}{b^2}a_{ij}$ therefore $\bar{a}_{ij}$ and a_{ij} are conformal metrics).

So we get

$$\left(\frac{1}{b^2}\, b_i\right)_{\|k} = 0 \Longleftrightarrow \partial_k b_i - \overline{\{ {j \atop ki} \}}\, b_j - b_i f_k = 0 \Longleftrightarrow$$

$$b_{i;k} = b_i f_k - b_k f_i + a_{ik} f^j b_j, \quad f^j := a^{ji} f_i$$

which also was obtained in [90], [77] by a different argument.

We can obtain other properties of Kropina spaces from those of Randers by using the $\mathcal{L}-$duality.

Theorem 7.6.2 i) *Kropina space is a Berwald space if and only if*

$$\nabla_k b_i = b_i f_k - b_k f_i + a_{ik} f^j b_j. \tag{6.2}$$

ii) *Kropina space is locally Minkowski space if and only if the condition* (6.2) *above holds and also*

$$\dot{R}_{ihjk} = \coprod_{(j,k)} \{a_{hk} f_{ij} + a_{ij} f_{hk} + f^m f_m a_{ik} a_{hj}\}, \tag{6.3}$$

where $f_k := \partial_k(\log b)$, $f^i := a^{ik} f_k$, $f_{ij} = \nabla_i f_j + f_i f_j$. ∇_k stands for the covariant derivative with respect to Levi-Cività connection of (M, a_{ij}) and $\dot{R}_{ihjk}$ is the Riemannian curvature *tensor.*

First of all we need the following

Lemma 7.6.1 *Let $(M, \bar{F})$ be Cartan space with Randers metric*

$$\bar{F}(x,p) = \sqrt{\bar{a}^{ij} p_i p_j} + \bar{b}^i p_i.$$

Then

(6.4)

(i) $(M, \bar{F})$ *is a Berwald space if and only if* $\bar{\nabla}_k \bar{b}_i = 0,$

(ii) $(M, \bar{F})$ *is Minkowski space if and only if* $\bar{\nabla}_k \bar{b}_i = 0, \quad \dot{\bar{R}}_i{}^r{}_{jk} = 0,$

where $\bar{\nabla}_k$ stands for Levi-Cività connection of $(M, \bar{a}_{ij})$ and $\circledast \circ \bar{R}$ is Riemannian curvature *tensor.*

Proof. The proof of this Lemma follows step by step the ideas of Kikuchi [77]. For example, here, to obtain that $(M, \bar{F})$ is Berwald space on the condition $\bar{\nabla} b_i =$

0, we use Theorem 7.4.3 and prove that the Cartan connection of this space is $\overline{C\Gamma} = (\overline{\{ {i \atop jk} \}} p_i, \overline{\{ {i \atop jk} \}}, V_i^{jk})$, where $\overline{\{ {i \atop jk} \}}$ are the coefficients of the Levi-Cività connection of $\bar{a}_{ij}$.

Proof of the Theorem 7.6.2. We take the $\mathcal{L}-$*dual* of the Kropina metric (4.14) and we get the Hamiltonian (4.18). We have

$$\bar{a}_{ij} = \frac{4}{b^2}\, a_{ij} = e^{2\sigma} a_{ij}, \quad \sigma := \log 2 - \log b.$$

Therefore, Riemannian manifolds (M, a_{ij}) and $(M, \bar{a}_{ij})$ are conformal and the coefficients of Levi-Cività connections are related as follows:

$$\overline{\{ {i \atop jk} \}} = \{ {i \atop jk} \} - (\delta^i_k f_j + \delta^i_j f_k - a_{jk} f^i).$$

The condition (6.4) is written as (6.2).

Also, for the conformal metrics, we have

$$\begin{aligned} \dot{\bar{R}}_i{}^h{}_{jk} = \dot{R}_i{}^h{}_{jk} \; & +\delta^h_k \sigma_{ij} - \delta^h_j \sigma_{ik} + a^{h\ell}(a_{ij}\sigma_{\ell k} - a_{ik}\sigma_{\ell j}) \\ & +(\delta^h_k a_{ij} - \delta^h_j a_{ik}) a^{mn} \sigma_m \sigma_n. \end{aligned}$$

For our position $\sigma_i = -f_i$, $\sigma_{ij} = -\nabla_i f_j - f_i f_j = -f_{ij}$ and using the above equality, we get (6.3). Finally, we apply the results of the previous section.

Remark. (i) The results of Theorem 7.6.2 were obtained by Kikuchi [77] in a different form and also by Matsumoto [90] in this form, by using of totally different ideas.

(ii) A Finsler space with a Kropina metric is Berwald space if and only if (6.2) is true. Indeed, it is easily checked that the Cartan space with Randers metric (4.15) is a Berwald space if and only if $\bar{\nabla}_k \bar{b}_i = 0$ (see [151] for Finsler spaces with Randers metric). Therefore, we follow the same idea as in Theorem 7.6.2.

Let us now give another example of using of $\mathcal{L}-$*duality* (see also [63]). If (M, F) is Finsler space with

$$F(x,y) = \{a_1(x)(y^1)^m + \ldots + a_n(x)(y^n)^m\}^{1/m}$$

($m-$th root metric [11], [152]), its $\mathcal{L}-$*dual* is Cartan space having fundamental function

$$\bar{F}(x,p) = \Big\{\frac{1}{(a_1(x))^{\ell-1}}\,(p_1)^\ell + \ldots + \frac{1}{(a_n(x))^{\ell-1}}\,(p_n)^\ell\Big\}^{1/\ell},$$

where $\frac{1}{\ell} + \frac{1}{m} = 1$.

In particular, if $a_1(x) = \ldots = a_n(x) = e^{m\phi(x)}$, $\phi(x) = \phi_1 x^1 + \ldots + \phi_n x^n$ ($\phi_i =$ const.), we get the ecological metric of Antonelli ([11], [13], [152]):

$$F(x, y) = e^{\phi(x)}\{(y^1)^m + \ldots + (y^n)^m\}^{1/m}$$

and its $\mathcal{L}-$dual is

$$\bar{F}(x, y) = e^{-\phi(x)}\{(p_1)^\ell + \ldots + (p_n)^\ell\}^{1/\ell}.$$

The geodesics of (M, F), parametrized by the arclength, are just the ecological equations (see [11], [12], [18], [152]). The $\mathcal{L}-$duals of these equations have a simpler form:

$$\frac{dx^i}{dt} = \dot{\partial}^i H, \qquad \frac{dp_i}{dt} = -\partial_i H, \tag{6.5}$$

(Hamilton-Jacobi equations), where $H = \frac{1}{2} F^2$. The solutions of (6.5) are $h-$paths of $(M, \bar{F})$ [66].

Chapter 8

Symplectic transformations of the differential geometry of T^*M

It is well-known that symplectic transformations preserve the form of the Hamilton-Jacobi equations. However, the natural metric tensor (kinetic energy matrix) is not generally invariant nor is its associated differential geometry. In this chapter we address precisely the question of how the geometry of the cotangent bundle changes under symplectic transformation. As a special case, we also consider the homogeneous contact transformations. The geometry of spaces admitting contact transformations was initiated and developed by Eisenhart [53], Eisenhart and Knebelman [55], where the first contact frame was introduced. Muto [122] and Doyle [52] introduced independently, the second contact frame and the geometry of the homogeneous contact manifolds was intensively studied by Yano and Muto [173, 174]. The chapter is based on [14].

8.1 Connection-pairs on cotangent bundle

Let M be a $n-$dimensional $C^\infty-$differentiable manifold and $\pi^* : T^*M \to M$ the cotangent bundle. As we have seen (Chapter 4) a nonlinear connection on T^*M is a supplementary distribution HT^*M of the vertical distribution $VT^*M = \text{Ker}\, \pi^*_*$. ($\pi^*_*$ is the tangent map of π^*). It is often more convenient to think of a nonlinear connection as an almost product structure Γ on T^*M such that $VT^*M = \text{Ker}\,(I+\Gamma)$ (Section 4.6).

If $f \in \text{Diff}\,(T^*M)$ and Γ is a connection on T^*M, the *push-forward* of Γ by f generally fails to be a connection. Because of this, we will now define a new geometrical structure which nevertheless is an extension of the above definition for connections.

Definition 8.1.1. A *connection-pair* ϕ on T^*M is an almost product structure on

T^*M such that Ker $(I - \phi)$ is supplementary to VT^*M.

$HT^*M =$ Ker $(I-\phi)$ will be called the *horizontal bundle* and $WT^*M =$ Ker $(I + \phi)$ the *oblique bundle.*

Remark. If ϕ is a connection-pair on T^*M, then a unique connection Γ can be associated to it, such that Ker $(I - \Gamma) =$ Ker $(I - \phi)$, therefore $\phi = \Gamma$ on Ker $(I - \phi)$. Conversely, if a connection Γ is given on T^*M, we can get a *connection-pair* on T^*M by taking a complementary subbundle of Ker $(I - \Gamma)$.

Let ϕ be a connection-pair on T^*M and Γ the associated connection. We will denote by h and v the projections induced by Γ :

$$h = \frac{1}{2}(I + \Gamma), \quad v = \frac{1}{2}(I - \Gamma) \tag{1.1}$$

and by h^{pr}, w those induced by ϕ

$$h' = \frac{1}{2}(I + \phi), \quad w = \frac{1}{2}(I - \phi). \tag{1.2}$$

The local expression of Γ is given by

$$\Gamma(\partial_i) = \partial_i + 2\Gamma_{ij}\dot{\partial}^j, \ \Gamma(\dot{\partial}^i) = -\dot{\partial}^i \tag{1.3}$$

and the local vector fields:

$$\delta_i := h(\partial_i) = \frac{1}{2}(\partial_i + \Gamma(\partial_i)) = \partial_i + \Gamma_{ij}\dot{\partial}^j \tag{1.4}$$

provide us with a frame for HT^*M at (x, p).

We also obtain:

$$\phi(\delta_i) = \Gamma(\delta_i) = \delta_i, \quad h'(\delta_i) = \delta_i. \tag{1.5}$$

On the other hand,

$$\phi(\dot{\partial}^i) = -\dot{\partial}^i + 2\Pi^{ij}\delta_j. \tag{1.6}$$

Indeed, from $\phi(\dot{\partial}^i) = a^i_j\dot{\partial}^j + b^{ij}\delta_j$ we get

$$\dot{\partial}^i = a^i_j(a^j_k\dot{\partial}^k + b^{jk}\delta_k) + b^{ij}\delta_j$$

and now (1.6) follows easily.

The local vector fields

$$\dot{\delta}^i := w(\dot{\partial}^i) = \frac{1}{2}(\dot{\partial}^i - \phi(\dot{\partial}^i)) = \dot{\partial}^i - \Pi^{ij}\delta_j \tag{1.7}$$

form a basis for WT^*M at (x,p) and

$$\phi(\dot{\delta}^i) = -\dot{\delta}^i. \tag{1.8}$$

Therefore, $(\delta_i, \dot{\delta}^i)$ is a frame for TT^*M at (x,p), *adapted to the connection-pair,* ϕ.

The dual of this *adapted frame* is $(\delta x^i, \delta p_i)$ where

$$\delta x^i = dx^i + \Pi^{ji}\delta p_j, \quad \delta p_i = dp_i - \Gamma_{ji}dx^j. \tag{1.9}$$

Using the notation above we have the following local expression of ϕ and its *associated connection* Γ :

$$\phi = \delta_i \otimes \delta x^i - \dot{\delta}^i \otimes \delta p_i, \quad \Gamma = \delta_i \otimes dx^i - \dot{\partial}^i \otimes \delta p_i. \tag{1.10}$$

With respect to natural frame, ϕ has the local form

(1.11)

$$\phi(\partial_i) = (\delta_i^k - 2\Gamma_{ij}\Pi^{jk})\partial_k + 2(\Gamma_{ik} - \Gamma_{ij}\Pi^{jh}\Gamma_{hk})\dot{\partial}^k, \phi(\dot{\partial}^i) = 2\Pi^{ik}\partial_k + (2\Pi^{ij}\Gamma_{jk} - \delta_k^i)\dot{\partial}^k.$$

From (1.4) and (1.7) we get:

Proposition 8.1.1. *The adapted basis $(\delta_i, \dot{\delta}^i)$ and its dual $(\delta x^i, \delta p_i)$ transform under a change of coordinates on T^*M as follows:*

$$\delta_{i'} = \partial_{i'}x^i\delta_i, \qquad \dot{\delta}^{i'} = \partial_i x^{i'}\dot{\delta}^i, \tag{1.12}$$

$$\delta x^{i'} = \partial_i x^{i'}\delta x^i, \quad \delta p_{i'} = \partial_{i'}x^i\delta p_i. \tag{1.13}$$

Proposition 8.1.2. *If a change of coordinates is performed on T^*M, then the coefficients of the connection-pair ϕ obey the following rules of transformation*

$$\Gamma_{i'j'}(x',p') = \partial_{i'}x^i\partial_{j'}x^j\Gamma_{ij}(x,p) + p_k\partial_{i'}\partial_{j'}x^k, \tag{1.14}$$

$$\Pi^{i'j'}(x',p') = \partial_i x^{i'}\partial_j x^{j'}\Pi^{ij}(x,p). \tag{1.15}$$

Remarks. 1. In spite of being an object on T^*M, Π^{ij} follows the same rule of transformation as a tensor of type $(1,1)$ on M, therefore Π^{ij} are the components of a d-tensor field.

2. If M is paracompact, there exists a connection-pair on T^*M if and only if, on the domain of each chart on T^*M there exists $2n^2$–differentiable functions Γ_{ij} and Π^{ij} satisfying (1.16) and (1.17) with respect to the transformation of coordinates on T^*M.

3. Explicit examples of connection-pair on T^*M. Let $g = (g_{ij})$ a Riemannian metric on M and $\{ {i \atop jk} \}$ the Christoffel symbols of g. We can define on every domain of a chart

$$\Gamma_{ij} = \{ {k \atop ij} \} p_k, \quad \Pi^{ij} = \frac{g^{ij}}{\sqrt{g^{ij}p_ip_j}}. \tag{1.16}$$

These are the local components of a connection pair [146].

More generally, if (M, H) is a Hamilton manifold, we can take Γ_{ij} as the coefficients of the canonical nonlinear connection and $\Pi^{ij} = \dot{\partial}^i\dot{\partial}^j H$.

Proposition 8.1.3. *We have the brackets:*

$$[\delta_i, \delta_j] = R_{ij\ell}\Pi^{\ell k}\delta_k + R_{ijk}\dot{\delta}^k, \tag{1.17}$$

$$\begin{aligned}[\dot{\delta}^j, \delta_i] = & (\delta_i\Pi^{jk} + \Pi^{j\ell}R_{i\ell h}\Pi^{hk} + \dot{\partial}^j\Gamma_{ih}\Pi^{hk})\delta_k \\ & +(\dot{\partial}^j\Gamma_{ik} + \Pi^{j\ell}R_{i\ell k})\dot{\delta}^k,\end{aligned} \tag{1.18}$$

$$[\dot{\delta}^i, \dot{\delta}^j] = R'^{ijk}\delta_k + (\Pi^{ir}\dot{\delta}^j\Gamma_{rk} - \Pi^{jr}\dot{\delta}^i\Gamma_{rk})\dot{\delta}^k, \tag{1.19}$$

where

$$R_{ijk} = \delta_i\Gamma_{jk} - \delta_j\Gamma_{ik}, \tag{1.20}$$

$$R'^{ijk} = \dot{\delta}^j\Pi^{ik} - \dot{\delta}^i\Pi^{jk} + (\Pi^{ir}\dot{\delta}^j\Gamma_{rs} - \Pi^{jr}\dot{\delta}^i\Gamma_{rs})\Pi^{sk}. \tag{1.21}$$

Let us put

$$\frac{1}{8}[\phi, \phi] = R + R',$$

where $[\phi, \phi]$ denotes the Nijenhuis bracket of ϕ and R, R' are given by

$$R(X, Y) = w[h'X, h'Y], \quad R'(X, Y) = h'[wX, wY].$$

We call R the *curvature* and R' the *cocurvature* of the *connection-pair* ϕ. *R and R' are obstructions to the integrability of HT^*M and WT^*M, respectively.*

Locally we have:

$$R = R_{ijk}\dot{\delta}^k \otimes \delta x^i \otimes \delta x^j, \quad R' = R'^{ijk}\delta_k \otimes \delta p_i \otimes \delta p_j. \tag{1.22}$$

HT^*M and WT^*M are integrable iff *ϕ is integrable, or equivalently, $R = R' = 0$.*

Definition 8.1.2. A *connection-pair ϕ* on T^*M is called *symmetric* if $\Gamma_{ij} = \Gamma_{ji}$ and $\Pi^{ij} = \Pi^{ji}$.

Let $\omega = p_i dx^i$ be *the canonical one form* of T^*M and $\theta = d\omega$ *the canonical symplectic* 2-form.

The Definition 8.1.2 above is invariant because of:

Proposition 8.1.4. *A connection-pair ϕ is symmetric if and only if*

$$\phi^*\theta = -\theta. \tag{1.23}$$

Proof. θ has the local expression

$$\theta = dp_i \wedge dx^i.$$

We obviously have $\phi^*\theta(\dot{\delta}^i, \delta_j) = -\theta(\dot{\delta}^i, \delta_j)$. From

$$\begin{aligned}
\phi^*\theta(\dot{\delta}^i, \dot{\delta}^j) &= -\theta(\dot{\delta}^i, \dot{\delta}^j) \Longleftrightarrow \theta(\dot{\delta}^i, \dot{\delta}^j) = 0,\\
\phi^*\theta(\delta_i, \delta_j) &= -\theta(\delta_i, \delta_j) \Longleftrightarrow \theta(\delta_i, \delta_j) = 0,
\end{aligned}$$

we get

$$\Pi^{ij} - \Pi^{ji} + \Pi^{is}\Pi^{jr}(\Gamma_{sr} - \Gamma_{rs}) = 0, \quad \Gamma_{sr} = \Gamma_{rs}.$$

We obtain therefore, $\Pi^{ij} = \Pi^{ji}$ and $\Gamma_{ij} = \Gamma_{ji}$.

Corollary 8.1.1. *The following statements are equivalent*

(i) *ϕ is symmetric.*

(ii) *WT^*M and HT^*M are Lagrangian (every subbundle is both isotropic and coisotropic with respect to θ).*

Proof. If ϕ is symmetric, using the proposition above we get $\theta(\delta_i, \delta_j) = 0, \quad \theta(\dot{\delta}^i, \dot{\delta}^j) = 0$ and on account of $\dim WT^*M = \dim HT^*M = \frac{1}{2}\dim TT^*M$ it follows that HT^*M and VT^*M are Lagrangian.

Now, if conversely HT^*M and VT^*M are Lagrangian, using again the proposition above, we get that ϕ is symmetric.

Remark. A connection-pair ϕ on T^*M induces two almost symplectic forms, globally defined on T^*M :

$$\theta' = \delta p_i \wedge dx^i, \quad \theta'' = \delta p_i \wedge \delta x^i. \tag{1.24}$$

ϕ is symmetric iff $\theta = \theta' = \theta''$.

Its associated connection Γ is symmetric (that is $\Gamma_{ij} = \Gamma_{ji}$, or equivalently $\Gamma^*\theta = -\theta$) iff $\theta = \theta'$.

Let $C = p_i\dot{\partial}^i$ be *the Liouville vector field,* globally defined on T^*M. We denote by $\widetilde{T}^*M$ the *slit cotangent bundle,* that is, the cotangent bundle with zero section removed.

Definition 8.1.3. A *connection-pair* on $\widetilde{T}^*M$ is called *homogeneous* if the Lie derivative of ϕ with respect to C vanishes, that is

$$L_C\phi = 0. \tag{1.25}$$

The following, characterize the property of *homogeneity for connection-pairs* in terms of homogeneity of its connectors.

Proposition 8.1.5. *A connection-pair ϕ is homogeneous iff Γ_{ij} and Π^{ij} are 1-homogeneous, respectively, $-1-$homogeneous, with respect to p.*

Proof. From $L_C\phi = 0$ and (1.5) we get

$$[C, \delta_k] - \phi[C, \delta_k] = 0.$$

But,

$$[C, \delta_k] = (-\Gamma_{kh} + p_i\dot{\partial}^i\Gamma_{kh})\dot{\partial}^h$$

and from (1.6)

$$\phi(\dot{\partial}^h) = -\dot{\partial}^h + 2\Pi^{hs}\delta_s.$$

Therefore,

$$(-\Gamma_{kh} + p_i\dot{\partial}^i\Gamma_{kh})\dot{\partial}^h + (\Gamma_{kh} - p_i\dot{\partial}^i\Gamma_{kh})\Pi^{hs}\delta_s = 0$$

and thus Γ_{ij} are $1-$homogeneous.

We also must have:

$$[C, \dot{\delta}^k] + \phi[C, \dot{\delta}^k] = 0,$$

but using the $1-$homogeneity of Γ_{ij} above, we get

$$[C, \dot{\delta}^k] = [p_i\dot{\partial}^i, \dot{\delta}^k] = (-\Pi^{ks} - p_i\dot{\partial}^i\Pi^{ks})\delta_s - \dot{\delta}^k$$

and thus

$$-\Pi^{ks} - p_i\dot{\partial}^i\Pi^{ks} = 0,$$

that is, Π^{ks} is $-1-$homogeneous relative to p.

8.2 Special Linear Connections on T^*M

Let ϕ be a fixed symmetrical connection-pair on T^*M and

$$TT^*M = HT^*M \oplus WT^*M,$$

the splitting generated by it. HT^*M is the *horizontal bundle* and WT^*M is the *oblique bundle.*

Every vector field $X \in \mathcal{X}(T^*M)$ has two components with respect to the above splitting

$$X = X^{H'} + X^W \tag{2.1}$$

where $X^{H'} = h'(X)$ is the horizontal component and $X^W = w(X)$ is the oblique component of X.

We can also introduce some special tensor fields, called $\phi-$*tensor* fields as objects in the algebra spanned by $\{1, \delta_i, \dot{\delta}^i\}$ over the ring of $\mathcal{F}(T^*M)$ of smooth real valued functions on T^*M. For instance

$$K = K^{ih}_{jk}\delta_i \otimes \dot{\delta}^j \otimes \delta x^k \otimes \delta p_h \tag{2.2}$$

is a $(2,2)$ ϕ-tensor field. For a change of coordinates given on T^*M the components of a $\phi-$tensor are transformed in exactly the same way as a tensor on M, in spite of p_i dependence, thus K is a d-tensor field.

Definition 8.2.1. Let ∇ be a linear connection of T^*M and ϕ a connection-pair. We say that ∇ is a $\phi-$*connection* if

$$\nabla\phi = 0 \quad \textit{and} \quad \nabla w = 0. \tag{2.3}$$

It can easily be proved that $\nabla\phi = 0$ is equivalent to $\nabla w = 0$ or $\nabla h' = 0$. This definition extends to a general setting, the definition of so called Finsler connection for Cartan space.

A $\phi-$connection can be characterized locally by a pair of coefficients (H^i_{jk}, V^{ik}_j) such that

$$\nabla_{\delta_k}\delta_j = H^i_{jk}\delta_i; \quad \nabla_{\delta_k}\dot{\delta}^i = -H^i_{jk}\dot{\delta}^j, \nabla_{\dot{\delta}^k}\dot{\delta}^i = -V^{ik}_j\dot{\delta}^j; \quad \nabla_{\dot{\delta}^k}\delta_j = V^{ik}_j\delta_i. \tag{2.4}$$

Proposition 8.2.1. *Under a change of coordinates on T^*M the coefficients of a $\phi-$connection ∇ change as follows:*

$$H^{i'}_{j'k'} = \partial_i x^{i'}\partial_{j'}x^j\partial_{k'}x^k H^i_{jk} + \partial_h x^{i'}\partial_{j'}\partial_{k'}x^h, \tag{2.5}$$

(2.6) $$V_{k'}^{i'j'} = \partial_i x^{i'} \partial_j x^{j'} \partial_{k'} x^k V_k^{ij} + \partial_k x^{i'} \partial_h x^{j'} \dot{\delta}^h (\partial_{k'} x^k).$$

Remarks.

1. (2.6) is equivalent to

(2.7) $$V_{k'}^{i'j'} = \partial_i x^{i'} \partial_j x^{j'} \partial_{k'} x^k V_k^{ij} + \Pi^{hs} \partial_h x^{j'} \partial_{k'} x^\ell \partial_\ell \partial_s x^{i'}.$$

2. A $\phi-$connection can be characterized by a couple of coefficients (H_{jk}^i, V_j^{ik}) which obey the transformation law of (2.5) and (2.6), if a change of coordinates on T^*M, is performed.

A ϕ -connection on T^*M induces two types of covariant derivative:
(a) the *h-covariant derivative*

(2.8) $$\nabla_X^{H'} Y := \nabla_{X^{H'}} Y \qquad \forall\, X, Y \in \mathcal{X}(T^*M)$$

(b) the *w-covariant derivative*

(2.9) $$\nabla_X^W Y := \nabla_{X^W} Y \qquad \forall\, X, Y \in \mathcal{X}(T^*M).$$

If K is the ϕ-tensor field of (2.2), then the local expressions of its $h-$ and $w-$covariant derivative have the following form:

$$\nabla_{\delta_k}^H K = K_{jj'|k}^{ii'} \delta_i \otimes \dot{\delta}^j \otimes x^{j'} \otimes \delta p_{i'},$$

$$\nabla_{\dot{\delta}^k}^W K = K_{jj'}^{ii'}|^k \delta_i \otimes \dot{\delta}^j \otimes \delta x^{j'} \otimes \delta p_{i'},$$

where,

(2.10) $$K_{jj'|k}^{ii'} = \delta_k K_{jj'}^{ii'} + H_{\ell k}^i K_{jj'}^{\ell i'} + H_{\ell k}^{i'} K_{jj'}^{i\ell} - H_{jk}^\ell K_{\ell j'}^{ii'} - H_{j'k}^\ell K_{j\ell}^{ii'},$$

(2.11) $$K_{jj'}^{ii'}|^k = \dot{\delta}^k K_{jj'}^{ii'} + V_\ell^{ik} K_{jj'}^{\ell i'} + V_\ell^{i'k} K_{jj'}^{i\ell} - V_j^{\ell k} K_{\ell j'}^{ii'} - V_{j'}^{\ell k} K_{j\ell}^{ii'}.$$

Let $C = p_i \dot{\partial}^i$ the Liouville vector field on T^*M.

Definition 8.2.2. A *$\phi-$connection* ∇ on T^*M is of *Cartan type* if $\nabla^H C = 0$ and $\nabla^W C = I$.

Proposition 8.2.2. *A $\phi-$connection is of Cartan type iff*

(2.12) $$p_{i|j} = 0 \Longleftrightarrow H^k_{ij}p_k - \Gamma_{ij} = 0$$

(2.13) $$p_i|^j = \delta^j_i \Longleftrightarrow V^{kj}_i p_k + \Pi^{j\ell}\Gamma_{\ell i} = 0.$$

Remarks. 1. If (2.12) is verified, we say that ∇ is *h-deflection free* and if (2.13) is true then ∇ will be called *v-deflection free.*

2. When $\Pi^{ij} = 0$, ϕ is just the connection Γ which arises as usual. We will denote this Γ-connection by $\overset{\circ}{\nabla}$.

Locally, we have

(2.14) $$\begin{aligned} \overset{\circ}{\nabla}_{\delta_k}\delta_j &= \overset{\circ}{H}{}^i_{jk}\delta_i, & \overset{\circ}{\nabla}_{\dot\partial^k}\dot\partial^i &= -\overset{\circ}{V}{}^{ik}_j\dot\partial^j, \\ \overset{\circ}{\nabla}_{\delta_k}\dot\partial^i &= -\overset{\circ}{H}{}^i_{jk}\dot\partial^j, & \overset{\circ}{\nabla}_{\dot\partial^k}\delta_j &= \overset{\circ}{V}{}^{ik}_j\delta_i. \end{aligned}$$

Theorem 8.2.1. *Let ϕ be a connection-pair and Γ its associated connection. Then a $\Gamma-$connection $\overset{\circ}{\nabla}$ induces a $\phi-$connection ∇ on T^*M, given by:*

(2.15) $$\nabla_X Y = w(\overset{\circ}{\nabla}_X Y^W) + h'(\overset{\circ}{\nabla}_X Y^{H'}), \quad X, Y \in \mathcal{X}(T^*M).$$

The local connectors of ∇ are the following:

(2.16) $$H^k_{ij} = \overset{\circ}{H}{}^k_{ij}; \quad V^{kj}_i = \overset{\circ}{V}{}^{kj}_i - \Pi^{j\ell}\overset{\circ}{H}{}^k_{i\ell}.$$

Proof. ∇ from (2.14) is clearly a linear connection. Let us find the local form of this connection.

We have

$$\begin{aligned}
\nabla_{\delta_j}\delta_i &= h'(\overset{\circ}{\nabla}_{\delta_j}\delta_i) = h'(\overset{\circ}{H}{}^k_{ij}\delta_k) = \overset{\circ}{H}{}^k_{ij}\delta_k,\\
\nabla_{\delta_i}\dot{\delta}^j &= w(\overset{\circ}{\nabla}_{\delta_i}\dot{\delta}^j) = w(\overset{\circ}{\nabla}_{\delta_i}(\dot{\partial}^j - \Pi^{js}\delta_s))\\
&= w(-\overset{\circ}{H}{}^j_{ki}\dot{\partial}^k - \delta_i(\Pi^{js})\delta_s - \Pi^{js}\overset{\circ}{H}{}^k_{si}\delta_k) = -\overset{\circ}{H}{}^j_{ki}\dot{\delta}^k,\\
\nabla_{\dot{\delta}^i}\delta_j &= h'(\overset{\circ}{\nabla}_{\dot{\delta}_i}\delta_j) = h'(\nabla_{(\dot{\partial}^i - \Pi^{is}\delta_s)}\delta_j)\\
&= h'(\overset{\circ}{V}{}^{ki}_j\delta_k - \Pi^{is}\overset{\circ}{H}{}^k_{js}\delta_k) = (\overset{\circ}{V}{}^{ki}_j - \Pi^{is}\overset{\circ}{H}{}^k_{js})\delta_k,\\
\nabla_{\dot{\delta}^i}\dot{\delta}^j &= w(\overset{\circ}{\nabla}_{\dot{\delta}^i}\dot{\delta}^j) = w(\nabla_{\dot{\delta}^i}(\dot{\partial}^j - \Pi^{jk}\delta_k) - \Pi^{is}\nabla_{\delta_s}(\dot{\partial}^j - \Pi^{jk}\delta_k))\\
&= w((\Pi^{is}\Pi^{j\ell}\overset{\circ}{H}{}^k_{s\ell} + \Pi^{i\ell}\delta_\ell(\Pi^{jk}) - \Pi^{j\ell}\overset{\circ}{V}{}^{ki}_\ell - \dot{\partial}^i(\Pi^{jk}))\delta_k\\
&\quad +(-\overset{\circ}{V}{}^{ji}_m + \Pi^{is}\overset{\circ}{H}{}^j_{ms})\dot{\partial}^m)\\
&= -(\overset{\circ}{V}{}^{ji}_m - \Pi^{is}\overset{\circ}{H}{}^j_{ms})\dot{\delta}^m.
\end{aligned}$$

Therefore, ∇ is a ϕ*−connection* and also (2.16) are verified.

Let ∇ be a ϕ -connection and

$$T(X,Y) = \nabla_X Y - \nabla_Y X - [X,Y], \tag{2.17}$$

its torsion.

Locally, with respect ot the frame $(\delta_i, \dot{\delta}^i)$, we have

$$\begin{aligned}
T(\delta_j,\delta_i) &= T^k_{\ ij}\delta_k + \widetilde{T}_{ijk}\dot{\delta}^k,\\
T(\dot{\delta}^i,\delta_j) &= \widetilde{P}_j^{\ ki}\delta_k + P^i_{\ kj}\dot{\delta}^k,\\
T(\dot{\delta}^j,\dot{\delta}^i) &= \widetilde{S}^{ijk}\delta_k + S_k^{\ ij}\dot{\delta}^k,
\end{aligned} \tag{2.18}$$

where

$$\begin{aligned}
T^k_{\ ij} &= H^k_{ij} - H^k_{ji} + R_{ij\ell}\Pi^{\ell k}; \quad \widetilde{T}_{ijk} = R_{ijk},\\
\widetilde{P}_j^{\ ki} &= V^{ki}_j - (\delta_j\Pi^{ik} + \Pi^{i\ell}R_{j\ell h}\Pi^{hk} + \dot{\partial}^i\Gamma_{jh}\Pi^{hk}),\\
P^i_{\ kj} &= H^i_{kj} - (\dot{\partial}^i\Gamma_{jk} + \Pi^{i\ell}R_{j\ell k}), \quad \widetilde{S}^{ijk} = -R'^{ijk},\\
S_k^{\ ij} &= V^{ji}_k - V^{ij}_k - (\Pi^{jr}\dot{\delta}^i\Gamma_{rk} - \Pi^{ir}\dot{\delta}^j\Gamma_{rk}).
\end{aligned} \tag{2.19}$$

Proposition 8.2.3. *The torsion connectors of* ∇ *and* $\overset{\circ}{\nabla}$ *are related as follows:*

$$(2.20)\quad \begin{aligned} T^k{}_{ij} &= \overset{\circ}{T}{}^k{}_{ij} - R_{ji\ell}\Pi^{\ell k}, \quad \widetilde{T}_{ijk} = \overset{\circ}{\widetilde{T}}_{ijk}, \\ \widetilde{P}_j{}^{ki} &= \overset{\circ}{\widetilde{P}}{}^{ki}{}_j - (\delta_j\Pi^{ik} + \Pi^{i\ell}R_{j\ell m}\Pi^{mk} + \dot{\partial}^i\Gamma_{j\ell}\Pi^{\ell k}), \\ P^i{}_{kj} &= \overset{\circ}{P}{}^i{}_{kj} - \Pi^{i\ell}R_{j\ell k}, \\ S_k{}^{ij} &= \overset{\circ}{S}{}_k{}^{ij} + (\Pi^{i\ell}\dot{\delta}^j\Gamma_{\ell k} - \Pi^{j\ell}\dot{\delta}^i\Gamma_{\ell k}). \end{aligned}$$

However, ∇ has an *extra torsion tensor* $\widetilde{S}^{ijk} = R'^{jik}$ which does not occur when $\Pi^{ij} = 0$. It is clear that ∇ is h-deflection free iff $\overset{\circ}{\nabla}$ is $h-$deflectionfree. The following result gives the relations between $v-$deflection free tensors:

Proposition 8.2.4. *Assume that* $\overset{\circ}{\nabla}$ *is* $h-$*deflection free. Then* ∇ *is* $v-$*deflection free iff* $\overset{\circ}{\nabla}$ *is* v*-deflection free.*

Proof.

$$\begin{aligned} p_i|^k &= \delta_i^k \iff V_i^{jk}p_j \\ &= -\Pi^{k\ell}\Gamma_{\ell i} \iff \overset{\circ}{V}{}_i^{jk}p_j - \Pi^{k\ell}\overset{\circ}{H}{}^j_{i\ell}p_j \\ &= -\Pi^{k\ell}\Gamma_{\ell i} \iff \overset{\circ}{V}{}_i^{jk}p_j = 0 \\ &\iff p_i\overset{\circ}{|}{}^k = \delta_i^k \end{aligned}$$

(here $\overset{\circ}{|}$ denote the $v-$covariant derivative induced by $\overset{\circ}{\nabla}$).

The *curvature tensor of a* $\phi-$connection ∇,

$$R(X,Y)Z = \nabla_X\nabla_Y Z - \nabla_Y\nabla_X Z - \nabla_{[X,Y]}Z$$

has three essential components.

We have:

$$(2.21)\quad \begin{aligned} R(\delta_h,\delta_k)\delta_j &= R^i{}_{jkh}\delta_i, & R(\delta_h,\delta_k)\dot{\delta}^i &= -R^i{}_{jkh}\dot{\delta}^j, \\ R(\dot{\delta}^h,\delta_k)\delta_j &= P^{ih}{}_{jk}\delta_i, & R(\dot{\delta}^h,\delta_k)\dot{\delta}^i &= -P^{ih}{}_{jk}\dot{\delta}^j, \\ R(\dot{\delta}^h,\dot{\delta}^k)\delta_j &= S_j{}^{ikh}\delta_i, & R(\dot{\delta}^h,\dot{\delta}^k)\dot{\delta}^i &= -S_j{}^{ikh}\dot{\delta}^j, \end{aligned}$$

where

(2.22)

$$
\begin{aligned}
R^i{}_{jkh} &= \bigcup_{(h,k)} \{\delta_h H^i_{jk} + H^m_{jk} H^i_{mh}\} - R_{hk\ell}(\Pi^{\ell m} H^i_{jm} + V^{i\ell}_j),\\
P^{ih}{}_{jk} &= \dot\delta^h(H^i_{jk}) - \delta_k(V^{ih}_j) + H^s_{jk} V^{ih}_s - H^i_{sk} V^{sh}_j\\
&\quad -(\delta_k \Pi^{h\ell} + \Pi^{hm} R_{kmr} \Pi^{r\ell} + \dot\partial^h \Gamma_{km} \Pi^{m\ell}) H^i_{j\ell} - (\dot\partial^h \Gamma_{ks} + \Pi^{hm} R_{kms}) V^{is}_j,\\
S_j{}^{ikh} &= \bigcup_{(h,k)} \{\dot\delta^h(V^{ik}_j) + V^{ih}_s V^{sk}_j\} - R'^{hks} H^i_{js} - (\Pi^{h\ell}\dot\delta^k \Gamma_{\ell s} - \Pi^{k\ell}\dot\delta^h \Gamma_{\ell s}) V^{is}_j,
\end{aligned}
$$

where $\bigcup_{(j,k)}\{\cdots\}$ indicates interchange of i and k for the terms in the brackets and *subtraction.*

Let us consider the diagonal lift metric tensor on T^*M

$$G = g_{ij}(x,p)\delta x^i \otimes \delta x^j + g^{ij}(x,p)\delta p_i \otimes \delta p_j, \tag{2.23}$$

where g_{ij} is a symmetric nondegenerate $d-tensorfield$.

Theorem 8.2.2. *Let ϕ be a fixed connection pair on T^*M. Then there exists only one $\phi-$connection such that the following properties are verified:*

$$
\begin{aligned}
&\text{(i)}\ \ g_{ij|k} = 0, && \text{(ii)}\ \ g_{ij}|^k = 0,\\
&\text{(iii)}\ \ T^i_{jk} = 0, && \text{(iv)}\ \ S^{jk}_i = 0.
\end{aligned}
$$

The coefficients of this $\phi-$connection are the following:

$$H^k_{ij} = \frac{1}{2} g^{km}(\delta_i g_{mj} + \delta_j g_{im} - \delta_m g_{ij}) - \frac{1}{2} g^{km}(A^s_{im} g_{sj} + A^s_{jm} g_{si} + A^s_{ij} g_{sm}), \tag{2.24}$$

$$V^{jk}_i = -\frac{1}{2} g_{im}(\dot\delta^j g^{mk} + \dot\delta^k g^{jm} - \dot\delta^m g^{jk}) - \frac{1}{2} g_{im}(g^{sj} B^{mk}_s + g^{sk} B^{mj}_s + g^{ms} B^{kj}_s), \tag{2.25}$$

where

$$A^i_{jk} := R_{kj\ell}\Pi^{\ell i}; \qquad B^{kj}_i := \Pi^{kr}\dot\delta^j \Gamma_{ri} - \Pi^{jr}\dot\delta^k \Gamma_{ri}. \tag{2.26}$$

Proof. (i) is written in the following from:

$$\delta_k g_{ij} = H^m_{ik} g_{mj} + H^m_{jk} g_{im} \tag{2.27}$$

and by using the same technique used to find the Christoffel symbols for Riemannian manifolds, and the first of (2.19), we get (2.24); similary for (ii), but using the last of (2.19). In particular, the $\Gamma-$*connection* $\overset{\circ}{\nabla}$ has the coefficients

$$\overset{\circ}{H}{}^k_{ij} = \frac{1}{2}\, g^{km}(\delta_i g_{mj} + \delta_j g_{im} - \delta_m g_{ij}), \tag{2.28}$$

$$\overset{\circ}{V}{}_i^{jk} = -\,\frac{1}{2}\, g_{im}(\dot\partial^j g^{mk} + \dot\partial^k g^{jm} - \dot\partial^m g^{jk}) \tag{2.29}$$

and verifies (i) - (iv); this connection is metrical with respect to

$$\overset{\circ}{G} = g_{ij}(x,p)dx^i \otimes dx^j + g^{ij}(x,p)\delta p_i \otimes \delta p_j. \tag{2.30}$$

Theorem 8.2.3. *Let* $\overset{\circ}{\nabla}$ *be the* $\overset{\circ}{G}-$*metrical connection above, and* ∇ *its induced* $\phi-$*connection* (2.15). *Then* ∇ *is* $G-$*metrical.*

Proof. We must show only that ∇ is $v-$metrical. By virtue of (2.16) we have:

$$\begin{aligned} g_{ij}|^k &= \dot\delta^k g_{ij} - V_i^{\ell k} g_{\ell j} - V_j^{\ell k} g_{i\ell} \\ &= \dot\partial^k g_{ij} - \Pi^{ks}\delta_s g_{ij} - (\overset{\circ}{V}{}_i^{\ell k} - \Pi^{ks}\overset{\circ}{H}{}^\ell_{is})g_{\ell j} - (\overset{\circ}{V}{}_j^{\ell k} - \Pi^{ks}\overset{\circ}{H}{}^\ell_{js})g_{i\ell} \\ &= \dot\partial^k g_{ij} - \overset{\circ}{V}{}_i^{\ell k} g_{j\ell} - \overset{\circ}{V}{}_j^{\ell k} g_{i\ell} - \Pi^{ks}(\delta_s g_{ij} - \overset{\circ}{H}{}^\ell_{is} g_{\ell j} - \overset{\circ}{H}{}^\ell_{js} g_{i\ell}) \\ &= g_{ij}\overset{\circ k}{|} - \Pi^{ks} g_{ij}\overset{\circ s}{|} = 0. \end{aligned}$$

Remark. This $\phi-$connection, induced by $\overset{\circ}{\nabla}$, is the appropriate one for studing the geometry of T^*M endowed with the metric tensor (2.23). Eisenhart [53] and also Yano-Davies [172] used a similar connection.

8.3 The homogeneous case

We specialize, here, the results of previous sections in the particular case when

$$g^{ij}(x,p) = \frac{1}{2}\,\dot\partial^i\dot\partial^j H(x,p) \tag{3.1}$$

and H is a real smooth function on a domain $D^* \subset \widetilde{T}^*M$, 2-homogeneous in p_i and such that the tensor $(g^{ij}(x,p))$ is everywhere nondegenerate on D^*.

$$\overset{c}{\Gamma}_{ij} = \gamma^0_{ij} - \frac{1}{2}\,\dot\partial^k g_{ij}\gamma^0_{k0}, \tag{3.2}$$

where we have put, as usual,

$$\gamma^k_{ij} := \tfrac{1}{2} g^{kh}(\partial_i g_{hj} + \partial_j g_{ih} - \partial_h g_{ij}), \quad \gamma^0_{ij} := \gamma^k_{ij} p_k,$$

$$\gamma^0_{i0} := \gamma^k_{ij} p_k p^j, \quad p^i := g^{ij} p_j.$$

Note that $\overset{c}{\Gamma}$ is deflection free, [97].

However, the geometry of Cartan manifolds as given, is *dramatically changed under a diffeomorphism which is not fiber-preserving.* In this case, the geometrical approach described in the previous section is the correct one to use.

Theorem 8.3.1. *Let ϕ be a homogeneous connection-pair on $\overset{\circ}{T}{}^*M$, such that $p_i \Pi^{ij} = 0$. If ∇ is the $\phi-$connection given by (2.24), (2.25) and g^{ij} are those of (3.1), then*

(i) *∇ is h-deflection free iff*

$$\overset{c}{\Gamma}_{ij} = \Gamma_{ij} + \frac{1}{2} p^m (R_{im\ell}\Pi^{\ell s} g_{sj} + R_{jm\ell}\Pi^{\ell s} g_{si}). \tag{3.3}$$

(ii) *If ∇ is h-deflection free, then $V_i^{jk} p_k = 0$.*

(iii) *If ∇ is h-deflection free, then ∇ is also $v-deflection free$.*

Proof. (i) By definition ∇ is $h-deflection$ free iff $\Gamma_{ij} = H^k_{ij} p_k$. From (3.1) we see that

$$\overset{\circ}{V}{}^{ijk} = \dot{\partial}^i g^{jk} \tag{3.4}$$

is completely symmetric and because g^{jk} is $0-homogeneous$ we get

$$p_i \dot{\partial}^i g^{jk} = p_i \dot{\partial}^j g^{ik} = p_i \dot{\partial}^k g^{ij} = 0,$$

and also

$$p^i \dot{\partial}^k g_{ij} = -p^i g_{i\ell} g_{jm} \dot{\partial}^k g^{\ell m} = -p_\ell \dot{\partial}^k g^{\ell m} g_{jm} = 0. \tag{3.5}$$

Taking into account that $p_i \Pi^{ij} = 0$ and using the above identities, from (2.24) transvecting by p_k, we get:

$$\Gamma_{ij} = \gamma^0_{ij} - \frac{1}{2} p^m \Gamma_{ms} \dot{\partial}^s g_{ij} - \frac{1}{2} p^m (R_{im\ell}\Pi^{\ell s} g_{sj} + R_{jm\ell}\Pi^{\ell s} g_{si}).$$

Transvecting again by p^j, we obtain $\Gamma_{i0} = \gamma^0_{i0}$ and thus

$$\Gamma_{ij} = \gamma^0_{ij} - \frac{1}{2}\dot{\partial}^s g_{ij}\gamma^0_{s0} - \frac{1}{2}p^m(R_{im\ell}\Pi^{\ell s}g_{sj} + R_{jm\ell}\Pi^{\ell s}g_{si}),$$

that is, (3.3) holds true.

To prove (ii) we need the following:

Lemma. *If ∇ is h-deflection free, then $R_{ij\ell}p^\ell = 0$.*

Proof. Because ∇ is $h-metrical$ and $h-deflection$ free and $H = g^{ij}p_ip_j$ we get $\delta_k H = 0$, that is,

$$\partial_i H = -\Gamma_{ij}\dot{\partial}^j H. \tag{3.6}$$

Therefore,

$$\begin{aligned} R_{sk\ell}p^\ell &= (\delta_s\Gamma_{k\ell} - \delta_k\Gamma_{s\ell})p^\ell = [(\partial_s\Gamma_{k\ell} - \partial_k\Gamma_{s\ell}) + (\Gamma_{si}\dot{\partial}^i\Gamma_{k\ell} - \Gamma_{ki}\dot{\partial}^i\Gamma_{s\ell})]\dot{\partial}^\ell H \\ &= \partial_s(\Gamma_{k\ell}\dot{\partial}^\ell H) - \partial_k(\Gamma_{s\ell}\dot{\partial}^\ell H) - \Gamma_{k\ell}\partial_s\dot{\partial}^\ell H + \Gamma_{s\ell}\partial_k\dot{\partial}^\ell H \\ &\quad + \Gamma_{si}\dot{\partial}^i(\Gamma_{k\ell}\dot{\partial}^\ell H) - \Gamma_{ki}\dot{\partial}^i(\Gamma_{s\ell}\dot{\partial}^\ell H) - \Gamma_{si}\Gamma_{k\ell}\dot{\partial}^i\dot{\partial}^\ell H + \Gamma_{ki}\Gamma_{s\ell}\dot{\partial}^i\dot{\partial}^\ell H \\ &= -\partial_s\partial_k H + \partial_k\partial_s H - \Gamma_{k\ell}\partial_s\dot{\partial}^\ell H + \Gamma_{s\ell}\partial_k\dot{\partial}^\ell H - \Gamma_{si}\dot{\partial}^i\partial_k H + \Gamma_{ki}\dot{\partial}^i\partial_s H \\ &= 0. \end{aligned}$$

Let us now prove (ii). We have

$$\begin{aligned} p_k\dot{\delta}^j g^{mk} &= \dot{\delta}^j(p_k g^{mk}) - \dot{\delta}^j p_k g^{mk} \\ &= \tfrac{1}{2}\dot{\delta}^j(\dot{\partial}^m H) - (\delta^j_k - \Pi^{js}\Gamma_{sk})g^{mk} \\ &= \tfrac{1}{2}\dot{\partial}^j\dot{\partial}^m H - \tfrac{1}{2}\Pi^{js}\partial_s\dot{\partial}^m H - \tfrac{1}{2}\Pi^{js}\Gamma_{s\ell}\dot{\partial}^\ell\dot{\partial}^m H - g^{jm} + \Pi^{js}\Gamma_{s\ell}g^{\ell m} \\ &= -\tfrac{1}{2}\Pi^{js}\partial_s\dot{\partial}^m H = \tfrac{1}{2}\Pi^{js}\dot{\partial}^m(\Gamma_{s\ell}\dot{\partial}^\ell H) \\ &= \tfrac{1}{2}\Pi^{js}\dot{\partial}^m\Gamma_{s\ell}\dot{\partial}^\ell H + \Pi^{js}\Gamma_{s\ell}g^{m\ell}. \end{aligned}$$

Thus,

$$p_k\dot{\delta}^j g^{mk} - p_k\dot{\delta}^m g^{jk} = \frac{1}{2}(\Pi^{js}\dot{\partial}^m\Gamma_{s\ell} - \Pi^{ms}\dot{\partial}^j\Gamma_{s\ell})p^\ell + (\Pi^{js}g^{m\ell} - \Pi^{ms}g^{j\ell})\Gamma_{s\ell},$$

and by using the above lemma we get,

$$p_k\dot{\delta}^j g^{mk} - p_k\dot{\delta}^m g^{jk} = \frac{1}{2}(\Pi^{jr}\dot{\delta}^m\Gamma_{rs} - \Pi^{mr}\dot{\delta}^j\Gamma_{rs})p^s + (\Pi^{js}g^{m\ell} - \Pi^{ms}g^{j\ell})\Gamma_{s\ell}.$$

On the other hand,

$$g^{sj}B_s^{mk}p_k = g^{\ell j}\Pi^{ms}\Gamma_{s\ell},$$
$$g^{ms}B_s^{kj}p_k = -g^{m\ell}\Pi^{js}\Gamma_{s\ell},$$

and

$$p_k g^{sk}B_s^{mj} = \frac{1}{2}(\Pi^{mr}\dot{\delta}^j\Gamma_{rs} - \Pi^{jr}\dot{\delta}^m\Gamma_{rs})p^s.$$

From the last four equalities, by again using (3.5) and $p_i\Pi^{ij} = 0$, we finally get, $V_i^{jk}p_k = 0$.

(iii) From the last of equation (2.19) we get

$$V_i^{jk}p_j = V_i^{kj}p_j + (\Pi^{js}\dot{\delta}^k\Gamma_{si} - \Pi^{ks}\dot{\delta}^j\Gamma_{si})p_j = -\Pi^{ks}\Gamma_{si}$$

and thus ∇ is $v - deflection\ free$.

Remarks. (1) The condition $p_i\Pi^{ij} = 0$ was used by Yano-Davies [172] and Yano-Muto [174] and it holds when WT^*M is the image of the vertical subbundle through a $\Gamma-$regular homogeneous contact transformation (see the next section). In fact, in this case the Liouville vector field $C = p_i\dot{\partial}^i$ belongs to WT^*M.

(2) Equation (3.3) shows how to select the connection Γ such that (2.24), (2.24) hold and the connection is of Cartan type.

8.4 f -related connection-pairs

Let ϕ be a connection-pair on T^*M and $WT^*M \to T^*M$, $HT^*M \to T^*M$, the oblique and horizontal bundles. We denote by Γ its *associated connection* (nonlinear).

Definition 8.4.1. $f \in \mathrm{Diff}\,(T^*M)$ is called $\Gamma-$regular if the restriction of the tangent map $(\pi f)_*$ to HT^*M

$$(\pi f)_* : HT^*M \to TM$$

is a diffeomorphism.

If f has the local expression $f(x,p) = (\overline{x}(x,p), \overline{p}(x,p))$, then it is $\Gamma-$regular iff $(\pi f)_*(\delta_i)$, $i \in \overline{1,n}$ are linearly independent, that is, the matrix with entries

$$\theta_i^k := \delta_i\overline{x}^k = \partial_i\overline{x}^k + \Gamma_{ij}\dot{\partial}^j\overline{x}^k \tag{4.1}$$

has maximal rank.

Theorem 8.4.1. *Let ϕ be a connection-pair on T^*M and $f \in$ Diff (T^*M). The following statements are equivalent*

(i) $\overline{\phi} = f_* \phi f_*^{-1}$ *is a connection-pair*

(ii) f *is* $\Gamma-$*regular.*

Proof. $\overline{\phi}$ is clearly an almost product structure on T^*M and it is a connection-pair iff Ker $(I - \overline{\phi})$ is transversal to VT^*M, or equivalently, $\pi_* :$ Ker $(I - \overline{\phi}) \to TM$ is an isomorphism. But this last condition is equivalent to (ii) because

$$f_*(\text{Ker}\,(I - \phi)) = \text{Ker}\,(I - \overline{\phi}).$$

Definition 8.4.2. The connection-pair $\overline{\phi}$ given by (i) above is called the *push-forward of* ϕ by f. The connection $\overline{\Gamma}$ associated to $\overline{\phi}$ will also be called the *push-forward of* Γ *by* f.

Also we will say that ϕ and $\overline{\phi}$ are $f -$ *related.*

Theorem 8.4.2. *The coefficients of two* $f-$*related connection-pairs* ϕ *and* $\overline{\phi}$ *are connected by the following equations:*

$$\theta_i^k \overline{\Gamma}_{kj} = \delta_i \overline{p}_j, \tag{4.2}$$

$$(\dot{\partial}^i \overline{p}_k - \dot{\partial}^i \overline{x}^\ell \overline{\Gamma}_{\ell k}) \overline{\Pi}^{kh} = \Pi^{ij} \theta_j^h - \dot{\partial}^i \overline{x}^h. \tag{4.3}$$

Proof. From $\overline{\phi} f_* = f_* \phi$ and (1.5) we get: $\overline{\phi} f_*(\delta_i) = f_*(\delta_i)$ that is $f_*(\delta_i) \in \overline{H}T^*M :=$ Ker $(I - \overline{\phi})$. On the other hand,

$$\begin{aligned} f_*(\delta_i) &= (\delta_i \overline{x}^k)\overline{\partial}_k + (\delta_i \overline{p}_k)\dot{\overline{\partial}}^k \\ &= (\delta_i \overline{x}^k)(\overline{\delta}_k - \overline{\Gamma}_{k\ell}\dot{\overline{\partial}}^\ell) + (\delta_i \overline{p}_k)\dot{\overline{\partial}}^k \\ &= \delta_i \overline{x}^k \overline{\delta}_k + (\delta_i \overline{p}_k - \delta_i \overline{x}^\ell \overline{\Gamma}_{\ell k})\dot{\overline{\partial}}^k. \end{aligned}$$

Therefore,

$$f_*(\delta_i) \in \overline{H}T^*M \Longleftrightarrow \delta_i \overline{p}_k = \delta_i \overline{x}^\ell \overline{\Gamma}_{\ell k},$$

that is (4.2).

Now, let us prove (4.3). We have

$$\overline{\phi} f_*(\dot{\delta}^i) = -f_*(\dot{\delta}^i) \Longleftrightarrow f_*(\dot{\delta}^i) \in \overline{W}T^*M := \text{Ker}\,(I + \overline{\phi}).$$

But,

$$\begin{aligned} f_*(\dot{\delta}^i) &= (\dot{\delta}^i\overline{x}^k)\overline{\partial}_k + (\dot{\delta}^i\overline{p}_k)\dot{\overline{\partial}}^k \\ &= \dot{\delta}^i\overline{x}^k(\overline{\delta}_k - \overline{\Gamma}_{ks}(\dot{\overline{\delta}}^s + \overline{\Pi}^{s\ell}\overline{\delta}_\ell)) + \dot{\delta}^i\overline{p}_k(\dot{\overline{\delta}}^k + \overline{\Pi}^{ks}\overline{\delta}_s) \\ &= (\dot{\delta}^i\overline{x}^k - \dot{\delta}^i\overline{x}^\ell\overline{\Gamma}_{\ell s}\overline{\Pi}^{sk} + \dot{\delta}^i\overline{p}_s\overline{\Pi}^{sk})\overline{\delta}_k + (\dot{\delta}^i\overline{p}_k - \dot{\delta}^i\overline{x}^s\overline{\Gamma}_{sk})\dot{\overline{\delta}}^k. \end{aligned}$$

Thus $f_*(\dot{\delta}^i) \in \overline{W}T^*M$ iff

$$\dot{\delta}^i\overline{x}^k - \dot{\delta}^i\overline{x}^\ell\overline{\Gamma}_{\ell s}\overline{\Pi}^{sk} + \dot{\delta}^i\overline{p}_s\overline{\Pi}^{sk} = 0.$$

By using (4.2) we get (4.3).

Corollary 8.4.1. *If* $(\delta_i, \dot{\delta}^i)$, $(\overline{\delta}_i, \dot{\overline{\delta}}^i)$ *are the adapted frames at* (x,p) *and* $(\overline{x},\overline{p})$ *induced by* ϕ *and* $\overline{\phi}$ *respectively, then:*

$$f_*(\delta_i) = \theta_i^k\overline{\delta}_k, \tag{4.4}$$

$$f_*(\dot{\delta}^i) = \widetilde{\theta}_j^i\dot{\overline{\delta}}^j, \tag{4.5}$$

where

$$\widetilde{\theta}_j^i := \dot{\partial}^i\overline{p}_j - \dot{\partial}^i\overline{x}^k\overline{\Gamma}_{kj}. \tag{4.6}$$

The regularity of $\widetilde{\theta}_j^i$ follows from (4.5) but is also a consequence of

Proposition 8.4.1. (i) *f is $\Gamma-$regular iff f^{-1} is $\overline{\Gamma}-$regular.*
If $f \in$ Diff (T^*M) is $\Gamma-$regular, then

(ii) $(\delta_i\overline{x}^k)(\overline{\delta}_k x^j) = \delta_i^j$; *and*

(iii) $(\dot{\partial}^i\overline{p}_k - \dot{\partial}^i\overline{x}^h\overline{\Gamma}_{hk})(\dot{\overline{\partial}}^k p_j - \dot{\overline{\partial}}^k x^s\Gamma_{sj}) = \delta_j^i$.

Proof. We have

$$\delta_i = f_*^{-1}(f_*(\delta_i)) = \delta_i\overline{x}^k f_*^{-1}(\overline{\delta}_k) = (\delta_i\overline{x}^k)(\overline{\delta}_k x^j)\delta_j$$

and (i), (ii) follow.

To prove (iii) we use the following equalities:

$$\begin{aligned} \dot{\delta}^i &= f_*^{-1}(f_*(\dot{\delta}^i)) = (\dot{\partial}^i\overline{p}_j - \dot{\partial}^i\overline{x}^k\overline{\Gamma}_{kj})f_*^{-1}(\dot{\overline{\delta}}^j) \\ &= (\dot{\partial}^i\overline{p}_j - \dot{\partial}^i\overline{x}^k\overline{\Gamma}_{kj})(\dot{\overline{\partial}}^j p_s - \dot{\overline{\partial}}^j x^k\Gamma_{ks})\dot{\delta}^s. \end{aligned}$$

Generally, if Γ is a connection on T^*M the push-forward of Γ is not a connection. Some consequences of Theorem 8.4.2 are the following.

Proposition 8.4.2. *Let Γ be a connection on T^*M and $f \in \mathrm{Diff}\,(T^*M)$ $\Gamma-$regular. Then $\overline{\phi} = f_*\Gamma f_*^{-1}$ is a connection-pair; the coefficients of the connection $\overline{\Gamma}$ associated to $\overline{\phi}$ are given by (4.2) while $\overline{\Pi}^{ij}$ has the following form:*

$$\overline{\Pi}^{ij} = (\dot{\overline{\partial}}^i x^j \Gamma_{sk} - \dot{\overline{\partial}}^i p_k)\dot{\partial}^k \overline{x}^j. \tag{4.7}$$

Proposition 8.4.3. *The push-forward of a connection-pair ϕ by a $\Gamma-$regular diffeomorphism is a connection if and only if*

$$\Pi^{ij} = \dot{\partial}^i \overline{x}^h \overline{\delta}_h x^j. \tag{4.8}$$

Corollary 8.4.2. *The push-forward of a connection Γ by a $\Gamma-$regular diffeomorphism is also a connection iff f is fiber preserving (that is, locally, $f(x,p) = (\overline{x}(x), \overline{p}(x,p))$.*

Now we will study when the push-forward of symmetric connection-pair by f is also symmetric.

Theorem 8.4.3. *Let ϕ be a symmetric connection pair on T^*M and $\overline{\phi}$ the push-forward of ϕ by a $\Gamma-$regular diffeomorphism f. The following statements are equivalent:*

(i) *$\overline{\phi}$ is symmetric*

(ii) *HT^*M and WT^*M are $f^*\theta-$Lagrangian.*

Proof. $\overline{\phi}$ is symmetric if and only if $\overline{HT^*M}$ and $\overline{WT^*M}$ are both Lagrangian (Corollary 8.4.2.). Therefore, we must have:

$$\theta_{(\overline{x},\overline{p})}(\overline{\delta}_i, \overline{\delta}_j) = 0, \quad \theta_{(\overline{x},\overline{p})}(\dot{\overline{\delta}}^i, \dot{\overline{\delta}}^j) = 0.$$

By using (4.4) and (4.5) these conditions are equivalent to

$$f^*\theta(\delta_i, \delta_j) = 0 \quad \textit{and} \quad f^*\theta(\dot{\delta}^i, \dot{\delta}^j) = 0,$$

therefore HT^*M and WT^*M are isotropic and thus Lagrangian with respect to $f^*\theta$. The converse statement is immediate.

Corollary 8.4.3. *$\overline{\phi}$ is a symmetric connection pair iff*

$$\delta_k\overline{p}_i\delta_h\overline{x}^i = \delta_h\overline{p}_i\delta_k\overline{x}^i, \tag{4.9}$$

$$\dot{\delta}^k\overline{p}_i\dot{\delta}^h\overline{x}^i = \dot{\delta}^h\overline{p}_i\dot{\delta}^k\overline{x}^i. \tag{4.10}$$

Proof. We have:

$$f^*\theta = d\overline{p}_i \wedge dx^i = (\delta_k\overline{p}_i\delta x^k + \dot{\delta}^s\overline{p}_i\delta p_s) \wedge (\delta_h\overline{x}^i\delta x^h + \dot{\delta}^r\overline{x}^i\delta p_r).$$

By using Theorem 8.4.3 we get the equalities above.

Note: $\overline{\Gamma}$ is symmetric iff (4.9) is verified.

Theorem 8.4.4. *Let f be a $\Gamma-$regular symplectomorphism. Then ϕ is symmetric iff $\overline{\phi}$ is symmetric.*

Proof. $\overline{\phi}$ is symmetric iff HT^*M and WT^*M are $f^*\theta-$Lagrangian, that is $\theta-$Lagrangian which is equivalent to ϕ being symmetric.

To summarise the results above we can state the following:

Theorem 8.4.5. *Let $f \in$ Diff (T^*M), $\Gamma-$regular, such that $f^*\theta(\dot{\delta}^i, \delta_i) = \delta^i_j$. Then each pair of the next statements implies the third:*

(i) *f is a symplectomorphism*

(ii) *ϕ is symmetric*

(iii) *$\overline{\phi}$ is symmetric.*

Remark. 1. The condition $f^*\theta(\dot{\delta}^i, \delta_j) = \delta^i_j$ is equivalent to

$$\dot{\delta}^i\overline{p}_k\delta_j\overline{x}^k - \dot{\delta}^i\overline{x}^k\delta_j\overline{p}_k = \delta^i_j \tag{4.11}$$

and it is obviously verified when f is a symplectomorphism.

2. If f is a $\Gamma-$regular symplectomorphism then

$$f_*(HT^*M \oplus WT^*M) = f_*(HT^*M) \oplus f_*(WT^*M).$$

Proposition 8.4.4. *Let f be a $\Gamma-$regular symplectomorphism of T^*M.*

Then

$$\widetilde{\theta}^i_j \theta^j_k = \delta^i_k, \quad \theta^i_j \widetilde{\theta}^j_k = \delta^i_k. \tag{4.12}$$

Proof. It follows from

$$f^*\theta(\dot{\delta}^i, \delta_j) = \delta^i_j$$

using (4.4) and (4.5).

Note: Under the conditions of the above result and Proposition 8.4.1 we have:

$$\theta^i_j = \partial_j \overline{x}^i + \Gamma_{jk} \dot{\partial}^k \overline{x}^i = \dot{\overline{\partial}}^i p_j - \Gamma_{sj} \dot{\overline{\partial}}^i x^s, \tag{4.13}$$

and its reciprocal,

$$\widetilde{\theta}^i_j = \dot{\partial}^i \overline{p}_j - \overline{\Gamma}_{kj} \dot{\partial}^i \overline{x}^k = \overline{\partial}_j x^i + \overline{\Gamma}_{ks} \dot{\overline{\partial}}^s x^i. \tag{4.14}$$

Proposition 8.4.5. *If $f \in \mathrm{Diff}\,(T^*M)$ is a Γ−regular symplectomorphism and $(\overline{\delta}\overline{x}^i, \overline{\delta}\overline{p}_i)$ is the dual of $(\overline{\delta}_i, \dot{\overline{\delta}}^i)$ then*

$$f_*(\delta x^i) = \widetilde{\theta}^i_j \overline{\delta}\overline{x}^j; \quad f_*(\delta p_i) = \theta^j_i \overline{\delta}\overline{p}_j. \tag{4.15}$$

Proof.

$$\begin{aligned} f_*(\delta x^i)(\overline{\delta}_k) &= \delta x^i(f_*^{-1}(\overline{\delta}_k)) = \delta x^i(\widetilde{\theta}^s_k \delta_s) \\ &= \widetilde{\theta}^s_k \delta x^i(\delta_s) = \widetilde{\theta}^s_k \delta^i_s = \widetilde{\theta}^i_k = \widetilde{\theta}^i_k \overline{\delta}\overline{x}^i(\overline{\delta}_k) \end{aligned}$$

and first equality (4.15) follows. A similar proof holds for the second.

Let us now study the connection between *curvature tensors R' and $\overline{R}'$ of* Γ and $\overline{\Gamma}$.

Proposition 8.4.6. *Let ϕ be a connection-pair and $\overline{\phi}$ its push-forward by a Γ−regular symplectomorphism. If R and $\overline{R}$ are the curvature tensors of Γ and $\overline{\Gamma}$ then:*

$$\overline{R}_{ijk} = R_{\ell ms} \widetilde{\theta}^\ell_i \widetilde{\theta}^m_j \widetilde{\theta}^s_k, \tag{4.16}$$

$$\overline{R}_{ijk} \overline{\Pi}^{k\ell} = \widetilde{\theta}^m_i \overline{\delta}_j(\theta^\ell_m) - \widetilde{\theta}^m_j \overline{\delta}_i(\theta^\ell_m) + \widetilde{\theta}^k_i \widetilde{\theta}^h_j \theta^\ell_m R_{khs} \Pi^{sm}. \tag{4.17}$$

Proof. From (1.17) we get:

$$f_*[\delta_i, \delta_j] = R_{ijk}\widetilde{\theta}^k_m \dot{\overline{\delta}}^m + R_{ijk}\Pi^{\ell k}\theta^m_k \overline{\delta}_m.$$

On the other hand,

$$\begin{aligned} f_*[\delta_i, \delta_j] &= [\theta^k_i \overline{\delta}_k, \theta^s_j \overline{\delta}_s] \\ &= \theta^k_i \overline{\delta}_k(\theta^m_j)\overline{\delta}_m - \theta^s_j \overline{\delta}_s(\theta^m_i)\overline{\delta}_m + \theta^k_i \theta^s_j [\overline{R}_{ksm}\dot{\delta}^m + \overline{R}_{ks\ell}\overline{\Pi}^{\ell m}\delta_m] \end{aligned}$$

and the relations above follow immediately.

Corollary 8.4.4. *If* $\Pi^{sm} = 0$ *then*

$$\overline{R}_{ijk}\overline{\Pi}^{k\ell} = \widetilde{\theta}^m_i \overline{\delta}_j(\theta^\ell_m) - \widetilde{\theta}^m_j \overline{\delta}_i(\theta^\ell_m). \tag{4.18}$$

8.5 f-related ϕ-connections

Let us now investigate the behaviour of geometrical objects described in Section 8.2 under *symplectomorphisms.*

If ϕ is a connection-pair on T^*M and $f : T^*M \to T^*M$ is a $\Gamma-$*regular* symplectomorphism, then the symmetry of the connection-pair $\overline{\phi} = f_*\phi f_*^{-1}$ is preserved and also

$$\begin{aligned} f_*(\delta_i) &= \theta^k_i \overline{\delta}_k, & f_*(\dot{\delta}^i) &= \widetilde{\theta}^i_k \dot{\overline{\delta}}^k, \\ f_*(\delta x^i) &= \widetilde{\theta}^i_j \overline{\delta}\,\overline{x}^j, & f_*(\delta p_i) &= \theta^j_i \overline{\delta}\overline{p}_j, \end{aligned}$$

where

$$\theta^i_k \widetilde{\theta}^k_j = \delta^i_j \quad and \quad \widetilde{\theta}^i_k \theta^k_j = \delta^i_j.$$

We can construct a new geometry on T^*M, generated by f, by pushing forward all geometrical objects described in Section 8.2, there by extending to a more general setting, the results of [55], [146], [147], [175].

For instance, if K is the tensor field, locally given by (2.2), then we can consider its push-forward:

$$\overline{K} = \overline{K}^{i'h'}_{j'k'} \overline{\delta}_{i'} \otimes \dot{\overline{\delta}}^{j'} \otimes \overline{\delta}\,\overline{x}^{k'} \otimes \overline{\delta}\,\overline{p}_{h'},$$

where

$$\overline{K}^{i'h'}_{j'k'} = \theta^{i'}_i \theta^{h'}_h \widetilde{\theta}^j_{j'} \widetilde{\theta}^k_{k'} K^{ih}_{jk} \circ f^{-1}. \tag{5.1}$$

In particular, the push-forward of G from (2.23) has the following local form:

$$\overline{G} = \overline{g}_{ij}\overline{\delta}\,\overline{x}^i \otimes \overline{\delta}\,\overline{x}^j + \overline{g}^{ij}\overline{\delta}\,\overline{p}_i \otimes \overline{\delta}\,\overline{p}_j, \tag{5.2}$$

where

$$\overline{g}^{ij} \circ f = \theta^i_k \theta^j_h g^{kh}. \tag{5.3}$$

If ∇ is a *linear connection on T^*M, we define its push-forward by f as follows*:

$$\overline{\nabla}_{\overline{X}} \overline{Y} := f_*(\nabla_X Y), \quad \overline{X} = f_*(X), \quad \overline{Y} = f_*(Y). \tag{5.4}$$

$\overline{\nabla}$ is clearly a linear connection on T^*M.

Proposition 8.5.1. (i) *∇ is a ϕ -connection iff $\overline{\nabla}$ is a $\overline{\phi}$ -connection.*
(ii) *∇ is G-metrical iff $\overline{\nabla}$ is $\overline{G}$–metrical.*

Proof. (i)

$$\begin{aligned} \overline{\nabla}\,\overline{\phi} = 0 \quad &\Longleftrightarrow \overline{\nabla}_{\overline{X}} \overline{\phi}(\overline{Y}) = \overline{\phi}(\overline{\nabla}_{\overline{X}} \overline{Y})) \\ &\Longleftrightarrow f_*(\nabla_X (\phi f_*^{-1}) f_* Y)) = (f_* \phi f_*^{-1}) f_*(\nabla_X Y) \Longleftrightarrow \nabla \phi = 0. \end{aligned}$$

On the other hand,

$$\overline{\nabla} \theta = \overline{\nabla} f_* \theta = f_*(\nabla \theta)$$

because f is symplectomorphism and thus

$$\nabla \theta = 0 \Longleftrightarrow \overline{\nabla} \theta = 0.$$

(ii) This follows from

$$\overline{\nabla}\,\overline{G} = f_*(\nabla G).$$

Proposition 8.5.2. *The coefficients of $\overline{\nabla}$ are related to those of ∇ by the following relations:*

$$\overline{H}^k_{ij} = \widetilde{\theta}^{i'}_i \widetilde{\theta}^{j'}_j \theta^k_{k'} H^{k'}_{i'j'} + \theta^k_{i'} \overline{\delta}_j \overline{\delta}_i x^{i'}, \tag{5.5}$$

$$\overline{V}^{ij}_k = \theta^i_{i'} \theta^j_{j'} \widetilde{\theta}^{k'}_k V^{i'j'}_{k'} + \theta^i_{i'} \dot{\overline{\delta}}^j \overline{\delta}_k x^{i'}. \tag{5.6}$$

Similar theorems to those of Section 8.2 (Theorems 8.2.2 and 8.2.3) hold when ∇ is replaced by $\overline{\nabla}$ and G by $\overline{G}$.

8.6 The geometry of a homogeneous contact transformation

In this section we will restrict our considerations to the slit tangent bundle $\widetilde{T}^*M$ (the cotangent bundle with zero section removed) instead of T^*M.

Let ω be the canonical one form of $\widetilde{T}^*M$, locally given by

$$\omega = p_i dx^i. \tag{6.1}$$

Definition 8.6.1. A diffeomorphism $f : \widetilde{T}^*M \to \widetilde{T}^*M$ is called a *homogeneous contact transformation* (*h.c.t.*) if ω is invariant under f, that is

$$f^*\omega = \omega. \tag{6.2}$$

Proposition 8.6.1. *If f is a* h.c.t. *then $f_*(C) = C$.*

Proof. We use the property of the Liouville vector field $C = p_i\dot{\partial}^i$, as the only one such that $i_C d\omega = \omega$ where "i" denotes the interior product of C and $d\omega$. We have

$$\begin{aligned} i_{f_*(C)}d\omega(X) &= d\omega(f_*(C), X) = d\omega(f_*(C), f_*(f^{-1})_*(X)) \\ &= (f^*d\omega)(C, (f^{-1})_*X) = d(f^*\omega)(C, (f^{-1})_*X) \\ &= d\omega(C, (f^{-1})_*X) = i_C d\omega((f^{-1})_*X) \\ &= \omega((f^{-1})_*X) = (f^{-1})^*\omega(X) = \omega(X) \end{aligned}$$

for every $X \in \mathcal{X}(\overset{\circ}{T}{}^*M)$.

Note: The set of *h.c.t.* is clearly a subgroup of the group of symplectomorphisms of $\overset{\circ}{T}{}^*M$.

Corollary 8.6.1. *If $f(x,p) = (\overline{x}(x,p), \overline{p}(x,p))$ is the local expression of a* h.c.t. *then $\overline{x} = \overline{x}(x,p)$ and $\overline{p} = \overline{p}(x,p)$ are homogeneous of degree* 0 *and* 1 *with respect to* p.

Proof.

$$\begin{aligned} f_*(C) = C &\iff p_i\dot{\partial}^i\overline{x}^k\overline{\partial}_k + p_i\dot{\overline{\partial}}{}^i\overline{p}_k\dot{\overline{\partial}}{}^k = \overline{p}_k\dot{\overline{\partial}}{}^k \\ &\iff p_i\dot{\partial}^i\overline{x}^k = 0, \quad p_i\dot{\overline{\partial}}{}^i\overline{p}_k = \overline{p}_k. \end{aligned}$$

Remarks.

(1) See also [53] for another proof of this result.

(2) A *h.c.t.* is a symplectomorphism, therefore we must have:

$$(6.3)\qquad \partial_i \overline{x}^k = \overline{\partial}^k p_i, \quad \partial_i \overline{p}_k = -\overline{\partial}_k p_i, \dot{\partial}^i \overline{x}^k = -\overline{\partial}^k x^i, \quad \dot{\partial}^i \overline{p}_k = \overline{\partial}_k x^i.$$

If $\overline{x} = \overline{x}(x,p)$, $\overline{p} = \overline{p}(x,p)$ are homogeneous of degree 0 and 1 with respect to p, Eq. (6.3) are also sufficient conditions for $f(x,p) = (\overline{x}, \overline{p})$ to be a *h.c.t.*

In [53] it is proved that f is a *h.c.t.* then

$$(6.4)\qquad \begin{aligned} \partial_i \overline{x}^k \partial_j p_k - \partial_j \overline{x}^k \partial_i \overline{p}_k &= 0,\\ \partial_j \overline{x}^k \dot{\partial}^i \overline{p}_k - \dot{\partial}^i \overline{x}^k \partial_j \overline{p}_k &= \delta^i_j,\\ \dot{\partial}^i \overline{x}^k \dot{\partial}^j \overline{p}_k - \dot{\partial}^j \overline{x}^k \dot{\partial}^i \overline{p}_k &= 0, \end{aligned}$$

which in fact results from (6.3).

(3) If $f_0 \in \text{Diff}\,(M)$ then the cotangent map induced by f_0 is a *h.c.t.* In fact, if $\overline{x} = \overline{x}(x)$ is the local form of f_0 then,

$$f(x,p) = (\overline{x}(x), \overline{p}(x,p)), \quad \overline{p}_k = p_i \overline{\partial}_k x^i.$$

In this case f is called *an extended point transformation.*

It can easily proved that every fibre preserving map which is also a *h.c.t.* is an extended point transformation (see also [146]).

(4) The reason to use the word "contact" in the name of this transformation is given by the property of preserving the tangency of some special submanifolds of $\overset{\circ}{T}{}^*M$. (See [53], [146].)

Proposition 8.6.2. *Let ϕ be a connection-pair, Γ its associate connection and f a* $\Gamma-$ regular *h.c.t.*

(i) *ϕ is homogeneous $\Longleftrightarrow$ $\overline{\phi}$ is homogeneous*

(ii) *If ∇ is a $\phi-$connection of $\overline{\nabla}$ is the $\overline{\phi}-$connection defined by* (5.4), *then ∇ is $h(v)-$deflection free iff $\overline{\nabla}$ is $h(v)$-deflection free.*

Proof. Straightforward consequence of Definition 8.2.2 and Proposition 8.6.1

Let $H : \overset{\circ}{T}{}^*M \to M$ be a 2-homogeneous regular Hamiltonian and $\overline{H}$ the push-forward of H by f,

$$(6.5)\qquad \overline{H} = H \circ f^{-1}.$$

$\overline{H}$ is also 2-homogeneous Hamiltonian, but the matrix with entries

$$\widetilde{g}^{ij} = \dot{\overline{\partial}}^i \dot{\overline{\partial}}^j \overline{H} \tag{6.6}$$

may not be regular. Assume also that f is $\overset{c}{\Gamma}-$*regular*, where $\overset{c}{\Gamma}$ is given by (3.2).

Using the homogeneity property of f we get

$$\overline{p}_i = p_k \dot{\partial}^k \overline{p}_i = p_k(\dot{\partial}^k \overline{p}_i - \dot{\partial}^k \overline{x}^s \overline{\Gamma}_{si}) = \widetilde{\theta}_i^k p_k.$$

Therefore,

$$\overline{p}_i = \widetilde{\theta}_i^k p_k \quad \textit{and} \quad p_k = \theta_k^i \overline{p}_i. \tag{6.7}$$

Let $\overline{G}$ be the metric tensor (5.2). The push-forward of G is given by (2.23), where

$$g^{ij} = \dot{\partial}^i \dot{\partial}^j H. \tag{6.8}$$

We have

$$\overline{g}^{ij}(\overline{x}, \overline{p})\overline{p}_i \overline{p}_j = \overline{p}_i \overline{p}_j \theta_k^i \theta_h^j g^{kh}(x, p) = p_k p_h g^{kh}(x, p).$$

Therefore,

$$\overline{H} = \frac{1}{2} \overline{g}^{ij} \overline{p}_i \overline{p}_j. \tag{6.9}$$

Of course, we also have

$$H = \frac{1}{2} \widetilde{g}^{ij} \overline{p}_i \overline{p}_j,$$

but $\widetilde{g}^{ij} \neq \overline{g}^{ij}$ may happen.

In fact, the metric tensor induced by f is $\overline{g}^{ij}$ and not $\widetilde{g}^{ij}$, in general.

The tensor $\overline{g}^{ij}$ is $0-$homogeneous with respect to $\overline{p}_i$ and nondegenerate, but it may lack the property

$$\dot{\overline{\partial}}^k \overline{g}^{ij} = \dot{\overline{\partial}}^i \overline{g}^{kj}$$

which assures that following Section 8.5, the geometry, as in Section 8.3, can be derived from it.

Therefore, it is from $\overline{g}^{ij}$ that we can derive the geometry described in Section 8.2.

Now let us find the relationship between g^{ij} and $\overline{g}^{ij}$.

Proposition 8.6.3. (i) $\dot{\overline{\partial}}^k \overline{H} = \theta_s^k \dot{\partial}^s H \circ f^{-1}$,

(ii) $\widetilde{g}^{ij} = \overline{g}^{ij} + (\dot{\partial}^m \theta_k^i \partial_m \overline{x}^j - (\partial_\ell \theta_k^i - \theta_h^i \dot{\partial}^h \Gamma_{\ell k}) \dot{\partial}^\ell \overline{x}^j) p^k.$

Proof. (i) $\dot{\overline{\partial}}^k\overline{H} = \partial_s H\dot{\overline{\partial}}^k x^s + \dot{\partial}^s H\dot{\overline{\partial}}^k p_s$.

But $\delta_s H = 0 \Longrightarrow \partial_s H = -\Gamma_{s\ell}\dot{\partial}^\ell H = -\Gamma_{s\ell}p^\ell$ and by using (3.2) we get, because of homogeneity of g_{ij},

$$\partial_s H = -\gamma^0_{s0} = -\overset{c}{\Gamma}_{si}p^i. \tag{6.10}$$

Therefore,

$$\begin{aligned}\dot{\overline{\partial}}^k\overline{H} &= -\gamma^0_{s0}\dot{\overline{\partial}}^k x^s + p^s\dot{\overline{\partial}}^k p_s\\ &= p^i(\dot{\overline{\partial}}^k p_i - (\gamma^0_{is} - \tfrac{1}{2}\,\dot{\partial}^m g_{is}\gamma^0_{0m})\dot{\overline{\partial}}^k x^s)\\ &= p^i(\dot{\overline{\partial}}^k p_i - \overset{c}{\Gamma}_{is}\dot{\overline{\partial}}^k x^s) = \dot{\partial}^i H\theta^k_i,\end{aligned}$$

where we also have used (4.13).

Therefore, we get (i).

Now,

$$\dot{\overline{\partial}}^i\dot{\overline{\partial}}^j H = \theta^i_m(\dot{\partial}^m\dot{\partial}^k H\dot{\overline{\partial}}^j p_k + \dot{\partial}^m\partial_k H\dot{\overline{\partial}}^j x^k) + \dot{\partial}^k H(\partial_m\theta^i_k\dot{\overline{\partial}}^j x^m + \dot{\partial}^m\theta^i_k\dot{\overline{\partial}}^j p_m).$$

Using equation. (6.9) we obtain

$$\dot{\partial}^m\partial_k H = -\dot{\partial}^m\Gamma_{k\ell}p^\ell - \Gamma_{k\ell}\dot{\partial}^m\dot{\partial}^\ell H$$

and this equality transforms into (ii) after a straightforward calculation.

Note: The relation (ii) above is just (3.19) combined with (3.20) of [54], if we start with a Riemannian metric $g^{ij} = \gamma^{ij}(x)$.

Remark. $\widetilde{g}^{ij} = \overline{g}^{ij} \Longleftrightarrow p^k(\dot{\partial}^m\theta^i_k\partial_m\overline{x}^j - (\partial_\ell\theta^i_k - \theta^i_h\dot{\partial}^h\Gamma_{\ell k})\dot{\partial}^\ell\overline{x}^j) = 0$. By using (6.3) we see that this equality is equivalent to

$$p^\ell\dot{\partial}^h\theta^i_\ell = A^i_k\dot{\partial}^h\overline{x}^k \quad \textit{and} \quad p^s(\partial_\ell\theta^i_s - \theta^i_m\dot{\partial}^m\Gamma_{\ell s}) = A^i_k\partial_\ell\overline{x}^k,$$

for some functions A^i_k.

But from these equalities we get

$$A^i_k = \widetilde{\theta}^\ell_k(\partial_\ell\theta^i_s - \theta^i_m\dot{\partial}^m\Gamma_{\ell s} + \Gamma_{\ell m}\dot{\partial}^m\theta^i_s)p^s$$

(see also [54], (3.26)).

Also note,

$$\begin{aligned}\widetilde{g}^{ij} &= \overline{g}^{ij} \quad \textit{holds}\\ &\Longleftrightarrow (\dot{\overline{\partial}}^k\overline{g}^{ij} - \dot{\overline{\partial}}^i\overline{g}^{kj})\overline{p}_i = 0\\ &\Longleftrightarrow \overline{g}^{kj}\overline{p}_j = \overline{p}^k = \widetilde{g}^{kj}\overline{p}_j.\end{aligned}$$

As a consequence of the discussion above and results of previous sections, we have the following summary:

(a) If we start with a Cartan manifold (M, H), we get the triple $(\overset{c}{\Gamma}, \overset{\circ}{\nabla}, \overset{\circ}{G})$ given by (3.2), (2.28), (2.29) and (2.30) $\overset{\circ}{\nabla}$ is a $\overset{c}{\Gamma}-$connection, $\overset{\circ}{G}-$metrical, of Cartan type and the torsion tensors $\overset{\circ}{T}{}^k_{ij}$, $\overset{\circ}{S}{}^{kj}_i$ vanish.

(b) Taking a $\overset{c}{\Gamma}$-regular *h.c.t.* we get a new triple $(\overline{\phi}, \overline{\nabla}, \overline{G})$. Here $\overline{\phi}$ is a homogeneous connection-pair those coefficients are given by (4.2) and (4.7); $\overline{\nabla}$ is the $\overline{\phi}-$connection of Theorem 8.2.2 (in (2.24) and (2.25) g_{ij} and g^{ij} are substituted by $\overline{g}_{ij}, \overline{g}^{ij}$ and $\delta_i, \dot{\delta}^i$ by $\overline{\delta}_i, \dot{\overline{\delta}}{}^i$) and $\overline{G}$ is given by (5.2). This linear connection is $\overline{G}-$metrical and of Cartan type. Also, the torsion tensors $\overline{T}^k{}_{ij}$ and $\overline{S}_i{}^{kj}$ vanish. In fact, $\overline{T}^k{}_{ij}$ and $\overline{S}_k{}^{ij}$ are contact transformation (as (5.1)) of $\overset{\circ}{T}{}^k{}_{ij}$, and $\overset{\circ}{S}{}_k{}^{ij}$.

We get a new function $\overline{H}$ as in (6.5) which may not be a regular Hamiltonian.

Also,

$$\overline{\delta}_i H = 0 \quad \text{and} \quad \overline{H}\,\overline{|}^i\overline{|}^j = \overline{H}\,\overline{|}^j\overline{|}^i = 2\overline{g}^{ij} \tag{6.11}$$

where $\overline{|}$ denotes the *v−covariant* derivative with respect to $\overline{\nabla}$.

(c) If $\overline{g}^{ij} = \tilde{g}^{ij}$, then $\overline{H}$ is a regular Hamiltonian and Theorem 8.3.1 is valid (baring all the coefficients). A simple consequence, for the deflection-free case, is :

$$\overline{\Gamma} = \overset{c}{\overline{\Gamma}} \quad \text{iff} \quad (\overline{R}_{im\ell}\overline{\Pi}^{\ell s}\overline{g}_{sj} + \overline{R}_{jm\ell}\overline{\Pi}^{\ell s}\overline{g}_{si})\overline{p}^m = 0. \tag{6.12}$$

The relation (6.11) can be also written by using Proposition 8.4.6 and (5.3) in terms of similar objects derived from H. ($\overline{R}_{im\ell}$ and $\overline{g}_{sj}$ are contact transformations of $R_{im\ell}$ and g_{sj}).

When (6.12) is verified, by virtue of Theorem 8.2.1, Proposition 8.2.2 and Theorem 8.2.3 we can pass to the triple $(\overset{c}{\overline{\Gamma}}, \overset{\circ}{\overline{\nabla}}{}', \overset{\circ}{\overline{G}})$ when $\overset{\circ}{\overline{\nabla}}{}'$ is a $\overset{c}{\overline{\Gamma}}-$connection, $\overset{\circ}{\overline{G}}-$metrical, h- and v-deflection free, but generally fails to have vanishing torsion tensor $\overset{\circ}{\overline{T}}{}^i_{jk}, \overset{\circ}{\overline{S}}{}^{ik}_j$. Therefore, it does not coincide with the Cartan linear connection for the Hamilton manifold $(M, \overline{H})$ [66], [97].

If f is an extended point transformation, then $\overline{\Pi} = 0$, $\overline{\Gamma} = \overset{c}{\overline{\Gamma}}$ and the push-forward of the geometry of Cartan manifold (M, H) is just the geometry of $(M, \overline{H})$ so this geometry is invariant.

8.7 Examples

We now construct a connection-pair on $T^*\mathbb{R}^2$ which is horizontally flat, but with complicated Π^{ij}. In fact, we construct a homogeneous contact transformation between $(\widetilde{T}^*\mathbb{R}^2, \overline{H})$, where $\overline{\Pi}^{ij} = 0$ and $\overline{\Gamma}_{ij} \neq 0$, and $(\widetilde{T}^*\mathbb{R}^2, H)$, where $\Pi^{ij} \neq 0$ and

$\Gamma_{ij} = 0$. Here, $\mathbb{H} = \frac{1}{2}(P_1^2 + P_2^2)$ is the Euclidean Hamiltonian in $\widetilde{T}^*\mathbb{R}^2$, spanned by $(\mathbb{Q}^1, \mathbb{Q}^2, P_1, P_2)$, and $g^{ij} = \delta^{ij}$.

Select, once and for all, a Finsler metric function $\lambda = \lambda(q^1, q^2, p_1, p_2)$ and the metric $\overline{g}^{ij}(q^i, p_i) = e^{-2\phi(q^1,q^2)} \cdot (\dot{\partial}^i\dot{\partial}^j\lambda^2)$ and set

$$P_i = \lambda\, \underset{\cdot}{\Omega}\mathbb{Q}^i\,, \quad p_i = -\lambda\, \underset{\cdot}{\Omega}q^i\,,$$

where,

$$\Omega(q^1, q^2, \mathbb{Q}^1, \mathbb{Q}^2) = f_1(q^1, q^2)\mathbb{Q}^1 + f_2(q^1, q^2)\mathbb{Q}^2 = A,$$

is defined in terms of C^∞ functions f_1, f_2 and A, some constant. Noting that

$$P_i d\mathbb{Q}^i - p_i dq^i = 0,$$

we have the possibility of constructing the desired contact transformation $(\mathbb{Q}^i, P_i) \mapsto (q^i, p_i)$ and its *inverse*, locally. But, two side conditions will be necessary for this.

Firstly,

$$\det{}_f = \det\left(\mathrm{f}_{\underset{\cdot}{i}}q^j\right) \neq 0,$$

must hold in some chart $(\overline{U}, \overline{h})$. Then

$$p_1 = -\lambda\, \mathrm{f}_{\underset{\cdot}{1}}q^1\; \mathbb{Q}^1 - \lambda\, \mathrm{f}_{\underset{\cdot}{2}}q^1\; \mathbb{Q}^2, \qquad p_2 = -\lambda\, \mathrm{f}_{\underset{\cdot}{1}}q^2\; \mathbb{Q}^1 - \lambda\, \mathrm{f}_{\underset{\cdot}{2}}q^2\; \mathbb{Q}^2$$

has a unique solution in $(\overline{U}, \overline{h})$,

$$\mathbb{Q}^i = \mathbb{Q}^i(p_1, p_2, q^1, q^2).$$

Of course,

$$P_i = \lambda f_i(q^1, q^2),$$

so that the transformation is determined by f_i and $\partial_j f_i$.

It is also required that the transformation by $\Gamma -$ *regular*. In this case, $\Gamma_{ij} = 0$, so this condition is merely

$$\det\left(\underset{\cdot}{\mathbb{Q}^i}q^j\right) \neq 0.$$

Now, the push-forward of $g^{ij} = \delta^{ij}$ is required to be (by results of Section 8.6)

$$\overline{g}^{ij} = \dot{\partial}^i\dot{\partial}^j(\mathbb{H} \circ \phi) = \frac{1}{2}\,(\dot{\partial}^i\dot{\partial}^j\lambda^2) \cdot e^{-2\phi(q^1,q^2)},$$

where,

$$\begin{aligned}\phi &= -\tfrac{1}{2}\,\ell n\left[\left(\underset{\cdot}{\Omega}\mathbb{Q}^1\right)^2 + \left(\underset{\cdot}{\Omega}\mathbb{Q}^2\right)^2\right]\\ &= -\tfrac{1}{2}\,\ell n\,[(f_1)^2 + (f_2)^2].\end{aligned}$$

We have supposed that $\phi(q^1, q^2)$ is known and defined in a chart $(\overline{U}, \overline{h})$.

Secondly, we must now select $f_i(q^1, q^2)$ so that the $\Gamma-$regularity condition holds, locally.

Set $\mathbb{R} = -(p_2\,\lambda)\,\partial_1 + (p_1\,\lambda)\,\partial_2$, to denote this linear operator and assume λ is independent of q^1, q^2. Thus, λ is a Minkowski metric function in the chart $(\overline{U}, \overline{h})$. Note that $\mathbb{Q}^1 = -1/\det_f \mathbb{R}(f_2)$ and $\mathbb{Q}^2 = -1/\det_f \mathbb{R}(f_1)$.

Proposition 8.7.1. *Under the condition* $\det_f \neq 0$ *in* $(\overline{U}, \overline{h})$ $\Gamma-$*regularity holds for* λ *Minkowski* $\Longleftrightarrow \partial_1(\ell n \det_f)[(\mathbb{R}f_2)(\mathbb{R}\circ\partial_2)(f_1) - (\mathbb{R}f_1)\cdot(\mathbb{R}\circ\partial_2)(f_2)] - \partial_2(\ell n \det_f) \times[(\mathbb{R}f_2)(\mathbb{R}\circ\partial_1))(f_1) - (\mathbb{R}f_1)(\mathbb{R}\circ\partial_1)(f_2)] \neq [(\mathbb{R}\circ\partial_1)(f_1)]\cdot[(\mathbb{R}\circ\partial_2)(f_2)] - [(\mathbb{R}\circ\partial_1)(f_2)][(\mathbb{R}\circ\partial_2)(f_1)]$ *in* $(\overline{U}, \overline{h})$.

Corollary 8.7.1. *In addition, assume* $f_2 = c\cdot q^2$, *where* $c > 0$ *is a sufficiently small constant. Then* $\Gamma-$*regularity holds in* $(\overline{V}, \overline{h}) \Longleftrightarrow \mathrm{Hess}\,(f_1) \neq 0$ *in* $(\overline{V}, \overline{h})$ *(Hessian determinant) and* $f_1 = \sqrt{e^{2\phi} - (c\cdot q^2)^2}$. *Here,* $\overline{V} \subseteq \overset{\circ}{\overline{B}} \subseteq \overline{U}$ ($\overset{\circ}{\overline{B}}$ *is interior of closed 2-disk).*

Proof. A short calculation shows that the condition of the proposition reduces to the non-zero Hessian condition. An easy continuity argument shows that f_1 above is well-defined in some closed 2-disk in $(\overline{U}, \overline{h})$. Merely note $m \leq \phi(q^1, q^2) \leq M$ holds in any closed disk $\overline{B} \subseteq (\overline{U}, \overline{h})$ and take the radius $r = \frac{1}{c}\,e^m$ so that $c\cdot q^2 < e^\phi$ in *this*, $\overline{B}$ (radius $= r$). Now choose a smaller chart $\overline{V}$ in the interior of $\overline{B}$. This completes the proof.

Also note that by linear adjustment, we can always suppose that ϕ (center of $\overline{B}$) $= 0$ in $\mathbb{R}^2$. We can now state the

Theorem 8.7.1. *If* ϕ *has a non-degenerate critical point* x *in* $(\overline{U}, \overline{h})$ *of* $\mathbb{R}^2$, *then* $\Gamma-$*regularity holds in some neighborhood of* x.

Consequently, $(T^*\mathbb{R}^2, \overline{H})$ is homogeneous contact equivalent to $(T^*\mathbb{R}^2, H)$ where $\overline{H} = \frac{1}{2}\,\overline{g}^{ij}p_ip_j = \frac{1}{2}\,e^{-2\phi}(\dot{\partial}^i\dot{\partial}^j\lambda^2)p_ip_j$. Moreover, Π^{ij} is by (4.7) not zero generally and is completely determined by $\overline{\Pi}^{ij} = 0$, $\Gamma_{ij} = 0$ and this transformation.

Similar results are possible even if ϕ has no nondegenerate critical points. For example, if $\phi = \omega_i q^i$, ω_i constants, the conclusion of the theorem above holds. It can be reformulated as

Theorem 8.7.2. *Any 2-dimensional constant Wagner space is the Legendre-dual of the homogeneous contact transformation of the flat Cartan space* $(\widetilde{T}^*\mathbb{R}^2, H)$ *with non-trivial oblique distribution* Π.

Similar reformulations can be made of the main theorem on $\Gamma-$*regularity*, as well, using the known result that Wagner spaces with vanishing $h-$*curvature* must have local metric functions of the form $e^\phi\cdot\lambda$, [11]. These have been found to be of fundamental importance in the ecology and evolution of colonial marine invertebrates (ibid.).

Chapter 9

The dual bundle of a k-osculator bundle

The cotangent bundle T^*M, dual of the tangent bundle carries some canonical geometric object field as: the Liouville vector field, a symplectic structure and a Poisson structure. They allow to construct a theory of Hamiltonian systems, and, via Legendre transformation, to transport this theory in that of Lagrangian systems on the tangent bundle. Therefore, the Lagrange spaces $L^n = (M, L(x,y))$ appear as dual of Hamilton spaces $H^n = (M, H(x,p))$, (cf. Ch.7).

In the theory of Lagrange spaces of order k, where the fundamental functions are Lagrangians which depend on point and higher order accelerations, we do not have a dual theory based on a good notion of higher order Hamiltonian, which depend on point, higher order accelerations and momentum (of order 1, only). This is because we have not find yet a differentiable bundle which have a canonical symplectic or presymplectic structure and a canonical Poisson structure and which is basic for a good theory of the higher order Hamiltonian systems.

The purpose of this chapter is to elliminate this incovenient. Starting from the k-osculator bundle $(\mathrm{Osc}^k M, \pi^k, M)$, identified with k-tangent bundle $(T^k M, \pi^k, M)$, we introduce a new differentiable bundle $(T^{*k}M, \pi^{*k}, M)$ called *dual* bundle of k-osculator bundle (or k-tangent bundle), where the total space $T^{*k}M$ is the fibered product:

$$T^{*k}M = T^{k-1}M \times_M T^*M.$$

We prove that on the manifold $T^{*k}M$ there exist a canonical Liouville 1–form, a canonical presymplectic structure and a canonical Poisson structure. Consequently we can develop a natural theory of Hamiltonian systems of order k, for the Hamilton functions which depend on point, accelerations of order $1, 2, ..., k-1$ and momenta.

These properties are fundamental for introducing the notion of Hamilton space of order k.

We will develop in the next chapters the geometry of second order Hamilton spaces and we remark that this is a natural extension of the geometry of Hamilton spaces, studied in the previous chapters.

All this theory is based on the paper [110] of the first author.

9.1 The $(T^{*k}M, \pi^{*k}, M)$ bundle

Let M be a real n–dimensional manifold and let (T^kM, π^k, M) be its k–osculator bundle. The canonical local coordinates of a point $u \in T^kM$ are $(x^i, y^{(1)i}, ..., y^{(k)i})$, $\pi^k(u) = x$ and the point u will be denoted by $u = (x, y^{(1)}, ..., y^{(k)})$.

The changes of coordinates on T^kM are given by [106]:

$$
\begin{aligned}
\widetilde{x}^i &= \widetilde{x}^i(x^1, ..., x^n),\ \det\left(\frac{\partial \widetilde{x}^i}{\partial x^j}\right) \neq 0\\
\widetilde{y}^{(1)i} &= \frac{\partial \widetilde{x}^i}{\partial x^j} y^j\\
2\widetilde{y}^{(2)i} &= \frac{\partial \widetilde{y}^{(1)i}}{\partial x^j} y^{(1)j} + 2\frac{\partial \widetilde{y}^{(1)i}}{\partial y^{(1)j}} y^{(2)j}\\
&\cdots\cdots\cdots\cdots\cdots\cdots\cdots\cdots\cdots\cdots\cdots\\
k\widetilde{y}^{(k)i} &= \frac{\partial \widetilde{y}^{(k-1)i}}{\partial x^j} y^{(1)j} + 2\frac{\partial \widetilde{y}^{(k-1)i}}{\partial y^{(1)j}} y^{(2)j} + \cdots +\\
&+ k\frac{\partial \widetilde{y}^{(k-1)i}}{\partial y^{(k-1)j}} y^{(k)j}.
\end{aligned}
\tag{1.1}
$$

For every point $u \in T^kM$, the natural basis

$$\left\{ \left.\frac{\partial}{\partial x^i}\right|_u, \left.\frac{\partial}{\partial y^{(1)i}}\right|_u, ..., \left.\frac{\partial}{\partial y^{(k)i}}\right|_u \right\}$$

in $T_u(T^kM)$ is transformed by (1.1), as follows:

(1.2)
$$\begin{aligned}
\left.\frac{\partial}{\partial x^i}\right|_u &= \frac{\partial \widetilde{x}^j}{\partial x^i}\left.\frac{\partial}{\partial \widetilde{x}^j}\right|_u + \frac{\partial \widetilde{y}^{(1)j}}{\partial x^i}\left.\frac{\partial}{\partial \widetilde{y}^{(1)j}}\right|_u + \cdots + \\
&+ \frac{\partial \widetilde{y}^{(k)j}}{\partial x^i}\left.\frac{\partial}{\partial \widetilde{y}^{(k)j}}\right|_u, \\
\left.\frac{\partial}{\partial y^{(1)i}}\right|_u &= \frac{\partial \widetilde{y}^{(1)j}}{\partial y^{(1)i}}\left.\frac{\partial}{\partial \widetilde{y}^{(1)j}}\right|_u + \cdots + \frac{\partial \widetilde{y}^{(k)j}}{\partial y^{(1)i}}\left.\frac{\partial}{\partial \widetilde{y}^{(k)j}}\right|_u, \\
&\cdots\cdots\cdots\cdots\cdots\cdots\cdots\cdots\cdots\cdots \\
\left.\frac{\partial}{\partial y^{(k)i}}\right|_u &= \frac{\partial \widetilde{y}^{(k)j}}{\partial y^{(k)i}}\left.\frac{\partial}{\partial \widetilde{y}^{(k)j}}\right|_u,
\end{aligned}$$

where

(1.3)
$$\frac{\partial \widetilde{y}^{(\alpha)i}}{\partial x^j} = \frac{\partial \widetilde{y}^{(\alpha+1)i}}{\partial y^{(1)j}} = \cdots = \frac{\partial \widetilde{y}^{(k)i}}{\partial y^{(k-\alpha)j}},\ (\alpha = 0, ..., k-1; y^{(0)i} = x^i).$$

For $\alpha \in \{0, 1, ..., k-1\}$ we denote by $\pi^k_\alpha : T^kM \longrightarrow T^\alpha M$ the canonical submersion, locally expressed by

$$\pi^k_\alpha(x, y^{(1)}, ..., y^{(k)}) = (x, y^{(1)}, ..., y^{(\alpha)}),\ \pi^k_0 = \pi^k.$$

Every of these submersions determines on T^kM a simple foliation, denoted by $\mathcal{F}_{\alpha+1}$. The sheafs of $\mathcal{F}_{\alpha+1}$ are embedding submanifolds of T^kM of dimension $(k-\alpha)n$ on which $(y^{(\alpha+1)i}, ..., y^{(k)i})$ are the local coordinates and $(x^i, y^{(1)i}, ..., y^{(\alpha)i})$ are the transverse coordinates.

Every foliation $\mathcal{F}_{\alpha+1}$ determines a tangent distribution

$$V_{\alpha+1} = T\mathcal{F}_{\alpha+1} = \mathrm{Ker} d\pi^k_\alpha, (\alpha = 0, ..., k-1),$$

where $d\pi^k_\alpha$ is the differential of the mapping π^k_α.

Therefore, we have a number of k distributions $V_1, ..., V_k$ which are integrable, of local dimension $kn, (k-1)n, ..., n$, respectively, and having the propertythat $\forall\, u \in T^kM$, $V_1(u) \supset V_2(u) \supset \cdots \supset V_k(u)$.

The manifold T^kM carries some others natural geometrical object fields [106], as the Liouville vector fields $\overset{1}{\Gamma}, ..., \overset{k}{\Gamma}$. $\overset{1}{\Gamma}$ belongs to the distribution V_k and is given by $\overset{1}{\Gamma} = y^{(1)i}\dfrac{\partial}{\partial y^{(k)i}}, \ \cdots$, $\overset{k}{\Gamma}$ belongs to V_1 and has the form $\overset{k}{\Gamma} = y^{(1)}\dfrac{\partial}{\partial y^{(1)i}} + 2y^{(2)i}\dfrac{\partial}{\partial y^{(2)i}} + \cdots + ky^{(k)i}\dfrac{\partial}{\partial y^{(k)i}}$.

Moreover, there is a tangent structure J defined on T^kM. It is given by

$$J = \frac{\partial}{\partial y^{(1)i}} \otimes dx^i + \frac{\partial}{\partial y^{(2)i}} \otimes dy^{(1)i} + \cdots + \frac{\partial}{\partial y^{(k)i}} \otimes dy^{(k-1)i}.$$

We have the properties

$$J\overset{k}{\Gamma} = \overset{k-1}{\Gamma},\ J\overset{k-1}{\Gamma} = \overset{k-2}{\Gamma}, ..., J\overset{2}{\Gamma} = \overset{1}{\Gamma},\ J\overset{1}{\Gamma} = 0.$$

However, it does not exist a canonical symplectic or presymplectic structure over the manifold T^kM.

We introduce the following differentiable bundle:

Definition 9.1.1. We call the *dual* of the k–osculator bundle (T^kM, π^kM) the differentiable bundle $(T^{*k}M, \pi^{*k}, M)$ whose total space is

$$T^{*k}M = T^{k-1}M \times_M T^*M. \tag{1.4}$$

The previous fibered product has a differentiable structure given by that of the $(k-1)$–osculator bundle $T^{k-1}M$ and the cotangent bundle T^*M. Of course, for $k = 1$, we have $T^{*1}M = T^*M$.

We will see that over the manifold $T^{*k}M$ there exist a natural presymplectic structure and a natural Poisson structure. Sometime we denote $(T^{*k}M, \pi^{*k}, M)$ by $T^{*k}M$. A point $u \in T^{*k}M$ will be denoted by $u = (x, y^{(1)}, ..., y^{(k-1)}, p)$. The canonical projection $\pi^{*k} : T^{*k}M \longrightarrow M$ is defined by $\pi^{*k}(x, y^{(1)}, ..., y^{(k-1)}, p) = x$. Of course, we take the projections on the factors of the fibered products (1.4):

$$\pi^{*k}_{k-1} : T^{*k}M \longrightarrow T^{k-1}M,\ \pi^* : T^*M \longrightarrow M$$

as being $\pi^{*k}_{k-1}(x, y^{(1)}, ..., y^{(k-1)}, p) = (x, y^{(1)}, ..., y^{(k-1)})$ and $\pi^*(x, p) = x$. It results the following commutative diagram:

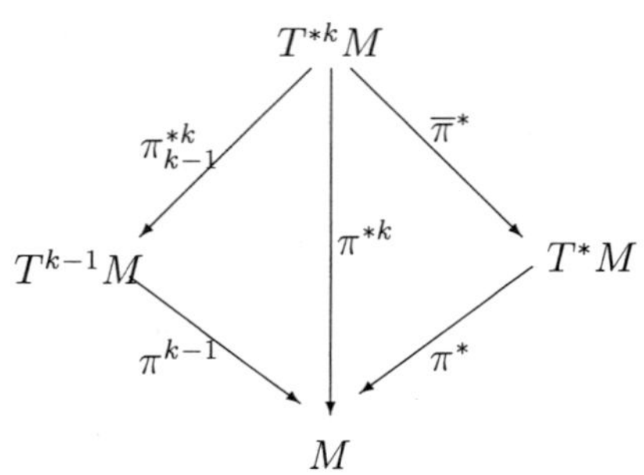

where $\overline{\pi}^*\ (x, y^{(1)}, .., y^{(k-1)}, p) = (x, p)$.

This diagram implies the existence of some natural geometrical object fields over the manifold $T^{*k}M$.

Let $(x^i, y^{(1)i}, ..., y^{(k-1)i}, p_i)$, $(i = 1, ..., n)$, be the coordinates of a point $u = (x, y^{(1)}, ..., y^{(k-1)}, p) \in T^{*k}M$ in a local chart $(\pi^{*-1}(U), \phi)$ on $T^{*k}M$.

The change of coordintes on the manifold $T^{*k}M$ is:

$$(1.5)\quad \begin{cases} \widetilde{x}^i = \widetilde{x}^i(x^1, ..., x^n), \ \det\left(\dfrac{\partial \widetilde{x}^i}{\partial x^j}\right) \neq 0 \\ \widetilde{y}^{(1)i} = \dfrac{\partial \widetilde{x}^i}{\partial x^j} y^{(1)j} \\ \cdots\cdots\cdots\cdots\cdots\cdots\cdots\cdots\cdots \\ (k-1)\widetilde{y}^{(k-1)i} = \dfrac{\partial \widetilde{y}^{(k-2)i}}{\partial x^j} y^{(1)j} + \cdots + (k-1)\dfrac{\partial \widetilde{y}^{(k-2)i}}{\partial y^{(k-2)j}} y^{(k-1)i} \\ \widetilde{p}_i = \dfrac{\partial x^j}{\partial \widetilde{x}^i} p_j \end{cases}$$

where the following relation holds:

$$(1.5)' \quad \frac{\partial \widetilde{y}^{(\alpha)i}}{\partial x^j} = \frac{\partial \widetilde{y}^{(\alpha+1)i}}{\partial y^{(1)j}} = \cdots = \frac{\partial \widetilde{y}^{(k-1)i}}{\partial y^{(k-1-\alpha)j}}, \quad (\alpha = 0, ..., k-2; y^{(0)} = x).$$

It follows that $T^{*k}M$ is a real differentiable manifold of dimension $(k+1)n$. It is the same dimension with that of the manifold T^kM.

With respect to (1.4) the natural basis

$$\left\{ \left.\frac{\partial}{\partial x^i}\right|_u, \left.\frac{\partial}{\partial y^{(1)i}}\right|_u, ..., \left.\frac{\partial}{\partial y^{(k-1)i}}\right|_u, \left.\frac{\partial}{\partial p_i}\right|_u \right\}$$

in $T_u(T^{*k}M)$ is transformed as follows:

$$
(1.6)\qquad \begin{cases}
\left.\dfrac{\partial}{\partial x^i}\right|_u = \left.\dfrac{\partial \widetilde{x}^j}{\partial x^i}\dfrac{\partial}{\partial \widetilde{x}^j}\right|_u + \left.\dfrac{\partial \widetilde{y}^{(1)j}}{\partial x^i}\dfrac{\partial}{\partial \widetilde{y}^{(1)j}}\right|_u + \cdots + \\
\qquad + \left.\dfrac{\partial \widetilde{y}^{(k-1)j}}{\partial x^i}\dfrac{\partial}{\partial \widetilde{y}^{(k+1)j}}\right|_u + \left.\dfrac{\partial \widetilde{p}_j}{\partial x^i}\dfrac{\partial}{\partial \widetilde{p}_j}\right|_u \\
\left.\dfrac{\partial}{\partial y^{(1)i}}\right|_u = \left.\dfrac{\partial \widetilde{y}^{(1)j}}{\partial y^{(1)i}}\dfrac{\partial}{\partial \widetilde{y}^{(1)j}}\right|_u + \cdots + \left.\dfrac{\partial \widetilde{y}^{(k-1)j}}{\partial y^{(1)i}}\dfrac{\partial}{\partial \widetilde{y}^{(k-1)j}}\right|_u \\
\cdots\cdots\cdots\cdots\cdots\cdots\cdots\cdots\cdots\cdots \\
\left.\dfrac{\partial}{\partial y^{(k-1)i}}\right|_u = \left.\dfrac{\partial \widetilde{y}^{(k-1)j}}{\partial y^{(k-1)i}}\dfrac{\partial}{\partial \widetilde{y}^{(k-1)j}}\right|_u \\
\dfrac{\partial}{\partial p_i} = \left.\dfrac{\partial x^i}{\partial \widetilde{x}^j}\dfrac{\partial}{\partial \widetilde{p}_j}\right|_u,
\end{cases}
$$

the conditions $(1.5)'$ being satisfied.

Consequently, the Jacobian matrix of the transformation of coordinate (1.5) is given by

$$
(1.6)'\qquad J_k(u) = \begin{pmatrix}
\dfrac{\partial \widetilde{x}^j}{\partial x^i}(u) & 0 & 0 & \cdots & 0 & 0 \\
\dfrac{\partial \widetilde{y}^{(1)j}}{\partial x^i}(u) & \dfrac{\partial \widetilde{x}^j}{\partial x^i}(u) & 0 & \cdots & 0 & 0 \\
\vdots & & & \ddots & & \\
\dfrac{\partial \widetilde{y}^{(k-1)j}}{\partial x^i}(u) & \dfrac{\partial \widetilde{y}^{k-1}j}{\partial y^{(1)i}}(u) & 0 & \cdots & \dfrac{\partial \widetilde{x}^j}{\partial x^i}(u) & 0 \\
\dfrac{\partial \widetilde{p}_j}{\partial x^i} & 0 & 0 & \cdots & 0 & \dfrac{\partial x^i}{\partial \widetilde{x}^j}(u)
\end{pmatrix}.
$$

It follows:

$$
(1.6)''\qquad \det J_k(u) = \left[\det\left(\frac{\partial \widetilde{x}^i}{\partial x^j}(u)\right)\right]^{k-1}.
$$

Theorem 9.1.1. *We have:*

a. *If k is an odd number, the manifold $T^{*k}M$ is orientable.*

b. *If k is an even number, the manifold $T^{*k}M$ is orientable if and only if the manifold M is orientable.*

Following the property of the bundle $(T^{*k}M, \pi^{*k}_{k-1}, T^{k-1}M)$ it is not difficult to prove:

Theorem 9.1.2. *If the differentiable manifold M is paracompact, then the differentiable manifold $T^{*k}M$ is paracompact, too.*

Let us introduce the following differential forms:

$$\omega = p_i dx^i, \tag{1.7}$$

$$\theta = d\omega = dp_i \wedge dx^i. \tag{1.7$'$}$$

Then, we have:

Theorem 9.1.3.

1° *The forms ω and θ are global defined on the manifold $T^{*k}M$.*

2° *θ is closed, i.e. $d\theta = 0$.*

3° *θ is a presymplectic structure of rank $2n$ on the manifold $T^{*k}M$.*

Proof. 1° The forms ω and θ are invariant with respect to (1.5).
2° $d\theta = d^2\omega = 0$.
3° θ is a 2–form of rank $2n$ and $2n \leq (k+1)n = \dim T^{*k}M$. **q.e.d.**

Now, let us consider the system of Poisson brackets:

$$\{\ \}_\alpha : (f, g) \in \mathcal{F}(T^{*k}M) \times \mathcal{F}(T^{*k}M) \longrightarrow \{f, g\}_\alpha \in \mathcal{F}(T^{*k}M), \tag{1.8}$$

defined by

$$\{f, g\}_\alpha = \frac{\partial f}{\partial y^{(\alpha)i}} \frac{\partial g}{\partial p_i} - \frac{\partial f}{\partial p_i} \frac{\partial g}{\partial y^{(\alpha)i}}, \ (\alpha = 0, 1, ..., k-1, y^{(0)} = x) \tag{1.8$'$}$$

We obtain:

Theorem 9.1.4. *Every bracket $\{\ \}_\alpha$, $(\alpha = 0, ..., k-1)$ defines a canonical Poisson structure on the manifold $T^{*k}M$.*
Proof. We prove that the Poisson bracket $\{f, g\}_\alpha$, $(\alpha = 1, ..., k-1)$ is invariant under the transformations of coordinates (1.5) on $T^{*k}M$. Indeed, by means of (1.6)

we have

$$\frac{\partial f}{\partial y^{(\alpha)i}} = \frac{\partial \widetilde{y}^{(\alpha)m}}{\partial y^{(\alpha)i}} \frac{\partial \widetilde{f}}{\partial \widetilde{y}^{(\alpha)m}} + \frac{\partial \widetilde{y}^{(\alpha+1)m}}{\partial y^{(\alpha)i}} \frac{\partial \widetilde{f}}{\partial \widetilde{y}^{(\alpha+1)m}} + \cdots +$$

$$+ \frac{\partial \widetilde{y}^{(k-1)m}}{\partial y^{(\alpha)i}} \frac{\partial \widetilde{f}}{\partial \widetilde{y}^{(k-1)m}} + \frac{\partial \widetilde{p}_m}{\partial y^{(\alpha)i}} \frac{\partial \widetilde{f}}{\partial \widetilde{p}_m},$$

$$\frac{\partial g}{\partial p_i} = \frac{\partial x^i}{\partial \widetilde{x}^s} \frac{\partial \widetilde{g}}{\partial \widetilde{p}_i}.$$

But the first formulae, by means of (1.5)′, can be written as follows:

$$\frac{\partial f}{\partial y^{(\alpha)i}} = \frac{\partial \widetilde{x}^m}{\partial x^i} \frac{\partial \widetilde{f}}{\partial \widetilde{y}^{(\alpha)m}} + \frac{\partial \widetilde{y}^{(1)m}}{\partial x^i} \frac{\partial \widetilde{f}}{\partial \widetilde{y}^{(\alpha+1)m}} + \cdots +$$

$$+ \frac{\partial \widetilde{y}^{k-1-\alpha)m}}{\partial x^i} \frac{\partial \widetilde{f}}{\partial \widetilde{y}^{(k-1)m}} + \frac{\partial \widetilde{p}_m}{\partial y^{(\alpha)i}} \frac{\partial \widetilde{f}}{\partial \widetilde{p}_m}.$$

And also, taking into account the identities:

$$\frac{\partial \widetilde{y}^{(\alpha)m}}{\partial x^i} \frac{\partial x^i}{\partial \widetilde{x}^s} = \frac{\partial \widetilde{y}^{(\alpha)m}}{\partial \widetilde{x}^s} = 0, \quad \text{for } \alpha = 1, 2, ..., k-1,$$

$$\frac{\partial \widetilde{p}_m}{\partial y^{(\alpha)i}} = 0, \ (\alpha = 1, ..., k-1), \ \frac{\partial \widetilde{p}_m}{\partial x^i} \frac{\partial x^i}{\partial \widetilde{x}^s} = 0,$$

we obtain:

$$\frac{\partial f}{\partial y^{(\alpha)i}} \frac{\partial f}{\partial p_i} = \frac{\partial \widetilde{f}}{\partial \widetilde{y}^{(\alpha)i}} \frac{\partial \widetilde{g}}{\partial \widetilde{p}_i} \quad (\alpha = 0, ..., k-1).$$

Consequently

$$\{f, g\}_\alpha = \widetilde{\{f, g\}}_{(\alpha)}, \ (\alpha = 0, ..., k-1).$$

It is clear that:

$$\begin{cases} \{f, g\}_\alpha = -\{g, f\}_\alpha; \ \{f, g\}_\alpha \text{ is } R\text{-linear in every argument} \\ \{f, g\}_\alpha = -\{g, f\}_\alpha \end{cases}$$

holds.

Finally, we prove that the Jacobi identities hold, i.e.:

$$\{\{f, g\}_\alpha, h\}_\alpha + \{\{g, h\}_\alpha, f\}_\alpha + \{\{h, f\}_\alpha, g\}_\alpha = 0, \tag{1.8)''}$$

$$(\alpha = 0, 1, ..., k-1).$$

Finally, by a direct calculus, it is not difficult to prove the Jacobi identities (1.8)$''$ for every Poisson bracket $\{f, g\}_\alpha$. **q.e.d.**

For more clarity, all these considerations will be applied in the case $k = 2$, in order to study the geometry of Hamilton spaces of order 2.

9.2 The dual of the 2–osculator bundle

The theory from the previous section can be particularized to the case $k = 2$. We obtain in this way the dual of the 2–tangent bundle (T^2M, π^2, M) given by $(T^{*2}M, \pi^{*2}, M)$, where $T^{*2}M = T^1M \times_M T^*M$ can be identified with $T^{*2}M = TM \times_M T^*M$. Notice that the last one is exactly the Whitney sum of vector bundles $TM \oplus T^*M$. We have

$$T^{*2}M = T^{*2}M = TM \oplus T^*M, \tag{2.1}$$

where (TM, π, M) is the tangent bundle of the manifold M and (T^*M, π^*, M) its cotangent bundle. A point $u \in T^{*2}M$ can be written in the form $u = (x, y, p)$, having the local coordinates (x^i, y^i, p_i). The projections $\pi^{*2}(u) = \pi^{*2}(x, y, p) = x$, $\pi^2_1 : T^{*2}M \longrightarrow TM$ are defined by $\pi^2_1(u) = \pi^2_1(x, y, p) = (x, y)$ and $\overline{\pi}^* : T^{*2}M \longrightarrow T^*M$ is given by $\overline{\pi}^*(u) = \overline{\pi}^*(x, y, p) = (x, p)$. Let us denote $\pi^2_1(u) = \overline{\pi}(u)$. We get the following commuttive diagram:

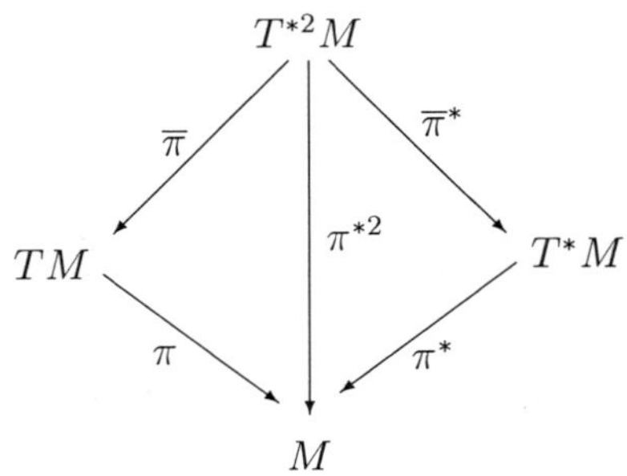

A change of local coordinates is given by

$$\begin{cases} \widetilde{x}^i &= \widetilde{x}^i(x^1, ..., x^n), \ \det\left(\dfrac{\partial \widetilde{x}^i}{\partial x^j}\right) \neq 0, \\ \widetilde{y}^i &= \dfrac{\partial \widetilde{x}^i}{\partial x^j} y^j, \\ \widetilde{p}_i &= \dfrac{\partial x^j}{\partial \widetilde{x}^i} p_j. \end{cases} \tag{2.2}$$

The dimension of the manifold $T^{*2}M$ is $3n$.

For every point $u \in T^{*2}M$ the natural basis

$$\left\{ \left.\frac{\partial}{\partial x^i}\right|_u, \left.\frac{\partial}{\partial y^i}\right|_u, \left.\frac{\partial}{\partial p_i}\right|_u \right\}$$

of the tangent space T_uT^{*2} transformes, with respect to (2.1), as follows:

$$\text{(2.3)}\qquad \begin{cases} \left.\dfrac{\partial}{\partial x^i}\right|_u = \dfrac{\partial \widetilde{x}^j}{\partial x^i} \left.\dfrac{\partial}{\partial \widetilde{x}^j}\right|_u + \dfrac{\partial \widetilde{y}^j}{\partial x^i} \left.\dfrac{\partial}{\partial \widetilde{y}^j}\right|_u + \dfrac{\partial \widetilde{p}_j}{\partial x^i} \left.\dfrac{\partial}{\partial \widetilde{p}_j}\right|_u \\[2ex] \left.\dfrac{\partial}{\partial y^i}\right|_u = \dfrac{\partial \widetilde{x}^j}{\partial x^i} \left.\dfrac{\partial}{\partial \widetilde{y}^j}\right|_u \\[2ex] \left.\dfrac{\partial}{\partial p_i}\right|_u = \dfrac{\partial x^i}{\partial \widetilde{x}^j} \left.\dfrac{\partial}{\partial \widetilde{p}^j}\right|_u . \end{cases}$$

The Jacobian matrix of the change of coordinates (2.1) is given by

$$\text{(2.4)}\qquad J(u) = \begin{pmatrix} \dfrac{\partial \widetilde{x}^j}{\partial x^i}(u) & 0 & 0 \\[2ex] \dfrac{\partial \widetilde{y}^j}{\partial x^i}(u) & \dfrac{\partial \widetilde{x}^j}{\partial x^i}(u) & 0 \\[2ex] \dfrac{\partial \widetilde{p}^j}{\partial x^i}(u) & 0 & \dfrac{\partial x^j}{\partial \widetilde{x}^i}(u) \end{pmatrix}.$$

Theorem 9.2.1. *The differentiable manifold $T^{*2}M$ is orientable if and only if the base manifold M is orientable.*

The nul section $0 : M \longrightarrow T^{*2}M$ of the projection π^{*2} is defined by $0 : (x) \in M \longrightarrow (x, 0, 0) \in T^{*2}M$ we denote by $\widetilde{T^{*2}M} = T^{*2}M \setminus \{0\}$.

Let us consider the tangent bundle of the differentiable manifold $T^{*2}M$. It is given by the triade $(TT^{*2}M, \tau^{*2}, T^{*2}M)$, where τ^{*2} is the canonical projection. Taking into account the kernel of the differential $d\tau^{*2}$ of the mapping τ^{*2} we get the vertical subbundle $VT^{*2}M$. This leads to the vertical distribution $V : u \in T^{*2}M \longrightarrow V(u) \subset T_uT^{*2}M$. The local dimension of the vertical distribution V is $2n$ and V is locally generated by the vector fields $\left\{ \left.\dfrac{\partial}{\partial y^i}\right|_u, \left.\dfrac{\partial}{\partial p_i}\right|_u \right\}_u$, $\forall u \in T^{*2}M$. As usually, let us denote

$$\text{(2.5)}\qquad \partial_i = \frac{\partial}{\partial x^i}, \; \dot{\partial}_i = \frac{\partial}{\partial y^i}, \; \dot{\partial}^i = \frac{\partial}{\partial p_i}.$$

It follows that the vertical distribution V is integrable. By means of the relation (2.3), we can consider the following subdistributions of V:

$$W_1 : u \in T^{*2}M \longrightarrow W_1(u) \subset T_u T^{*2}M \tag{2.6}$$

locally generated by the vector fields $\left\{ \dot{\partial}_i\big|_u ,\ u \in T^{*2}M \right\}$. It is an integrable distribution of local dimension n

Let us consider also the subdistribution

$$W_2 : u \in T^{*2}M \longrightarrow W_2(u) \subset T_u T^{*2}M \tag{2.6$'$}$$

locally generated by the vector fields $\{ \dot{\partial}^i\big|_u ,\ u \in T^{*2}M\}$. Of course, W_2 is also an integrable distribution of local dimension n

Clearly, we have

Proposition 9.2.1. *The vertical distribution V has the property*

$$V(u) = W_1(u) \oplus W_2(u),\ \forall\, u \in T^{*2}M. \tag{2.7}$$

Now, some important geometrical object fields can be introduced:

(i) the Liouville vector field on $T^{*2}M$:

$$\mathbb{C}(u) = y^i\, \dot{\partial}_i\big|_u ,\ \forall\, u \in T^{*2}M, \tag{2.8}$$

(ii) the Hamilton vector field on $T^{*2}M$:

$$\mathbb{C}^*(u) = p_i\, \dot{\partial}^i\big|_u ,\ \forall\, u \in T^{*2}M, \tag{2.9}$$

(iii) the scalar field

$$\varphi = p_i y^i. \tag{2.10}$$

We remark that $\mathbb{C} \in W_1$ and $\mathbb{C}^* \in W_2$.

Also, let us consider the following forms

$$\omega = p_i dx^i, \qquad \text{(Liouville 1-form)} \tag{2.11}$$

$$\theta = d\omega = dp_i \wedge dx^i. \tag{2.12}$$

Then, Theorem 9.1.3 leads to the following result:

Theorem 9.2.2.

1° *The differential forms ω and θ are globally defined on the manifold $T^{*2}M$.*

2° *The 2-form θ is closed and rank θ is $2n$.*

3° *θ is a presymplectic structure on $T^{*2}M$.*

The Poisson brackets $\{\ \}_0, \{\ \}_1$ can be defined on the manifold $T^{*2}M$ by:

$$\{f,g\}_0 = \frac{\partial f}{\partial x^i}\frac{\partial g}{\partial p_i} - \frac{\partial f}{\partial p_i}\frac{\partial g}{\partial x^i} \tag{2.13}$$
$$\{f,g\}_1 = \frac{\partial f}{\partial y^i}\frac{\partial g}{\partial p_i} - \frac{\partial f}{\partial p_i}\frac{\partial g}{\partial y^i}$$

Therefore, Theorem 9.1.4 can be particularized in:

Theorem 9.2.3. *Every bracket $\{\ \}_0$ and $\{\ \}_1$ defines a canonical Poisson structure on the manifold $T^{*2}M$.*

Now, it is not difficult to prove that the following $\mathcal{F}(T^{*2}M)$–linear mapping

$$J : \mathcal{X}(T^{*2}M) \longrightarrow \mathcal{X}(T^{*2}M)$$

defined by

$$J(\partial_i) = \dot{\partial}_i,\ J(\dot{\partial}_i) = 0,\ J(\dot{\partial}^i) = 0,\ u \in T^{*2}M \tag{2.14}$$

has geometrical meaning.

It is not difficult to prove:

Theorem 9.2.4. *The following properties hold:*

1°. *J is a tensor field of type $(1,1)$ on the manifold $T^{*2}M$.*

2°. *J is a tangent structure on $T^{*2}M$, i.e. $J \circ J = 0$.*

3°. *J is an integrable structure.*

4°. $\mathrm{Ker}J = W_1 \oplus W_2,\ \mathrm{Im}J = W_1$.

We are going to use these object fields to construct the geometry of the manifold $T^{*2}M$.

9.3 Dual semisprays on $T^{*2}M$

An important notion suggested by the geometrical theory of the Lagrange spaces of order 2 is that of dual semisprays.

Definition 9.3.1. A dual semispray on $T^{*2}M$ is a vector field S on $T^{*2}M$ with the property:

$$JS = \mathbb{C}. \tag{3.1}$$

Taking into account (2.6) and (2.11) it follows:

Proposition 9.3.1. *A dual semispray S on $T^{*2}M$ can be represented locally by*

$$S = y^i \partial_i + 2\xi^i \dot{\partial}_i + f_i \dot{\partial}^i. \tag{3.2}$$

The system of functions $(\xi^i(x,y,p), f_i(x,y,p))$ is called *the coefficients of the dual semispray S*. However, they are not any arbitrary functions. In fact, $\{\xi^i\}$ and $\{f_i\}$ are important geometrical object fields.

Theorem 9.3.1. *With respect to the transformation law (2.2) on $T^{*2}M$, the functions $\{\xi^i\}$ and $\{f_i\}$ transform as follows:*

$$2\widetilde{\xi}^i = \frac{\partial \widetilde{x}^i}{\partial x^j} 2\xi^j + \frac{\partial \widetilde{y}^i}{\partial x^j} y^j \tag{3.3}$$

$$\widetilde{f}_i = \frac{\partial x^j}{\partial \widetilde{x}^i} f_j + \frac{\partial \widetilde{p}_i}{\partial x^j} y^j \tag{3.3$'$}$$

Conversely, if on every domain of local chart on $T^{*2}M$ are given the systems of functions $\{\xi^i\}$ and $\{f_i\}$ $(i = 1, ..., n)$ such that, with respect to (2.2), *the formula* (3.3) *and* (3.3)′ *hold, then S given by* (3.2) *is a dual semispray on $T^{*2}M$.*

The proof is not difficult. It is similar with the proof given by semisprays on the osculator bundle T^2M.

Two immediate properties are the following:

Proposition 9.3.2. *The integral curves of the dual semispray S, from* (3.2), *are given by the solution curves of the system of differential equations:*

$$\frac{dx^i}{dt} = y^i, \quad \frac{d^2x^i}{dt^2} = 2\xi^i(x,y,p), \quad \frac{dp_i}{dt} = f_i(x,y,p). \tag{3.4}$$

Proposition 9.3.3. *Every dual semispray S on the manifold $T^{*2}M$ with the coefficients (ξ^i, f_i) determines a bundle morphism*

$$\xi : (x, y, p) \in T^{*2}M \to (x, y^{(1)i}, y^{(2)i}) \in T^{(2)}M, \tag{3.5}$$

defined by

$$x^i = x^i,\ \ y^{(1)i} = y^i,\ \ y^{(2)i} = \xi^i(x, y, p). \tag{3.5)'$$

Moreover, it is a local diffeomorphism if and only if $\text{rank}\left\|\dot{\partial}^i \xi^j\right\| = n$.

We shall see in Chapter 10 that the bundle morphism ξ, defined in (3.5)′, is uniquely determined by the Legendre transformation between a Lagrange space of order two, $L^{(2)n} = (M, L(x, y^{(1)}, y^{(2)})$, and a Hamilton space of order two, $H^{(2)n} = (M, H(x, y, p))$.

Consequently, if the bundle morphism ξ, defined in (3.5)′ is apriori given, the dual spray S, denoted by

$$S_\xi = y^i \partial_i + 2\xi^i \dot{\partial}_i + f_i \dot{\partial}^i \tag{3.2)'$$

is characterized only by the coefficients $f_i(x, y, p)$.

We have, also:

Proposition 9.3.4. *The formula:*

$$\omega(S_\xi) = \varphi, \tag{3.6}$$

holds.

An important problem is the existence of the dual semisprays on $T^{*2}M$.

Theorem 9.3.2. *If the base manifold M is paracompact, then on $T^{*2}M$ there exist dual semisprays S_ξ with apriori given bundle morphism ξ.*

Proof. Assuming that the manifold M is paracompact by means of Theorem 9.1.2, it follows that the manifold $T^{*2}M$ is paracompact, too. We shall see (Ch.10) that a bundle morphism ξ, defined in (3.5)′ exists. Now, let $\gamma_{ij}(x)$, $x \in M$ be a Riemannian metric on M and $\gamma^i_{jk}(x)$ its Christoffel symbols.

Setting

$$f_j = \gamma^i_{jh}(x) p_i y^h \tag{3.7}$$

we can prove that the rule of transformations of the systems of functions $\{f_i\}$ with respect to the transformation of local coordinates (2.2) are given by (3.3)′.

Indeed, we have

$$\tilde{\gamma}^i_{rs}\frac{\partial\tilde{x}^r}{\partial x^j}\frac{\partial\tilde{x}^s}{\partial x^h}=\gamma^s_{jh}\frac{\partial\tilde{x}^i}{\partial x^s}-\frac{\partial^2\tilde{x}^i}{\partial x^j\partial x^h}. \tag{3.8}$$

The contractions with y^h and p_i lead to

$$\tilde{f}_i=\frac{\partial x^j}{\partial\tilde{x}^i}f_j+\frac{\partial\tilde{p}_i}{\partial x^j}y^j$$

which is exactly (3.3)$'$.

Therefore f_i, from (3.7) are the coefficients of a dual semispray.

As a consequence, we have:

Theorem 9.3.3. *The following systems of functions*

$$N^i{}_j=\dot{\partial}^i f_j,\ N_{ij}=\dot{\partial}_j f_i \tag{3.9}$$

are geometrical object fields on $T^{*2}M$, having the following rules of transformations, with respect to the changing of local coordinates (2.2)*:*

$$\begin{aligned}\widetilde{N}^i{}_s\frac{\partial\tilde{x}^s}{\partial x^j}&=N^s{}_j\frac{\partial\tilde{x}^i}{\partial x^s}-\frac{\partial\tilde{y}^i}{\partial x^j},\\ \widetilde{N}_{ij}&=\frac{\partial x^s}{\partial\tilde{x}^i}\frac{\partial x^r}{\partial\tilde{x}^j}N_{sr}+p_r\frac{\partial^2x^r}{\partial\tilde{x}^i\partial\tilde{x}^j}.\end{aligned} \tag{3.10}$$

These properties can be proved by a direct computation starting from the formulae (3.3)$'$.

Remarks.

1°. We will see that the system of functions $(N^i{}_j, N_{ij})$ gives us the coefficients of a nonlinear connection on $T^{*2}M$.

2°. With respect to (2.2), the system of functions

$$\tau_{ij}=N_{ij}-N_{ji}$$

is transformed like a skew symmetric, covariant d–tensor field.

9.4 Homogeneity

The notion of homogeneity for the functions $f(x, y, p)$, defined on the manifold $T^{*2}M$ can be defined with respect to y^i, as well as, with respect to p_i respectively. Indeed, any homothety

$$h_a : (x, y, p) \longrightarrow (x, ay, p), \ a \in \mathbb{R}^+$$

is preserved by the transformation of local coordinates (2.2).

Let H_y be the group of transformation on $T^{*2}M$,

$$H_y = \{h_a : (x, y, p) \longrightarrow (x, ay, p) \mid a \in \mathbb{R}^+\}.$$

The orbit of a point $u_0 = (x_0, y_0, p_0)$ by H_y is given by

$$x^i = x^i{}_0, \ y^i = ay^i{}_0, \ p_i = p^0{}_i, \ \forall\, a \in \mathbb{R}^+.$$

The tangent vector in the point $u_0 = h_1(u_0)$ is the Liouville vector field $\mathbb{C}(u_0) = y^i{}_0\dot{\partial}_i|_{u_0}$.

Definition 9.4.1. A function $f : T^{*2}M \longrightarrow \mathbb{R}$ differentiable on $T^{*2}M \setminus \{0\} = \widetilde{T^{*2}M}$ and continuous on the null section of the projection $\pi^{*2} : T^{*2}M \longrightarrow M$ is called *homogeneous of degree* $r \in \mathbb{Z}$ with respect to y^i, if

$$f \circ h_a = a^r f, \ \forall\, a \in \mathbb{R}^+. \tag{4.1}$$

It follows[106]:

Theorem 9.4.1. *A function $f \in \mathcal{F}(T^{*2}M)$ differentiable on $\widetilde{T^{*2}M}$ and continuous of the null section is r-homogeneous with respect to y^i if and only if*

$$\mathcal{L}_{\mathbb{C}} f = rf, \tag{4.1$'$}$$

where $\mathcal{L}_{\mathbb{C}}$ is the Lie derivation with respect to the Liouville vector field $\mathbb{C}$.

We notice that (4.1) can be written in the form:

$$y^i \frac{\partial f}{\partial y^i} = rf. \tag{4.1$''$}$$

The entire theory of homogeneity, with respect to y^i, exposed in the book [106], can be applied.

However, in our case it is important to define the notion of homogeneity with respect to variables p_i.

Let H_p be the group of homotheties

$$H_p = \{h'_a : (x, y, p) \in T^{*2}M \longrightarrow (x, y, ap) \in T^{*2}M \mid a \in \mathbb{R}^+\}.$$

The orbit of a point $u_0 = (x_0, y_0, p_0)$ by H_p is given by

$$x^i = x^i{}_0,\ y^i = y^i{}_0,\ p_i = ap^0{}_i,\ \forall\, a \in \mathbb{R}^+.$$

Its tangent vector in the point $u_0 = h'_1(u)$ is the Hamilton vector field $\mathbb{C}^*(u_0)$.

A function $f : T^{*2}M \longrightarrow \mathbb{R}$ differentiable on $\widetilde{T^{*2}M}$ and continuous on the null section is called homogeneous of degree r, $(r \in \mathbb{Z})$ with respect to the variables p_i if

$$f \circ h'_a = a^r f,\ \forall\, a \in \mathbb{R}^+. \tag{4.2}$$

In other words:

$$f(x, y, ap) = a^r f(x, y, p).$$

It follows

Theorem 9.4.2. *A function $f \in \mathcal{F}(T^{*2}M)$, differentiable on $\widetilde{T^{*2}M}$ and continuous on the null section is r–homogeneous with respect to p_i if and only if we have*

$$\mathcal{L}_{\mathbb{C}^*} f = rf. \tag{4.2)'$$

Of course, (4.2)′ is given by

$$p_i \frac{\partial f}{\partial p_i} = rf. \tag{4.2)''$$

A vector field $X \in \mathcal{X}(\widetilde{T^{*2}M})$ is r–homogeneous with respect to p_i if

$$X \circ h'_a = a^{r-1} h'^*_a \circ X,\ \forall\, a \in \mathbb{R}^+.$$

We have:

Theorem 9.4.3. *A vector field $X \in \mathcal{X}(\widetilde{T^{*2}M})$ is r–homogeneous with respect to p_i if and only if*

$$\mathcal{L}_{\mathbb{C}^*} X = (r - 1)X. \tag{4.3}$$

Corollary 9.4.1. *The vector fields $\dfrac{\partial}{\partial x^i}, \dfrac{\partial}{\partial y^i}, \dfrac{\partial}{\partial p_i}$ are $1, 1, 0$–homogeneous with respect to p_i, respectively.*

Corollary 9.4.2. *A vector field*

$$X = \overset{(0)}{X}{}^i \frac{\partial}{\partial x^i} + \overset{(1)}{X}{}^i \frac{\partial}{\partial y^i} + \overset{(2)}{X}{}_i \frac{\partial}{\partial p_i}, \tag{4.3)'$$

is r—homogeneous with respect to p_i, if and only if

$$\overset{(0)}{X}{}^i \text{ is } r-1 \text{ homogeneous,}$$

$$\overset{(1)}{X}{}^i \text{ is } r-1 \text{ homogeneous,}$$

$$\overset{(2)}{X}_i \text{ is } r \text{ homogeneous.}$$

It results:

Proposition 9.4.1. *If $f \in \mathcal{F}(\widetilde{T^{*2}M})$, $X \in \mathcal{X}(\widetilde{T^{*2}M})$ are $r-$ and s–homogeneous with respect to p_i, respectively, then fX is $r+s$–homogeneous.*

In particular:

1°. *The Hamilton vector field $\mathbb{C}^*$ is 1–homogeneous.*

2°. *A dual semispray*

$$S_\xi = y^i\frac{\partial}{\partial x^i} + 2\xi^i\frac{\partial}{\partial y^i} + f_i\frac{\partial}{\partial p_i} \tag{4.4}$$

is 1–homogeneous with respect to p_i if and only if its coefficients ξ^i are 0–homogeneous and f_i are 1–homogeneous with respect to p_i.

Proposition 9.4.2. *If $X \in \mathcal{X}(\widetilde{T^{*2}M})$ is r–homogeneous and f is s–homogeneous with respect to p_i, then Xf is $r+s-1$–homogeneous with respect to p_i.*

Corollary 9.4.3. *If $f \in \mathcal{F}(\widetilde{T^{*2}M})$ is r–homogeneous with respect to p_i and differentiable on $\widetilde{T^{*2}M}$, then*

1°. $\dfrac{\partial f}{\partial p_i}$ *are $(r-1)$–homogeneous.*

2°. $\dfrac{\partial^2 f}{\partial p_i \partial p_j}$ *is $(r-2)$–homogeneous.*

A q–form $\omega \in \Lambda^q(\widetilde{T^{*2}M})$ is called s–homogeneous with respect to p_i if

$$\omega \circ h_a^{'*} = a^s\omega, \ \forall\, a \in \mathbb{R}^+.$$

Corollary 9.4.4. *If the functions $f, g \in \mathcal{F}(T^{*2}M)$ are $r-$ and $s-$ homogeneous with respect to p_i, respectively, then the functions given by the Poisson brackets $\{f,g\}_0$ and $\{f,g\}_1$ are $(r+s-1)$–homogeneous, respectively.*

The following result holds:

Theorem 9.4.4. *A q–form* $\omega \in \Lambda^q(\widetilde{T^{*2}M})$ *is s–homogeneous with respect to* p_i *if and only if*

$$\mathcal{L}_{\mathbb{C}^*}\omega = s\omega. \tag{4.5}$$

It follows:

Proposition 9.4.3. *The 1–forms* dx^i, dy^i, dp_i $(i = 1, ..., n)$ *are* $0, 0, 1$ *homogeneous with respect to* p_i*, respectively.*

In the next section we will apply these considerations for study the notion of the Hamilton spaces of order 2.

Finally, we remark:

Proposition 9.4.4. *A dual semispray* S_ξ *is 2–homogeneous with respect to* y^i *if and only if the coefficients* ξ^i *are 2–homogeneous and* f_i *are 1–homogeneous with respect to* y^i.

A dual semispray S_ξ which is 2–homogeneous with respect to y^i is called a dual spray.

9.5 Nonlinear connections

We extend the classical definition [97] of the nonlinear connection on the total space of the dual bundle $(T^{*2}M, \pi^{*2}, M)$.

Definition 9.5.1. A nonlinear connection on the manifold $T^{*2}M$ is a regular distribution N on $T^{*2}M$ supplementary to the vertical distribution V, i.e.

$$T_uT^{*2}M = N(u) \oplus V(u), \ \forall\, u \in T^{*2}M. \tag{5.1}$$

Taking into account Proposition 9.2.7 it follows that the distribution N has the property:

$$T_uT^{*2}M = N(u) \oplus W_1(u) \oplus W_2(u). \tag{5.2}$$

Therefore, the main geometrical objects on $T^{*2}M$ will be reported to the direct sum (5.2) of vector spaces.

We denote by

$$\left(\frac{\delta}{\delta x^i}, \frac{\partial}{\partial y^i}, \frac{\partial}{\partial p_i}\right) \ (i = 1, ..., n) \tag{5.3}$$

a local adapted basis to N, W_1, W_2. Clearly, we have

$$\frac{\delta}{\delta x^i} = \frac{\partial}{\partial x^i} - N^j{}_i \frac{\partial}{\partial y^j} + N_{ij}\frac{\partial}{\partial p_j}. \tag{5.4}$$

The systems of functions $(N^j{}_i(x,y,p), N_{ij}(x,y,p))$ are the *coefficients* of the nonlinear connection N.

With respect to the coordinate transformations (2.2), $\frac{\delta}{\delta x^i}, \frac{\partial}{\partial y^i}, \frac{\partial}{\partial p_i}$ are transformed by the rule:

$$\frac{\delta}{\delta x^i} = \frac{\partial \tilde{x}^j}{\partial x^i}\frac{\delta}{\delta \tilde{x}^j}; \ \frac{\partial}{\partial y^i} = \frac{\partial \tilde{x}^j}{\partial x^i}\frac{\partial}{\partial \tilde{y}^j}; \ \frac{\partial}{\partial p_i} = \frac{\partial x^i}{\partial \tilde{x}^j}\frac{\partial}{\partial \tilde{p}_j}. \tag{5.4'}$$

It is not difficult to prove the following property

Theorem 9.5.1. *The coefficients $(N^i{}_j, N_{ij})$ of a nonlinear connection N on $T^{*2}M$ obey the rule of transformations* (3.10) *with respect to a changing of local coordinates* (2.2). *Conversely, if the systems of functions $(N^i{}_j, N_{ij})$ are given on the every domain of local chart of the manifold $T^{*2}M$ such that the first two equations* (3.10) *hold, then $(N^i{}_j, N_{ij})$ are the coefficients of a nonlinear connection on $T^{*2}M$.*

It is convenient then to use the basis (5.3), if the coefficients $N^i{}_j$ and N_{ij} are determined only by the coefficients f_i of the semispray S_ξ.

It is not difficult to prove the following theorem:

Theorem 9.5.2. *If S_ξ is a dual semispray with the coefficients f_i, then the systems of functions*

$$N^i{}_j = \frac{\partial f_j}{\partial p_i}, \ N_{ij} = \frac{\partial f_i}{\partial y^j} \tag{5.5}$$

are the coefficients of a nonlinear connection.

Conversely:

Theorem 9.5.3. *If $(N^i{}_j, N_{ij})$ are coefficients of a nonlinear connection N, then the following systems of functions*

$$\check{f}_i = N_{ij}y^j$$

are the coefficients of a dual semispray S_ξ, where ξ_i are apriori given.

Taking into account Theorems 9.3.2 and 9.5.2, we can affirm:

Theorem 9.5.4. *If the base manifold M is paracompact, then there exist nonlinear connections on the manifold $T^{*2}M$.*

From now on we denote the basis (5.3) by:

$$(\delta_i, \dot{\partial}_i, \dot{\partial}^i). \tag{5.3)'$$

The dual basis of the adapted basis (5.3) is given by

$$(\delta x^i, \delta y^i, \delta p_i) \tag{5.6}$$

where

$$\delta x^i = dx^i,\ \delta y^i = dy^i + N^i{}_j dx^j,\ \delta p_i = dp_i - N_{ji}dx^j. \tag{5.6)'$$

With respect to (2.2), the covector fields (5.6) are transformed by the rules:

$$\delta \widetilde{x}^i = \frac{\partial \widetilde{x}^i}{\partial x^j}\delta x^j,\ \delta \widetilde{y}^i = \frac{\partial \widetilde{x}^i}{\partial x^j}\delta y^j,\ \delta \widetilde{p}_i = \frac{\partial x^j}{\partial \widetilde{x}^i}\delta p_j. \tag{5.6)'$$

Also, we remark that the differential of a function $f \in \mathcal{F}(\widetilde{T^{*2}M})$ can be written in the form

$$df = \frac{\delta f}{\delta x^i}dx^i + \frac{\partial f}{\partial y^i}\delta y^i + \frac{\partial f}{\partial p_i}\delta p_i. \tag{5.7}$$

9.6 Distinguished vector and covector fields

Let N be a nonlinear connection. Then, it gives rise to the direct decomposition (5.2). Let h, w_1, w_2 be the projectors defined by the distributions N, W_1, W_2. They have the following properties:

$$\begin{gathered} h + w_1 + w_2 = I,\ h^2 = h,\ w_1^2 = w_1,\ w_2^2 = w_2, \\ h \circ w_1 = w_1 \circ h = 0,\ h \circ w_2 = w_2 \circ h = 0,\ w_1 \circ w_2 = w_2 \circ w_1 = 0. \end{gathered} \tag{6.1}$$

If $X \in \mathcal{X}(\widetilde{T^{*2}M})$ we denote:

$$X^H = hX,\ X^{W_1} = w_1 X,\ X^{W_2} = w_2 X. \tag{6.2}$$

Therefore we have the unique decomposition:

$$X = X^H + X^{W_1} + X^{W_2}. \tag{6.3}$$

Each of the components X^H, X^{W_1}, X^{W_2} is called a d–vector field on $\widetilde{T^{*2}M}$.

In the adapted basis (5.3) we get

$$X^H = X^{(0)i}\delta_i,\ X^{W_1} = X^{(1)i}\dot{\partial}_i,\ X^{W_2} = X^{(2)}_i\dot{\partial}^i. \tag{6.3)'$$

By means of (5.4) we have

$$\widetilde{X}^{(0)i} = \frac{\partial\widetilde{x}^i}{\partial x^j}X^{(0)j};\ \widetilde{X}^{(1)i} = \frac{\partial\widetilde{x}^i}{\partial x^j}X^{(1)j};\ \widetilde{X}^{(2)}_i = \frac{\partial x^j}{\partial\widetilde{x}^i}X^{(2)}_j. \tag{6.4}$$

But, these are the classical rules of the transformations of the local coordinates of vector and covector fields on the base manifold M [97]. Therefore $X^{(0)i}, X^{(1)i}$ are called d–vector fields and $X^{(2)}_i$ is called a d–covector field.

For instance, the Liouville vector field $\mathbb{C}$ and Hamilton vector field $\mathbb{C}^*$ have the properties:

$$\mathbb{C} = \mathbb{C}^{W_1} = y^i\dot{\partial}_i, \qquad \mathbb{C}^H = 0, \quad \mathbb{C}^{W_2} = 0,$$

$$\mathbb{C}^* = \mathbb{C}^{*W_2} = p_i\dot{\partial}^i, \quad \mathbb{C}^{*H} = 0, \quad \mathbb{C}^{*W_1} = 0.$$

A dual semispray S_ξ, from (4.4), in the adapted basis (5.2), has the decomposition

$$S_\xi = S_\xi^H + S_\xi^{W_1} + S_\xi^{W_2} = y^i\delta_i + 2k_i\dot{\partial}_i - h_i\dot{\partial}^i, \tag{6.5}$$

where:

$$2k^i = 2\xi^i + N^i{}_j y^j,\ h_i = f_i - N_{ji}y^j, \tag{6.6}$$

k^i being a d–vector field and h_i a d–covector field.

Assuming that the nonlinear connection N provides from a dual semispray S_ξ with the coefficients f_i, we get

$$N^i{}_j = \dot{\partial}^i f_j,\ N_{ij} = \dot{\partial}_j f_i. \tag{6.6)'$$

It follows that the vector k^i and covector h_i are given by

$$2k^i = 2\xi^i + (\dot{\partial}^i f_j)y^j,\ h_i = f_i - (\dot{\partial}_i f_j)y^j. \tag{6.6)''$$

A similar theory can be done for distinguished 1-forms.

With respect to the direct decomposition (5.2) a 1-form $\omega \in \mathcal{X}^*(T^{*2}M)$ can be uniquely written in the form:

$$\omega = \omega^H + \omega^{W_1} + \omega^{W_2}, \tag{6.7}$$

where

$$\omega^H = \omega \circ h,\ \omega^{W_1} = \omega \circ w_1,\ \omega^{W_2} = \omega \circ w_2. \tag{6.7)'$$

In the adapted cobasis (5.6) and (5.6)′, we have

$$\omega = \overset{(0)}{\omega}_i \delta x^i + \overset{(1)}{\omega}_i \delta y^i + \overset{(2)}{\omega}{}^i \delta p_i \tag{6.8}$$

The quantities $\omega^H, \omega^{W_1}, \omega^{W_2}$ are called d-1-forms.

The coefficients $\overset{(0)}{\omega}_i, \overset{(1)}{\omega}_i$ and $\overset{(2)}{\omega}{}^i$ are transformed by (2.2) as follows:

$$\overset{(0)}{\omega}_i = \frac{\partial \widetilde{x}^j}{\partial x^i} \overset{(0)}{\widetilde{\omega}}_j, \quad \overset{(1)}{\omega}_i = \frac{\partial \widetilde{x}^j}{\partial x^i} \overset{(1)}{\widetilde{\omega}}_j, \quad \overset{(2)}{\widetilde{\omega}}{}^i = \frac{\partial \widetilde{x}^i}{\partial x^j} \overset{(2)}{\omega}{}^j .$$

Hence $\overset{(0)}{\omega}_i$ and $\overset{(1)}{\omega}_i$ are called d-covector fields and $\overset{(2)}{\omega}{}^i$ will be called d-vector field.

If the nonlinear connection N is apriori given, then some remarkable d-1-forms can be associated in a natural way. Namely, let us consider:

$$\begin{aligned} \omega &= \omega^H = p_i dx^i, \\ \alpha &= \alpha^{W_1} = p_i \delta y^i, \\ \beta &= \beta^{W_2} = y^i \delta p_i. \end{aligned} \tag{6.9}$$

We will use these d-forms for studying the Hamilton geometry of order 2 on $T^{*2}M$.

Proposition 9.6.1. *The following properties hold:*

If S_ξ is a dual semispray, as in (6.5), *and nonlinear connection N is determined by S_ξ, as in* (6.6)′, *then we have*

$$\omega(S_\xi) = p_i y^i, \ \alpha(S_\xi) = 2p_i k^i, \ \beta(S_\xi) = y^i h_i. \tag{6.10}$$

Now, let us consider a function f on $T^{*2}M$. Its differential can be written in the form (5.7). Therefore

$$\begin{cases} df = (df)^H + (df)^{W_1} + (df)^{W_2}, \\ \text{where} \\ (df)^H = \delta_i f \, dx^i, \ (df)^{W_1} = \dot{\partial}_i f \, \delta y^i, \ (df)^{W_2} = \dot{\partial}^i f \, \delta p_i. \end{cases} \tag{6.11}$$

Let us consider a smooth parametrized curve $\gamma : I \subset \mathbb{R} \longrightarrow \widetilde{T^{*2}M}$ such that $\text{Im}\gamma \subset (\pi^{*2})^{-1}(U)$. It can be analytical represented by:

$$x^i = x^i(t), \ y^i = y^i(t), \ p_i = p_i(t), \ t \in I. \tag{6.12}$$

The tangent vector $\dfrac{d\gamma}{dt}$, in a point of the curve γ, can be written in the form:

$$\frac{d\gamma}{dt} = \left(\frac{d\gamma}{dt}\right)^{H} + \left(\frac{d\gamma}{dt}\right)^{W_1} + \left(\frac{d\gamma}{dt}\right)^{W_2} = \frac{dx^i}{dt}\delta_i + \frac{\delta y^i}{dt}\dot{\partial}_i + \frac{\delta p_i}{dt}\dot{\partial}^i, \tag{6.13}$$

where

$$\frac{\delta y^i}{dt} = \frac{dy^i}{dt} + N^i{}_j\frac{dx^j}{dt}, \ \frac{\delta p_i}{dt} = \frac{dp_i}{dt} - N_{ji}\frac{dx^j}{dt}. \tag{6.14}$$

The curve in (6.3) is called horizontal if $\dfrac{d\gamma}{dt} = \left(\dfrac{d\gamma}{dt}\right)^{H}$ in every point of the curve γ.

Proposition 9.6.2. *An horizontal curve on $\widetilde{T^{*2}M}$ is characterized by the following system of differential equations:*

$$x^i = x^i(t), \ \frac{\delta y^i}{dt} = 0, \ \frac{\delta p_i}{dt} = 0, \ t \in I. \tag{6.15}$$

Clearly, the system of differential equations (6.15) has local solutions, if the initial points $x^i_0 = x^i(t_0)$, y^i_0, p^0_i on $T^{*2}M$ are given, $t_0 \in I$.

9.7 Lie brackets. Exterior differentials

In applications, the Lie brackets of the vector fields $\{\delta_i, \dot{\partial}_i, \dot{\partial}^i\}$, from the adapted basis to the direct decomposition (5.2), are important.

Proposition 9.7.1. *The Lie brackets of the vector fields of the adapted basis are given by*

$$\begin{aligned}
[\delta_j, \delta_h] &= \underset{(1)}{R}{}^i{}_{jh}\dot{\partial}_i + \underset{(2)}{R}{}_{ijh}\dot{\partial}^i, \\
\left[\delta_j, \dot{\partial}_h\right] &= \underset{(1)}{B}{}^i{}_{jh}\dot{\partial}_i + \underset{(2)}{B}{}_{ijh}\dot{\partial}^i, \\
\left[\delta_j, \dot{\partial}^h\right] &= \underset{(1)}{B}{}_j^{ih}\dot{\partial}_i + \underset{(2)}{B}{}^h_{ij}\dot{\partial}^i, \\
\left[\dot{\partial}_j, \dot{\partial}_h\right] &= \left[\dot{\partial}_j, \dot{\partial}^h\right] = \left[\dot{\partial}^j \dot{\partial}^h\right] = 0,
\end{aligned} \tag{7.1}$$

where

$$\underset{(1)}{R}{}^i{}_{jh} = \delta_h N^i{}_j - \delta_j N^i{}_h, \quad \underset{(2)}{R}{}_{ijh} = \delta_j N_{hi} - \delta_h N_{ji},$$

(7.2) $$\underset{(1)}{B}{}^i{}_{jh} = \dot{\partial}_h N^i{}_j, \qquad \underset{(2)}{B}{}_{ijh} = -\dot{\partial}_h N_{ji},$$

$$\underset{(1)}{B}{}^i{}_j{}^h = \dot{\partial}^h N^i{}_j, \qquad \underset{(2)}{B}{}^h{}_{ij} = -\dot{\partial}^h N_{ji}.$$

The proof of this relations can be done by a direct calculus.

Now we can establish:

Proposition 9.7.2. *The exterior differentials of the* 1*-forms* $(\delta x^i, \delta y^i, \delta p_i)$*, which determine the adapted cobasis* $(5.6)'$*, are given by*

$$d(\delta x^i) = 0,$$

(7.3) $$d(\delta y^i) = \left\{ \frac{1}{2} \underset{(1)}{R}{}^i{}_{jm} dx^m + \underset{(1)}{B}{}^i{}_{jm} \delta y^m + \underset{(1)}{B}{}^{im}{}_j \delta p_m \right\} \wedge dx^j,$$

$$d(\delta p_i) = \left\{ \frac{1}{2} \underset{(2)}{R}{}_{ijm} dx^m + \underset{(2)}{B}{}_{ijm} \delta y^m + \underset{(2)}{B}{}^m{}_{ji} \delta p_m \right\} \wedge dx^j.$$

Indeed, from $(5.6)'$ we deduce

$$d(\delta y^i) = dN^i{}_j \wedge dx^j, \; d(\delta p_i) = -dN_{ji} \wedge dx^j.$$

Using (6.11) for $dN^i{}_j$ and dN_{ji} we have the formula (7.3).

Now, the exterior differentials of the ω, α, β, from (6.9), can be easily determined.

Let us consider the following coefficients from (7.1):

(7.4) $$\underset{(1)}{B}{}^i{}_{jh} = \dot{\partial}_h N^i{}_j; \; - \underset{(2)}{B}{}^i{}_{jh} = \dot{\partial}^i N_{hj}.$$

By means of (3.10) it follows:

Proposition 9.7.3. *The coefficients* $\underset{(1)}{B}{}^i{}_{jh}, - \underset{(2)}{B}{}^i{}_{jh}$ *have the same rule of transformation with respect to the local changing of coordinates on* $T^{*2}M$*. This is*

(7.5) $$\widetilde{B}^i{}_{rs} \frac{\partial \widetilde{x}^r}{\partial x^j} \frac{\partial \widetilde{x}^s}{\partial x^k} = \frac{\partial \widetilde{x}^i}{\partial x^r} B^r{}_{jk} - \frac{\partial^2 \widetilde{x}^i}{\partial x^j \partial x^k}.$$

We will see that these coefficients are the horizontal coefficients of an N–linear connection.

We obtain also:

Proposition 9.7.4. *The coefficients:*

$$B_{(1)}{}^{i}{}_{j}{}^{h} = \dot{\partial}^{h} N^{i}{}_{j}, \quad B_{(2)}{}_{ijh} = -\dot{\partial}_{h} N_{ji} \tag{7.6}$$

are d-tensor fields.

9.8 The almost product structure $\mathbb{P}$. The almost contact structure $\mathbb{F}$.

Assuming that a nonlinear connection N is given, we define a $\mathcal{F}(T^{*2}M)$–linear mapping

$$\mathbb{P} : \mathcal{X}(T^{*2}M) \longrightarrow \mathcal{X}(T^{*2}M),$$

by defined

$$\mathbb{P}(X^{H}) = X^{H}, \ \mathbb{P}(X^{W_1}) = -X^{W_1}, \ \mathbb{P}(X^{W_2}) = -X^{W_2}, \quad \forall X \in \mathcal{X}(T^{*2}M). \tag{8.1}$$

We have also,

$$\begin{cases} \mathbb{P} \circ \mathbb{P} = I, \\ \mathbb{P} = I - 2(w_1 + w_2) = 2h - I, \\ \text{rank}\mathbb{P} = 3n. \end{cases} \tag{8.2}$$

Theorem 9.8.1. *A nonlinear connection N on $T^{*2}M$ is characterized by the existence of an almost product structure $\mathbb{P}$ on $T^{*2}M$ whose eigenspaces corresponding to the eigenvalue -1 coincide with the linear spaces of the vertical distribution V on E.*

Proof. If N is given, then we have the direct sum (5.1). Denoting by h and v the supplementary projectors determined by (5.1) we have $\mathbb{P} = I - 2v$ with the properties $\mathbb{P}(X^{H}) = X^{H}$, $\mathbb{P}(X^{V}) = -X^{V}$. So, the imposed condition is verified. Conversely, if $\mathbb{P}^2 = I$ and $\mathbb{P}(X^{V}) = -X^{V}$, then let $v = \frac{1}{2}(I - \mathbb{P})$ and $h = \frac{1}{2}(I + \mathbb{P})$. We verify easy that $h + v = I$. So $N = \text{Ker}v$. It follows $N \oplus V = TT^{*2}M$. **q.e.d.**

Proposition 9.8.1. *The almost product structure* $\mathbb{P}$ *is integrable if, and only if, the horizontal distribution* N *is integrable.*

Proof. The Nijenhuis tensor of the structure $\mathbb{P}$

$$N_{\mathbb{P}}(X,Y) = \mathbb{P}^2[X,Y] + [\mathbb{P}X, \mathbb{P}Y] - \mathbb{P}[\mathbb{P}X, Y] - \mathbb{P}[X, \mathbb{P}Y] \tag{8.3}$$

gives us for $X = X^H,\ Y = Y^H$

$$\begin{aligned}
N_{\mathbb{P}}(X^H, Y^H) &= 2[X^H, Y^H] - 2\mathbb{P}[X^H, Y^H] = 2(I - \mathbb{P})[X^H, Y^H] = 4v[X^H, Y^H] \\
N_{\mathbb{P}}(X^H, Y^V) &= [X^H, Y^V] - [X^H, Y^H] - \mathbb{P}[X^H, Y^V] + \mathbb{P}[X^H, Y^V] = 0 \\
N_{\mathbb{P}}(X^V, Y^V) &= [X^V, Y^V] + [X^V, Y^V] + \mathbb{P}[X^V, Y^V] + \mathbb{P}[X^V, Y^V] = \\
&= 2(I + \mathbb{P})[X^V, Y^V] = 4h[X^V, Y^V] = 0.
\end{aligned}$$

Therefore $N_{\mathbb{P}} = 0$ if and only if $[X^H, Y^H]^V = 0$. But $[X^H, Y^H]^V = 0$, $\forall\, X, Y \in \mathcal{X}(\widetilde{T^{*2}M})$ allows to say that the horizontal distribution N is integrable.

The nonlinear connection N being fixed we have the direct decomposition (5.1), (5.2) and the corresponding adapted basis (5.4).

Let us consider the $\mathcal{F}(\widetilde{T^{*2}M})$–linear mapping:

$$\mathbb{F} : \mathcal{X}(\widetilde{T^{*2}M}) \longrightarrow \mathcal{X}(\widetilde{T^{*2}M}),$$

determined by

$$\mathbb{F}(\delta_i) = -\dot{\partial}_i,\ \ \mathbb{F}(\dot{\partial}_i) = \delta_i,\ \ \mathbb{F}(\dot{\partial}^i) = 0. \tag{8.4}$$

Then, we deduce:

Theorem 9.8.2. *The mapping* $\mathbb{F}$ *has the following properties:*

1°. *It is globally defined on* $\widetilde{T^{*2}M}$.

2°. $\mathbb{F}$ *is a tensor field of type* $(1,1)$.

3°. $\mathrm{Ker}\mathbb{F} = W_2,\ \mathrm{Im}\mathbb{F} = N \oplus W_1$.

4°. $\mathrm{rank}||\mathbb{F}|| = 2n$.

5°. $\mathbb{F}^3 + \mathbb{F} = 0$.

Proof.

1°. Taking into account (5.4) we have $\frac{\delta\widetilde{x}^i}{\delta x^j}\mathbb{F}\left(\frac{\delta}{\delta x^i}\right) = -\frac{\partial\widetilde{x}^i}{\partial x^j}\frac{\partial}{\partial y^i}$ implies $\mathbb{F}\left(\frac{\delta}{\delta\widetilde{x}^j}\right) = -\frac{\partial}{\partial\widetilde{y}^j}\cdot$ Also, $\frac{\partial\widetilde{x}^i}{\partial x^j}\mathbb{F}\left(\frac{\partial}{\partial y^i}\right) = \frac{\partial\widetilde{x}^i}{\partial x^j}\frac{\delta}{\delta x^i}$ and $\frac{\partial\widetilde{x}^j}{\partial x^i}\mathbb{F}\left(\frac{\partial}{\partial p_i}\right) = 0$, lead to $\mathbb{F}\left(\frac{\partial}{\partial\widetilde{y}^j}\right) = \frac{\delta}{\delta\widetilde{x}^j}$ and $\mathbb{F}\left(\frac{\partial}{\partial\widetilde{p}_i}\right) = 0$.

2°. $\mathbb{F}$ is $\mathcal{F}(\widetilde{T^{*2}M})$–linear mapping from $\mathcal{X}(\widetilde{T^{*2}M})$ to $\mathcal{X}(\widetilde{T^{*2}M})$.

3°. $\mathbb{F}\left(\frac{\partial}{\partial p_i}\right) = 0$ implies $\mathbb{F}_{|W_2}$ is trivial and $\mathbb{F}(N \oplus W_1 \oplus W_2) = N \oplus W_1$.

4°. Evidently, by means of 3°.

5°. $\mathbb{F}^2(X^H) = \mathbb{F}(-X^{W_1}) = -X^H$; $\mathbb{F}^3(X^H) = X^{W_1}$ and $\mathbb{F}(X^H) = -X^{W_1}$. So $(\mathbb{F}^3 + \mathbb{F})X^H = 0$, $\forall\, X^H \in N$ and $(\mathbb{F}^3 + \mathbb{F})(X^{W_1}) = 0$, $(\mathbb{F}^3 + F)(X^{W_2}) = 0$.

We can say that $\mathbb{F}$ is a *natural almost contact structure* determined by the nonlinear connection N.

The Nijenhuis tensor of the structure $\mathbb{F}$ is given by:

$$\mathcal{N}_{\mathbb{F}}(X,Y) = \mathbb{F}^2[X,Y] + [\mathbb{F}X, \mathbb{F}Y] - \mathbb{F}[\mathbb{F}X, Y] - \mathbb{F}[X, \mathbb{F}Y]$$

and the normality condition of $\mathbb{F}$ reads as follows:

$$\mathcal{N}_{\mathbb{F}}(X,Y) + \sum_{i=1}^{n} d(\delta p_i)(X,Y) = 0,\ \forall\, X, Y \in \mathcal{X}(\widetilde{T^{*2}M}). \tag{8.5}$$

Of course, in the adapted basis, using the formula (7.3) we can obtainthe explicit form of the equation (8.5).

9.9 The Riemannian structures on $\widetilde{T^{*2}M}$

Let us consider a Riemannian structure $\mathbb{G}$ on the manifold $\widetilde{T^{*2}M}$.

The following problem is arises: *Can the Riemannian structure* $\mathbb{G}$*determine a nonlinear connection* N *on* $\widetilde{T^{*2}M}$*? Also, can* $\mathbb{G}$ *determine a dual semispray* S_ξ *on* $\widetilde{T^{*2}M}$ *?*

In order to determine a nonlinear connection on $\widetilde{T^{*2}M}$ by means of $\mathbb{G}$, it is sufficient to determine a distribution N orthogonal to the vertical distribution V. The solution is immediate. Namely, it is important to determine the coefficients ${N^i}_j$ and N_{ij} of N.

In the natural basis, $\mathbb{G}$ is given locally by

$$\mathbb{G} = \overset{(11)}{g}_{ij} dx^i \otimes dx^j + \overset{(12)}{g}_{ij} dx^i \otimes dy^j + \overset{(13)}{g}_i{}^j dx^i \otimes dp_j + \tag{9.1}$$
$$+ \cdots + \overset{(33)}{g}{}^{ij} dp_i \otimes dp_j,$$

where the matrix $\left\| \overset{(\alpha\beta)}{g} \right\|$ is positively defined.

Let $\{\delta_i\}$, $(i = 1, ..., n)$, be the adapted basis of N :

$$\frac{\delta}{\delta x^i} = \partial_i - N^j{}_i \dot{\partial}_j + N_{ij} \dot{\partial}^j. \tag{9.2}$$

The following conditions of orthogonality between N and V:

$$\mathbb{G}(\delta_i, \dot{\partial}_i) = 0, \ \mathbb{G}(\delta_i, \dot{\partial}^j) = 0, \ (i, j = 1, ..., n) \tag{9.3}$$

give us the following system of equations for determining the coefficients $N^i{}_j$ and N_{ij}:

$$\overset{(22)}{g}_{mj} N^m{}_i + \overset{(32)}{g}{}^m_j N_{im} = \overset{(12)}{g}_{ij}, \tag{9.4}$$
$$\overset{(23)}{g}_m{}^j N^m{}_i + \overset{(33)}{g}{}^{mj} N_{im} = \overset{(12)}{g}_i{}^j,$$

where, the matrix

$$\begin{pmatrix} \overset{(22)}{g}_{mj} & \overset{(32)}{g}{}^{mj} \\ \overset{(23)}{g}_m{}^j & \overset{(33)}{g}{}^{mj} \end{pmatrix} \tag{9.4$'$}$$

is nonsingular.

Therefore the system (9.4) has an unique solution.

Whether, take into account the rule of transformation of the coefficients $\overset{(\alpha\beta)}{g}_{ij}$ from $\mathbb{G}$ we can prove that the solution $(N^i{}_j, N_{ij})$ of (9.4) has the rule of transformation (3.10), by means of the transformations of local coordinates on $\widetilde{T^{*2}M}$. Consequently, we have:

Theorem 9.9.1. *A Riemannian structure* $\mathbb{G}$ *on* $\widetilde{T^{*2}M}$ *determines uniquely a nonlinear connection* N*, if the distribution of* N *is orthogonal to the vertical distribution* V*. The coefficients* $N^i{}_j, N_{ij}$ *of* N *are given by the system of equations* (9.4).

Remarking that $f_i = N_{ij} y^j$ are the coefficients of a dual spray S_ξ, we have:

Theorem 9.9.2. *A Riemannian structure* $\mathbb{G}$ *on* $\widetilde{T^{*2}M}$ *determines a dual semispray* S_ξ *with the coefficients*

$$f_i = N_{ij}y^j,$$

N_{ij} *being determined by the system* (9.4).

Let $\mathbb{F}$ be the natural almost contact structure determined by the previous nonlinear connection N.

The following problem arises: *When will the pair* $(\mathbb{G}, \mathbb{F})$ *is a Riemannian almost contact structure?*

Of course, it is necessary to have:

$$\mathbb{G}(\mathbb{F}X, Y) = -\mathbb{G}(X, \mathbb{F}Y), \ \forall X, Y \in \mathcal{X}(\widetilde{T^{*2}M}).$$

Consequently, we get:

Theorem 9.9.3. *The pair* $(\mathbb{G}, \mathbb{F})$ *is a Riemannian almost contact structure if and only if in the adapted basis determined by* N *and* V *the tensor* $\mathbb{G}$ *has the form*

$$\mathbb{G} = g_{ij}dx^i \otimes dx^j + g_{ij}\delta y^i \otimes \delta y^j + h^{ij}\delta p_i \otimes \delta p_j. \tag{9.5}$$

Corollary 9.9.1. *With respect to the Riemannian structure* (9.5) *the distributions* N, W_1, W_2 *are orthogonal respectively.*

Remarks.

1° The form (9.5) will be used to define a lift to $\widetilde{T^{*2}M}$ of a metric structure given only by a nonsingular and symmetric d–tensor field g_{ij}. Namely, we have

$$\mathbb{G} = g_{ij}dx^i \otimes dx^j + g_{ij}\delta y^i \otimes \delta y^j + g^{ij}\delta p_i \otimes \delta p_j. \tag{9.5$'$}$$

These problemes will be studied in a next chapter.

2° Using the metric $\mathbb{G}$ on $\widetilde{T^{*2}M}$ we can introduce a new almost contact structure $\check{\mathbb{F}}$, defined by

$$\check{\mathbb{F}}(\delta_i) = -g_{ij}\dot{\partial}^j, \ \check{\mathbb{F}}(\dot{\partial}_i) = 0, \ \check{\mathbb{F}}(g_{ij}\dot{\partial}^j) = \delta_i \tag{9.6}$$

We will prove that $(\mathbb{G}, \check{\mathbb{F}})$ is a Riemannian almost contact structure and its associated 2-form $\hat{\theta}$ is given by

$$\hat{\theta} = \delta p_i \wedge dx^i.$$

The pair $(\mathbb{G}, \check{\mathbb{F}})$ will be studied in the Chapter 11 about the generalized Hamilton spaces of order 2.

Chapter 10

Linear connections on the manifold $T^{*2}M$

The main topics of this chapter is to show that there are the linear connection compatible to the direct decomposition (5.2) determined by a nonlinear connection N, on the total space of the dual bundle $(T^{*2}M, \pi^{*2}, M)$.

We are going to study the distinguished Tensor Algebra (or d–Tensor Algebra), N-linear connections, torsions and curvatures, structure equations, autoparallel curves, etc.

10.1 The d–Tensor Algebra

Let N be a nonlinear connection on $T^{*2}M$. Then N determines the direct decomposition (5.2), Ch.9. With respect to (5.2), Ch.9, a vector field X and an one form ω can be uniquely written in the form (6.3) and (6.7), Ch.9, respectively, i.e.

$$X = X^H + X^{W_1} + X^{W_2},$$
$$\omega = \omega^H + \omega^{W_1} + \omega^{W_2}. \tag{1.1}$$

Definition 10.1.1. A distinguished tensor field (briefly: d-tensor field) on $T^{*2}M$ of type (r, s) is a tensor field T of type (r, s) on $T^{*2}M$ with the property:

$$T(\overset{1}{\omega}, ..., \overset{r}{\omega}, \underset{1}{X}, ..., \underset{s}{X}) = T(\overset{1}{\omega}^H, ..., \overset{r}{\omega}^{W_2}, \underset{1}{X}^H, ..., \underset{s}{X}^{W_2}), \tag{1.2}$$

for any $(\overset{1}{\omega}, ..., \overset{r}{\omega}) \in \mathcal{X}^*(T^{*2}M)$ and for any $(\underset{1}{X}, ..., \underset{s}{X}) \in \mathcal{X}(T^{*2}M)$.

For instance, every component X^H, X^{W_1} and X^{W_2} of a vector field $X \in \mathcal{X}(T^{*2}M)$ is a d-vector field.

Also, every component ω^H, ω^{W_1} and ω^{W_2} of the 1-form $\omega \in \mathcal{X}^*(T^{*2}M)$ is a d-1-form.

In the adapted basis $\left(\delta_i, \dot{\partial}_i, \dot{\partial}^i\right)$ and cobasis $(dx^i, \delta y^i, \delta p_i)$ to the direct decomposition (5.2), a d-tensor field T of type (r,s) can be written in the form:

$$T = T^{i_1 \dots i_r}_{j_1 \dots j_s}(x,y,p)\delta_{i_1} \otimes \cdots \dot{\partial}^{j_s} \otimes dx^{j_1} \otimes \cdots \otimes \delta p_{i_r}. \tag{1.3}$$

It follows that the set $\left\{1, \delta_i, \dot{\partial}_i, \dot{\partial}^i\right\}$ generates the algebra of the d-tensor fields over the ring of functions $\mathcal{F}(T^{*2}M)$.

For example, if $f \in \mathcal{F}(\widetilde{T^{*2}M})$, then $\dfrac{\delta f}{\delta x^i} = \delta_i f$, $\dfrac{\partial f}{\partial y^i} = \dot{\partial}_i f$ are d-1-covectors and $\dfrac{\partial f}{\partial p_i} = \dot{\partial}^i f$ is a d-vector field.

Clearly, with respect to a local transformation a coordinates on $T^{*2}M$, the coefficients $T^{i_1 \dots i_r}_{j_1 \dots j_s}(x,y,p)$ of a d-tensor fields are transformed by the classical rule:

$$\widetilde{T}^{i_1 \dots i_r}_{j_1 \dots j_s} = \frac{\partial \widetilde{x}^{i_1}}{\partial x^{h_1}} \cdots \frac{\partial \widetilde{x}^{i_r}}{\partial x^{h_r}} \frac{\partial x^{k_1}}{\partial \widetilde{x}^{j_1}} \cdots \frac{\partial x^{k_r}}{\partial \widetilde{x}^{j_s}} T^{h_1 \dots h_r}_{k_1 \dots k_s}. \tag{1.3$'$}$$

10.2 $\mathcal{N}$-linear connections

The notion of N-linear connection will be defined in the known manner [97]:

Definition 10.2.1. A linear connection D on $T^{*2}M$ is called an N-linear connection, if:

(1) D preserves by parallelism distributions N, W_1 and W_2.

(2) The 2-tangent structure J is absolute parallel with respect to D.

(3) The presymplectic structure θ is absolute parallel with respect to D.

Starting from this definition, any N-linear connection is characterized by the following:

Theorem 10.2.1. *A linear connection D is an N–linear connection on $T^{*2}M$ if and only if:*

(1) *D preserves by parallelism every of distributions N, W_1, W_2.*

(2) $D_X(JY^H) = J(D_XY^H)$, $D_X(JY^{W_\alpha}) = J(D_XY^{W_\alpha})$, $(\alpha = 1,2)$, $\forall X, Y \in \mathcal{X}(T^{*2}M)$.

(3) $D\theta = 0$.

The proof is similar with the case given in the book [106].

We remark that $D_X(JY^{W_\alpha}) = J(D_XY^{W_\alpha})$ is trivial, because $JY^{W_\alpha} = 0$, $\forall Y \in \mathcal{X}(T^{*2}M)$ and that by means of the property (1), it follows $J(D_XY^{W_\alpha}) = 0$.

We obtain also:

Theorem 10.2.2. *For any N–linear connection$D$ we have*

$$D_Xh = D_Xw_1 = D_Xw_2 = 0, \tag{2.1}$$

$$D_X\mathbb{P} = 0, \;\; D_X\mathbb{F} = 0. \tag{2.2}$$

Indeed, from $(D_Xh)Y = D_X(hY)-h(D_XY)$ if $Y = Y^H$, and $Y = Y^{W_\alpha}$, $(\alpha = 1,2)$ we obtain $D_Xh = 0$. Similarly, we get $D_Xw_1 = 0$, $D_Xw_2 = 0$.

Now, taking into account the expression (8.2), Ch.IX, of $\mathbb{P}$ it follows $D_X\mathbb{P} = 0$. The last equality $D_X\mathbb{F} = 0$ can be proved in a similar way.

Let us consider a vector field $X \in \mathcal{X}(T^{*2}M)$, written in the form (1.1). It follows, from the property of an N–linear connection that

$$D_XY = D_{X^H}Y + D_{X^{W_1}}Y + D_{X^{W_2}}Y, \; \forall X, Y \in \mathcal{X}(T^{*2}M). \tag{2.3}$$

We can introduce new operators of derivation in the d-tensor algebra, defined by:

$$D_X^H = D_{X^H}, \; D_X^{W_1} = D_{X^{W_1}}, \; D_X^{W_2} = D_{X^{W_2}}. \tag{2.4}$$

These operators are not the covariant derivations in the d-tensor algebra, since $D_X^Hf = X^Hf \neq Xf$ (etc.). However they have similar properties with the covariant derivatives.

From (2.3) and (2.4) we deduce

$$D_XY = D_X^HY + D_X^{W_1}Y + D_X^{W_2}Y, \; \forall X, Y \in \mathcal{X}(T^{*2}M). \tag{2.5}$$

By means of Theorem 10.2.2, the action of the operator D_X^H on the d-vector fields Y^{W_1}, Y^{W_2} is the same as its action on the d-vectors Y^H. This property holds for the operators $D_X^{W_1}$ and $D_X^{W_2}$, too.

Theorem 10.2.3. *The operators $D_X^H, D_X^{W_1}, D_X^{W_2}$ have the following properties:*

1) *Every $D_X^H, D_X^{W_1}, D_X^{W_2}$ maps a vector field belonging to one of distributions N, W_1, W_2 in a vector field belonging to the same distribution.*

2) $D_X^Hf = X^Hf, \; D_X^{W_1}f = X^{W_1}f, \; D_X^{W_2}f = X^{W_2}f.$

3) $D_X^H(fY) = X^HfY + fD_X^HY; \; D_X^{W_\alpha}fY = X^{W_\alpha}fY + D_X^{W_\alpha}Y.$

4) $D_X^H(Y+Z) = D_X^HY + D_X^HZ; \; D_X^{W_\alpha}(Y+Z) = D_X^{W_\alpha}Y + D_X^{W_\alpha}Z.$

5) $D^H_{X+Y} = D^H_X + D^H_Y;\ D^{W_\alpha}_{X+Y} = D^{W_\alpha}_X + D^{W_\alpha}_Y$.

6) $D^H_{fX} = fD^H_X;\ D^{W_\alpha}_{fX} = fD^{W_\alpha}_X$.

7) $D^H_X(JY) = JD^H_X Y;\ D^{W_\alpha}_X(JY) = JD^{W_\alpha}_X Y,\ (\forall\, \alpha = 1,2)$.

8) $D^H_X\theta = 0,\ D^{W_\alpha}_X\theta = 0,\ (\alpha = 1,2)$, *$\theta$ being the presymplectic structure from Theorem 6.7.1.*

9) *The operators $D^H_X, D^{W_1}_X, D^{W_2}_X$ have the property of localization on the manifold $T^{*2}M$, i.e. $(D^H_X Y)_{|U} = D^H_{X_{|U}} Y_{|U}$ etc. for any open set $U \subset T^{*2}M$.*

The proof of the previous theorem can be done by the classical methods [106].

The operators $D^H_X, D^{W_1}_X, D^{W_2}_X$ will be called the operators of $h-$, w_1- and w_2-covariant derivation.

The actions of these operators over the 1-form fields ω, on $T^{*2}M$, are given by

$$(2.6)\qquad \begin{aligned} (D^H_X\omega)(Y) &= X^H\omega(Y) - \omega(D^H_X Y), \\ (D^{W_1}_X\omega)(Y) &= X^{W_1}\omega(Y) - \omega(D^{W_1}_X Y), \\ (D^{W_2}_X\omega)(Y) &= X^{W_2}\omega(Y) - \omega(D^{W_2}_X Y). \end{aligned}$$

Of course, the action of the previous operators can be extended to any tensor field, particularly to any d-tensor field on $T^{*2}M$.

Now, let us consider a parametrized smooth curve $\gamma : t \subset I \longrightarrow \gamma(t) \in \widetilde{T^{*2}M}$, having the image in a domain of a local chart.

Its tangent vector field $\dot\gamma = \dfrac{d\gamma}{dt}$ can be uniquely written in the form

$$(2.7)\qquad \dot\gamma = \dot\gamma^H + \dot\gamma^{W_1} + \dot\gamma^{W_2}$$

In the case when γ is analytically given by the equation (6.12), Ch.9, then $\dot\gamma^H, \dot\gamma^{W_1}, \dot\gamma^{W_2}$ are given by (6.13), Ch.9. And we can define the horizontal curve.

A vector field Y defined along the curve γ has the covariant derivative

$$D_{\dot\gamma}Y = D^H_{\dot\gamma}Y + D^{W_1}_{\dot\gamma}Y + D^{W_2}_{\dot\gamma}Y.$$

The vector field $Y(u(\gamma))$ is called parallel along the curve γ if

$$D_{\dot\gamma}Y = 0.$$

In particular, the curve γ is autoparallel with respect to an N–linear connection D if $D_{\dot\gamma}\dot\gamma = 0$.

In a next section we will study these notions by means of adapted basis.

10.3 Torsion and curvature

The torsion $\mathbb{T}$ of an N-linear connection D is expressed, as usually, by

$$\mathbb{T}(X,Y) = D_X Y - D_Y X - [X,Y]. \tag{3.1}$$

It can be characterized by the vector fields

$$\mathbb{T}(X^H, Y^H), \mathbb{T}(X^H, Y^{W_a}), \mathbb{T}(X^{W_a}, Y^{W_b}), \ (a,b=1,2).$$

Taking the $h-$ and w_α–components we obtain the torsion *d-tensors*

$$\begin{aligned} \mathbb{T}(X^H,Y^H) &= h\mathbb{T}(X^H,Y^H) + \sum_{a=1}^{2} w_a \mathbb{T}(X^H,Y^H), \\ \mathbb{T}(X^H,Y^{W_a}) &= h\mathbb{T}(X^H,Y^{W_a}) + \sum_{b=1}^{2} w_b \mathbb{T}(X^H,Y^{W_a}), \\ \mathbb{T}(X^{W_a},Y^{W_b}) &= h\mathbb{T}(X^{W_a},Y^{W_b}) + \sum_{c=1}^{2} w_c \mathbb{T}(X^{W_a},Y^{W_b}). \end{aligned} \tag{3.2}$$

Since D preserves by parallelism the distributions H, W_1, W_2 and the distributions W_1, W_2 are integrable it follows

Proposition 10.3.1. *The following property of the torsion $\mathbb{T}$ holds:*

$$h\mathbb{T}(X^{W_a}, Y^{W_b}) = 0, \ (a,b=1,2). \tag{3.3}$$

Now we can express, without difficulties, the torsion d-tensors by means of the formula (3.1).

The curvature of D is given by

$$\mathbb{R}(X,Y)Z = (D_X D_Y - D_Y D_X)Z - D_{[X,Y]}Z, \ \forall X,Y,Z \in \mathcal{X}(T^{*2}M). \tag{3.4}$$

We will express $\mathbb{R}$ by means of the components (2.5), taking into account the decomposition (1.1) for the vector fields on $T^{*2}M$.

Proposition 10.3.2. *For any vector fields $X,Y,Z \in \mathcal{X}(T^{*2}M)$ the following properties holds:*

$$J(\mathbb{R}(X,Y)Z) = \mathbb{R}(X,Y)JZ; \ D_X\theta = 0. \tag{3.5}$$

The previous properties have an important consequence:

Corollary 10.3.1.

1° *The essential components of the curvature tensor field* $\mathbb{R}$ *are* $\mathbb{R}(X,Y)Z^H$, $\mathbb{R}(X,Y)Z^{W_2}$ *and* $R(X,Y)Z^{W_2}$.

2° *The vector field* $\mathbb{R}(X,Y)Z^H$ *is horizontal.*

3° *The vector field* $\mathbb{R}(X,Y)Z^{W_a}$ $(a=1,2)$ *belongs to the distribution* W_a.

4° *The following properties hold*

$$w_a[\mathbb{R}(X,Y)Z^H]=0, \quad h[\mathbb{R}(X,Y)Z^{W_a}]=0,$$
$$w_b[\mathbb{R}(X,Y)Z^{W_a}]=0, \quad (a\neq b,\ a,b=1,2). \tag{3.6}$$

Of course, we can express the *d-tensors* of curvature by means of the operators of $h-$, w_1-, w_2-covariant derivatives $(2.5)''$.

From (3.4) we get the following Ricci identities

$$\begin{aligned}[D_X,D_Y]Z^H &= \mathbb{R}(X,Y)Z^H+D_{[X,Y]}Z^H\\ [D_X,D_Y]Z^{W_a} &= \mathbb{R}(X,Y)Z^{W_a}+D_{[X,Y]}Z^{W_a}\ (a=1,2).\end{aligned}\tag{3.7}$$

As a consequence, we obtain:

Theorem 10.3.1. *For any N-linear connection D there are the following identities*

$$\begin{aligned}[D_X,D_Y]\mathbb{C} &= \mathbb{R}(X,Y)\mathbb{C}-D_{[X,Y]}\mathbb{C}\\ [D_X,D_Y]\mathbb{C}^* &= \mathbb{R}(X,Y)\mathbb{C}^*-D_{[X,Y]}\mathbb{C}^*\end{aligned}\tag{3.8}$$

where $\mathbb{C}$ *is the Liouville vector field, and* $\mathbb{C}^*$ *is the Hamilton vector field on the manifold* $T^{*2}M$.

Using the previous considerations we can express the Bianchi identities of the N-linear connection D, by means of the operators $D^H_X, D^{W_1}_X, D^{W_2}_X$ taking into account the classical Bianchi identities

$$\sum_{(XYZ)}\{(D_X\mathbb{T})(Y,Z)-\mathbb{R}(X,Y)Z+\mathbb{T}(\mathbb{T}(X,Y),Z\}=0,$$

$$\sum_{(XYZ)}\{(D_X\mathbb{R})(U,Y,Z)-\mathbb{R}(\mathbb{T}(X,Y),Z)U\}=0,$$

where $\sum\limits_{(XYZ)}$ means the cyclic sum.

10.4 The coefficients of an N-linear connection

An N–linear connection is characterized by its coefficientsin the adapted basis

$$\delta_i = \frac{\delta}{\delta x^i},\ \dot{\partial}_i = \frac{\partial}{\partial y^i},\ \dot{\partial}^i = \frac{\partial}{\partial p_i}.$$

These coefficients obey particular rules of transformation with respect to the changing of local coordinates on the manifold $T^{*2}M$.

Taking into account Proposition 10.3.2, we can prove the following theorem:

Theorem 10.4.1.

1° *An N-linear connection D can be uniquely represented, in the adapted basis $(\delta_i, \dot{\partial}_i, \dot{\partial}^i)$ in the following form:*

$$(4.1)\qquad \begin{cases} D_{\delta_j}\delta_i = H^k_{ij}\delta_k, & D_{\delta_j}\dot{\partial}_i = H^k_{ij}\dot{\partial}_k, \quad D_{\delta_j}\dot{\partial}^i = -H^i_{kj}\dot{\partial}^k,\\ D_{\dot{\partial}_j}\delta_i = C^k_{ij}\delta_k, & D_{\dot{\partial}_j}\dot{\partial}_i = C^k_{ij}\dot{\partial}_k, \quad D_{\dot{\partial}_j}\dot{\partial}^i = -C^i_{kj}\dot{\partial}^k,\\ D_{\dot{\partial}^j}\delta_i = C^{kj}_i\delta_k, & D_{\dot{\partial}^j}\dot{\partial}_i = C^{kj}_i\dot{\partial}_k, \quad D_{\dot{\partial}^j}\dot{\partial}^j = -C^{ij}_k\dot{\partial}^k. \end{cases}$$

2° *With respect to the coordinate transformation* (1.1), *Ch.6, the coefficients $H^i_{jk}(x, y, p)$ obey the rule of transformation:*

$$(4.2)\qquad \widetilde{H}^i_{rs}\frac{\partial x^r}{\partial \widetilde{x}^j}\frac{\partial x^s}{\partial \widetilde{x}^k} = \frac{\partial \widetilde{x}^i}{\partial x^r}H^r_{jk} - \frac{\partial^2 \widetilde{x}^i}{\partial x^j \partial x^k}.$$

3° *The coefficients $C^i_{jk}(x, y, p)$ and $C^{jk}_i(x, y, p)$ are d-tensor fields of type $(1,2)$ and $(2,1)$, respectively.*

Indeed, putting

$$D_{\delta_j}\delta_i = \underset{(0)}{H}{}^k{}_{ij}\delta_k,\ D_{\delta_j}\dot{\partial}_i = \underset{(1)}{H}{}^k{}_{ij}\dot{\partial}_k,\ D_{\delta_j}\dot{\partial}^i = -\underset{(2)}{H}{}^i{}_{kj}\dot{\partial}^k,$$

and taking into account Theorem 10.2.3, it follows

$$\underset{(0)}{H}{}^k{}_{ij} = \underset{(1)}{H}{}^k{}_{ij} = \underset{(2)}{H}{}^k{}_{ij}.$$

The statements 2° and 3° can be proved by a direct calculus, taking into account the rule of transformations (5.4)', Ch.9, for $\delta_i, \dot{\partial}_i$ and $\dot{\partial}^i$.

The system of functions

$$(4.3)\qquad D\Gamma(N) = \{H^i_{jk}, C^i_{jk}, C^{jk}_i\}$$

are called the coefficients of the N–linear connection D.

The inverse statement of the previous theorem holds also.

Theorem 10.4.2. *If the systems of functions (4.3) are apriori given over every domain of local chart on the manifold $T^{*2}M$, having the rule of transformation mentioned in the previous theorem, then there exists a unique N–linear connection D whose coefficients are just the systems of given functions.*

Corollary 10.4.1. *The following formula hold:*

$$D_{\delta_j} dx^i = -H^i_{kj} dx^k, \quad D_{\delta_j} \delta y^i = -H^i_{kj} \delta y^k, \quad D_{\delta_j} \delta p_i = H^k_{ij} \delta p_k,$$

$$D_{\dot{\partial}_j} dx^i = -C^i_{kj} dx^k, \quad D_{\dot{\partial}_j} \delta y^i = -C^i_{kj} \delta y^k, \quad D_{\dot{\partial}_j} \delta p_i = C^k_{ij} \delta p_k, \tag{4.4}$$

$$D_{\dot{\partial}^j} dx^i = -C^{ij}_k dx^k, \quad D_{\dot{\partial}^j} \delta y^i = -C^{ij}_k \delta y^k, \quad D_{\dot{\partial}^j} \delta p_i = C^{kj}_i \delta p_k.$$

Indeed, the formula (4.1), the condition of duality between $(\delta_j, \dot{\partial}_j, \dot{\partial}^j)$ and $(dx^i, \delta y^i, \delta p_i)$ leads to the formula (4.4).

10.5 The h–, w_1–, w_2–covariant derivatives in local adapted basis

Let us consider a d–tensor field T, of type (r, s) in the adapted basis $(\delta_j, \dot{\partial}_j, \dot{\partial}^j)$ and its dual (see (1.3)):

$$T = T^{i_1 \dots i_r}_{j_1 \dots j_s} \delta_{i_1} \otimes \dots \otimes \dot{\partial}^{j_s} \otimes dx^{j_1} \otimes \dots \otimes \delta p_{i_r}. \tag{5.1}$$

For $X = X^H = X^i \delta_i$, applying (4.1), (4.4) and using the properties of the operator D^H_X we deduce:

$$D^H_X T = X^m T^{i_1 \dots i_r}_{j_1 \dots j_s | m} \delta_{i_1} \otimes \dots \otimes \dot{\partial}^{j_s} \otimes dx^{j_1} \otimes \dots \otimes \delta p_{i_r} \tag{5.2}$$

where

$$\begin{aligned} T^{i_1 \dots i_r}_{j_1 \dots j_s | m} &= \delta_m T^{i_1 \dots i_r}_{j_1 \dots j_s} + T^{k i_2 \dots i_r}_{j_1 \dots j_s} H^{i_1}_{km} + \dots + T^{i_1 \dots i_{r-1} k}_{j_1 \dots j_s} H^{i_r}_{km} - \\ &- T^{i_1 \dots i_r}_{k j_2 \dots j_s} H^k_{j_1 m} - \dots - T^{i_1 \dots i_r}_{j_1 \dots j_{s-1} k} H^k_{j_s m}. \end{aligned} \tag{5.2'}$$

The operator "$_{|m}$" is called h–covariant derivative with respect to $D\Gamma(N)$.

Now, putting $X = X^{W_1} = X^i \dot{\partial}_i$ we obtain for the d-tensor field T from (5.1), the formula:

$$D^{W_1}_X T = X^m T^{i_1 \dots i_r}_{j_1 \dots j_s} |_m \delta_{i_1} \otimes \dots \otimes \dot{\partial}^{j_s} \otimes dx^{j_1} \otimes \dots \otimes \delta p_{i_r}, \tag{5.3}$$

where

$$(5.3)' \quad \begin{aligned} T^{i_1\dots i_r}_{j_1\dots j_s}\big|_m &= \dot{\partial}_m T^{i_1\dots i_r}_{j_1\dots j_s} + T^{ki_2\dots i_r}_{j_1\dots j_s} C^{i_1}_{km} + \dots + T^{i_1\dots k}_{j_1\dots j_s} C^{i_r}_{km} - \\ &- T^{i_1\dots i_r}_{kj_2\dots j_s} C^k_{j_1m} - \dots - T^{i_1\dots i_r}_{j_1\dots k} C^k_{j_sm}. \end{aligned}$$

The operator "$|_m$" will be called "the w_1–covariant derivative" with respect to $D\Gamma(N)$.

Finally, taking $X = X^{W_2} = X_i \dot{\partial}^i$, then $D^{W_1}_X T$ has the following form

$$(5.4) \qquad D^{W_2}_X T = X_m T^{i_1\dots i_r}_{j_1\dots j_s}|^m \delta_{i_1} \otimes \dots \otimes \dot{\partial}^{j_s} \otimes dx^{j_1} \otimes \dots \otimes \delta p_{i_r}.$$

where

$$(5.4)' \quad \begin{aligned} T^{i_1\dots i_r}_{j_1\dots j_s}|^m &= \dot{\partial}^m T^{i_1\dots i_r}_{j_1\dots j_s} + T^{ki_2\dots i_r}_{j_1\dots j_s} C^{i_1m}_k + \dots + T^{i_1\dots k}_{j_1\dots j_s} C^{i_rm}_k - \\ &- T^{i_1\dots i_r}_{kj_2\dots j_s} C^{km}_{j_1} - \dots - T^{i_1\dots i_r}_{j_1\dots k} C^{km}_{j_s}. \end{aligned}$$

The operator "$|^m$" will be called *the w_2–covariant derivative.*

It is not difficult to prove:

Proposition 10.5.1. *The following properties hold:*

$$T^{i_1\dots i_r}_{j_1\dots j_s|m},\ T^{i_1\dots i_r}_{j_1\dots j_s}|_m T^{i_1\dots i_r}_{j_1\dots j_s}|^m$$

are d–tensor fields. The first two are of type $(r, s+1)$ and the last one is of type $(r+1, s)$.

Proposition 10.5.2. *The operators "$_{|m}$", "$|_m$", and "$|^m$" have the properties:*

1°. $f_{|m} = \delta_m f$, $f|_m = \dot{\partial}_m f$, $f|^m = \dot{\partial}^m f$, $\forall f \in \mathcal{F}(T^{*2}M)$.

2°. *They are distributive with respect to the addition of the d–tensor of the same type.*

3°. *They commute with the operation of contraction.*

4°. *They verify the Leibniz rule with respect to the tensor product.*

As an application let us consider "*the (y)–deflection tensor fields*"

$$(5.5) \qquad D^i{}_j = y^i{}_{|j},\ d^i{}_j = y^i|_j,\ d^{ij} = y^i|^j.$$

Proposition 10.5.3. *The (y)-deflection tensor field have the expression*

$$(5.5)' \qquad D^i{}_j = -N^i{}_j + y^m H_m{}^i{}_j,\ d^i{}_j = \delta^i{}_j + y^m C_m{}^i{}_j,\ d^{ij} = y^m C^{ij}_m.$$

These equalities are easy to prove, if one notice

$$y^i{}_{|j} = \delta_j y^i + y^m H_m{}^i{}_j,\ y^i|_j = \dot{\partial}_j y^i + y^m C_m{}^i{}_j,\ y^i|^j = \dot{\partial}^j y^i + y^m C^{ij}_m.$$

Now, we consider the so called "(p)-deflection tensor fields":

$$\Delta_{ij} = p_{i|j},\ \delta_{ij} = p_i|_j,\ \delta_i{}^j = p_i|^j. \tag{5.6}$$

Proposition 10.5.4. *The (p)-deflection tensors are given by*

$$\Delta_{ij} = N_{ij} - p_m H^m{}_{ij},\ \delta_{ij} = -p_m C^m{}_{ij},\ \delta_i{}^j = \delta^i{}_j - p_m C_i^{mj}. \tag{5.6'}$$

A particular class of the N–connection with the coefficients $D\Gamma(N)$ is given by the Berwald connectioin.

Definition 10.5.1. An N–linear connection D with the coefficients (4.3) is called a Berwald connection if its coefficients are:

$$H^i_{jk} = \dot{\partial}_j N^i{}_k,\ C^i_{jk} = 0,\ C_i{}^{jk} = 0. \tag{5.7}$$

This definition has a geometrical meaning if we take into account Proposition 9.7.4.

The existence of the Berwald connection is an interesting example of N–linear connection.

Remarking that the Berwald connection is uniquely determined by the nonlinear connection N and the fact that the nonlinear connection exists over a paracompact manifold $T^{*2}M$ (cf. Theorem 9.3.2), we can state:

Theorem 10.5.1. *If the base manifold M is paracompact, then on the manifold $T^{*2}M$ there exists the N–linear connections.*

Of course, the (y)–deflections and (p)–deflection tensor fields of the Berwald connection

$$B\Gamma(N) = (\dot{\partial}_j N^i{}_k, 0, 0) \tag{5.8}$$

are very particular.

We get

$$\begin{aligned} &D^i{}_j = -N^i{}_j + y^m \dot{\partial}_m N^i{}_j,\quad d^i{}_j = \delta^i{}_j,\quad d^{ij} = 0,\\ &\Delta_{ij} = N_{ij} - p_m \dot{\partial}_i N^m{}_j,\qquad \delta_{ij} = 0,\qquad \delta_i{}^j = \delta_i{}^j. \end{aligned} \tag{5.9}$$

Hence, $D^i{}_j = 0$ if and only if the coefficients $N^i{}_j$ are 1-homogeneous with respect to y^i.

10.6 Ricci identities. The local expressions of curvature and torsion.

In order to determine the local expressions of d–tensors of torsion and curvature of an N–linear connection we establish the Ricci identities applied to a d–vector field, using the covariant derivatives (4.5)′, (4.6)′ and (4.7)′.

Theorem 10.6.1. *For any N-linear connection D the following Ricci identities hold:*

$$(6.1)\quad \begin{cases} X^i{}_{|j|k} - X^i{}_{|k|j} = X^h R_h{}^i{}_{jk} - X^i{}_{|h} T^h{}_{jk} - X^i|_h \underset{(1)}{R}{}^h{}_{jk} - X^i|^h \underset{(2)}{R}{}_{hjk}, \\ X^i{}_{|j}|_k - X^i|_{k|j} = X^h P_h{}^i{}_{jk} - X^i_{|h} C^h{}_{jk} - X^i|_h \underset{(1)}{P}{}^h{}_{jk} - X^i|^h \underset{(2)}{B}{}_{hjk}, \\ X^i{}_{|j}|^k - X^i|^k{}_{|j} = X^h P_h{}^i{}_j{}^k - X^i{}_{|h} C_j{}^{hk} - X^i|_h \underset{(1)}{B}{}_j{}^{hk} - X^i|^h \underset{(2)}{P}{}^k{}_{hj}, \end{cases}$$

and

$$(6.2)\quad \begin{cases} X^i|_j|_k - X^i|_k|_j = X^h S_h{}^i{}_{jk} - X^i|_h S^h{}_{jk}, \\ X^i|_j|^k - X^i|^k|_j = X^h S_h{}^i{}_j{}^k - X^i|_h C^h{}_j{}^k - X^i|^h C_h{}^k{}_j, \\ X^i|^j|^k - X^i|^k|^j = X^h S_h{}^{ijk} - X^i|^h S_h{}^{jk}, \end{cases}$$

where the following tensors

$$(6.3)\quad \underset{(1)}{R}{}^i{}_{jk} = \delta_k N^i{}_j - \delta_j N^i{}_k,\ \underset{(2)}{R}{}_{ijk} = \delta_j N_{ki} - \delta_k N_{ji},\ C^i{}_{jk},\ C_i^{jk},\ \underset{(1)}{B}{}_j{}^{ik},\ \underset{(2)}{B}{}_{ijk}$$

and

$$(6.3)'\quad T^i{}_{jk} = H^i{}_{jk} - H^i{}_{kj},\ S^i{}_{jk} = C^i{}_{jk} - C^i{}_{kj},\ S_i^{jk} = C_i^{jk} - C_i^{kj}$$

and

$$(6.3)''\quad \underset{(1)}{P}{}^i{}_{jk} = \dot{\partial}_k N^i{}_j - H^i{}_{kj},\ \underset{(2)}{P}{}^i{}_{jk} = H^i{}_{jk} - \dot{\partial}^i N_{kj}$$

are torsion d-tensors.

The d-tensors of curvature are given by

$$
(6.4)\quad \begin{cases}
R_h{}^i{}_{jk} = \delta_k H^i_{hj} - \delta_j H^i_{hk} + H^m_{hj} H^i_{mk} - H^m_{hk} H^i_{mj} - X^i{}_{|m} T^m{}_{jk} + \\
\qquad + C^i_{hs} \underset{(1)}{R}{}^s{}_{jk} + C_h{}^{is} \underset{(2)}{R}{}_{sjk}, \\
P_h{}^i{}_{jk} = \dot{\partial}_k H^i{}_{hj} - C^i{}_{hk|j} + C^i{}_{hs} \underset{(1)}{P}{}^s{}_{jk} + C^{im}_s \underset{(2)}{B}{}_{mjk}, \\
P_h{}^i{}_j{}^k = \dot{\partial}_k H^i{}_{hj} - C_h{}^{ik}{}_{|j} + C^{is}{}_h \underset{(2)}{P}{}^i{}_{sj} + C^i{}_{hs} \underset{(1)}{B}{}^s{}_j{}^k,
\end{cases}
$$

and

$$
(6.4)'\quad \begin{cases}
S_h{}^i{}_{jk} = \dot{\partial}_k C^i_{hj} - \dot{\partial}_j C^i_{hk} + C^m_{hj} C^i_{mk} - C^m_{hk} C^i_{mj}, \\
S_h{}^i{}_j{}^k = \dot{\partial}^k C^i_{hj} - \dot{\partial}_j C^{ik}_h + C^m_{hj} C^{ik}_m - C^{mk}_h C^i_{mj}, \\
S_h{}^{ijk} = \dot{\partial}^k C^{ij}_h - \dot{\partial}^j C^{ik}_h + C^{mj}_h C^{ik}_m - C^m{}_h{}^k C^{ij}_m.
\end{cases}
$$

Proof. By a direct calculus we have

$$
\begin{aligned}
X^i{}_{|j|k} &= \delta_k (X^i{}_{|j}) + X^m{}_{|j} H^i_{mk} - X^i{}_{|m} H^m_{jk} = \\
&= \delta_k \delta_j X^i + \delta_k X^m H^i_{mj} + X^s \delta_k H^i_{sj} + \\
&+ (\delta_j X^m + X^s H^m_{sj}) H^i{}_{mk} - X^i{}_{|m} H^m{}_{jk}.
\end{aligned}
$$

Interchanging j and k and subtracting, we get

$$
\begin{aligned}
&X^i{}_{|j|k} - X^i{}_{|k|j} = [\delta_k, \delta_j] X^i + \\
&+ X^s (\delta_k H^i_{sj} - \delta_j H^i_{sk} + H^m_{sj} H^i_{mk} - H^m_{sk} H^i_{mj}) - X^i{}_{|m} T^m_{jk}.
\end{aligned}
$$

And since the Lie brackets $[\delta_k, \delta_j]$ are given by (7.1), Ch.9., the previous equalities give us the first identity (6.1), the coefficients being given by (6.3), (6.3)$'$ and (6.4).

The identities (6.1) and (6.2) can be proved in the same manner. **q.e.d.**

Remark. Cf. Proposition 9.7.4, the d-torsions $\underset{(1)}{B}{}_j{}^{ik}, \underset{(2)}{B}{}_{ijk}$ are d-tensors.

As usually, we extend the Ricci identities for any d-tensor field, given by (1.3).

As a first application let us consider a Riemann d-metric g_{ij}, which is covariant constant, i.e.:

$$g_{ij|k} = 0,\ g_{ij}|_k = 0,\ g_{ij}|^k = 0. \tag{6.5}$$

Then we have:

Theorem 10.6.2. *If the Riemann d-metric g_{ij} verifies the condition* (6.5), *then the following d-tensors*

$$(6.5)' \quad \begin{cases} R_{ijkh} = g_{jm}R_i{}^m{}_{kh},\ P_{ijkh} = g_{jm}P_i{}^m{}_{kh},\ P_{ijk}{}^h = g_{jm}P_i{}^m{}_k{}^h, \\ S_{ijkh} = g_{jm}S_i{}^m{}_{kh},\ S_{ijk}{}^h = g_{jm}S_i{}^m{}_k{}^h,\ S_{ij}{}^{kh} = g_{jm}S_i{}^{mkh} \end{cases}$$

are skew-symmetrics in the first two indices (ij).

Indeed, writing the Ricci identities for d-tensor g_{ij} and taking into account by the equations (6.5) we deduce

$$g_{im}R_j{}^m{}_{kh} + g_{jm}R_i{}^m{}_{kh} = 0, \ldots$$

And using (6.5)′, we get $R_{jikh} + R_{ijkh} = 0$, etc.

The Ricci identities (6.1), (6.2) applied to the Liouville d-vector field y^i, and to the Hamilton d-covector field p_i lead to the same fundamental identities.

Theorem 10.6.3. *Any canonical N-linear connection D satisfies the following identities:*

$$(6.6) \quad \begin{cases} D^i{}_{j|k} - D^i{}_{k|j} = y^h R_h{}^i{}_{jk} - D^i{}_h T^h{}_{jk} - d^i{}_h \underset{(1)}{R}{}^h{}_{jk} - d^{ih} \underset{(2)}{R}{}_{hjk} \\ D^i{}_j|_k - d^i{}_{k|j} = y^h P_h{}^i{}_{jk} - D^i{}_h C^h{}_{jk} - d^i{}_h \underset{(1)}{P}{}^h{}_{jk} - d^{ih} \underset{(2)}{B}{}_{hjk} \\ D^i_j|^k - d^{ik}{}_{|j} = y^h P_h{}^i{}_j{}^k - D^i{}_h C_j{}^{hk} - d^{ih} \underset{(2)}{P}{}^k{}_{hj} \end{cases}$$

and

$$(6.6)' \quad \begin{cases} d^i{}_j|_k - d^i{}_k|_j = y^h S_h{}^i{}_{jk} - D^i{}_h S^h{}_{jk} \\ d^i{}_j|^k - d^{ik}|_j = y^h S_h{}^i{}_j{}^k - D^i{}_h C_j{}^{hk} - d^{ih} C^k{}_{hj}, \\ d^{ij}|^k - d^{ik}|^j = y^h S_h{}^{ijk} - d^{ih} S_h{}^{jk}, \end{cases}$$

as well as

$$(6.7)\qquad \begin{cases} \Delta_{ij|k} - \Delta_{ik|j} = -p_h R_i{}^h{}_{jk} - \Delta_{ih} T^h{}_{jk} - \delta_{ih} R^h{}_{jk} - \delta_i{}^h R_{hjk}, \\[2ex] \Delta_{ij}|_k - \delta_{ik|j} = -p_h P_i{}^h{}_{jk} - \Delta_{ih} C^h{}_{jk} - \delta_{ih} \underset{(1)}{P}{}^h{}_{jk} - \delta_i{}^h \underset{(2)}{B}{}_{hjk}, \\[2ex] \Delta_{ij}|^k - \delta_i{}^k|_j = -p_h P_i{}^h{}_j{}^k - \Delta_{ih} C_j{}^{kh} - \delta_{ih} \underset{(1)}{B}{}_j{}^{hk} - \delta_i{}^h \underset{(2)}{P}{}^k{}_{hj}, \end{cases}$$

and

$$(6.7)'\qquad \begin{cases} \delta_{ij}|_k - \delta_{ik}|_j = -p_h S_i{}^h{}_{jk} - \delta_{ih} S^h{}_{jk}, \\[2ex] \delta_{ij}|^k - \delta_i{}^k|_j = -p_h S_i{}^h{}_j{}^k - \delta_{ih} C_j{}^{hk} - \delta_i{}^h C^k{}_{hj}, \\[2ex] \delta_i{}^h|^k - \delta_i{}^k|^j = -p_h S_i{}^{hjk} - \delta_i{}^h S_h{}^{jk}. \end{cases}$$

In the case of Berwald connection $B\Gamma(N)$, the previous theory is a very simple one.

Also, if the (y)–deflection tensors and (p)–deflection tensors have the following particular form

$$(6.8)\qquad \begin{cases} D^i{}_j = 0, \quad d^i{}_j = \delta^i{}_j, \quad d^{ij} = 0, \\[2ex] \Delta_{ij} = 0, \quad \delta_{ij} = 0, \quad \delta^i{}_j = \delta^i{}_j, \end{cases}$$

then, the fundamental identities from (6.6), (6.6)′ and (6.7), (6.7)′ are very important, especially for applications.

Proposition 10.6.1. *If the deflection tensors are given by* (6.8), *then the following identities hold:*

$$(6.8)'\qquad \begin{cases} y^h R_h{}^i{}_{jk} = \underset{(1)}{R}{}^i{}_{jk}; \; y^h P_h{}^i{}_{jk} = \underset{(1)}{P}{}^i{}_{jk}, \; y^h P_h{}^i{}_j{}^k = \underset{(1)}{B}{}_j{}^{ik} \\[2ex] y^h S_h{}^i{}_{jk} = S^i{}_{jk}; \; y^h S_h{}^i{}_j{}^k = C^i{}_h{}^k, \; y^h S_h{}^{ijk} = 0 \end{cases}$$

and

$$(6.8)''\qquad \begin{cases} p_h R_i{}^h{}_{jk} = -\underset{(1)}{R}{}_{ijk}; \; p_h P_i{}^h{}_{jk} = -\underset{(1)}{B}{}_{ijk}, \; p_h P_i{}^h{}_j{}^k = \underset{(1)}{P}{}^k{}_{ij} \\[2ex] p_h S_i{}^h{}_{jk} = 0, \; p_h S_i{}^h{}_j{}^k = -C^k{}_{jk}; \; p_h S_i{}^{hjk} = -S_i{}^{jk}. \end{cases}$$

By means of this analytical aparatus we will study the notion of parallelism on the manifold $T^{*2}M$.

10.7 Parallelism of the vector fields on the manifold $T^{*2}M$

Let D be an N–linear connection with the coefficients $D\Gamma(N) = (H^i_{jk}, C^i_{jk}, C_i^{jk})$ in the adapted basis $(\delta_i, \dot{\partial}_i, \dot{\partial}^i)$.

Let us consider a smooth parametrized curve $\gamma : I \longrightarrow T^{*2}M$ having the image in a domain of a chart of $T^{*2}M$.

Thus, γ has an analytical expression of the form:

$$x^i = x^i(t),\ y^i = y^i(t),\ p_i = p_i(t),\ t \in I. \tag{7.1}$$

The tangent vector field $\dot{\gamma} = \dfrac{d\gamma}{dt}$, by means of (6.13) and §6, Ch.9, can be written as follows:

$$\dot{\gamma} = \frac{dx^i}{dt}\delta_i + \frac{\delta y^i}{dt}\dot{\partial}_i + \frac{\delta p_i}{dt}\dot{\partial}^i \tag{7.2}$$

where

$$\frac{\delta y^i}{dt} = \frac{dy^i}{dt} + N^i_j\frac{dx^j}{dt};\ \frac{\delta p_i}{dt} = \frac{dp_i}{dt} - N_{ji}\frac{dx^j}{dt}. \tag{7.3}$$

Let us denote

$$D_{\dot{\gamma}}X = \frac{DX}{dt},\ DX = \frac{DX}{dt}dt,\ \forall X \in \mathcal{X}(T^{*2}M). \tag{7.4}$$

The quantity DX is the *covariant differential* of the vector X and $\dfrac{DX}{dt}$ is the *covariant differential* along the curve γ.

If X is written in the form

$$X = X^H + X^{W_1} + X^{W_2} = \overset{0}{X}{}^i\ \delta_i + \overset{1}{X}{}^i\ \dot{\partial}_i + X_i\dot{\partial}^i$$

and we put

$$D_{\dot{\gamma}} = D_{\dot{\gamma}^H} + D_{\dot{\gamma}^{W_1}} + D_{\dot{\gamma}^{W_2}} = D^H_{\dot{\gamma}} + D^{W_1}_{\dot{\gamma}} + D^{W_2}_{\dot{\gamma}} =$$

$$= \frac{dx^i}{dt}D_{\delta_i} + \frac{\delta y^i}{dt}D_{\dot{\partial}_i} + \frac{\delta p_i}{dt}D_{\dot{\partial}^i}$$

then, after a straighforward calculus, we have

$$DX = \left(d\overset{0}{X}{}^i + \overset{0}{X}{}^s\ \omega^i_s\right)\delta_i + \left(d\overset{1}{X}{}^i + \overset{1}{X}{}^s\ \omega^i_s\right)\dot{\partial}_s + \left(dX_i - X_s\omega^s_i\right)\dot{\partial}^i, \tag{7.5}$$

where

$$\omega^i{}_j = H^i_{jk}dx^k + C^i_{jk}\delta y^k + C^{ik}_j \delta p_k. \tag{7.6}$$

Here, $\omega^i{}_j$ are called 1–*forms of connection* of D.

Putting

$$\frac{\omega^i{}_j}{dt} = H^i_{jk}\frac{dx^k}{dt} + C^i_{jk}\frac{\delta y^k}{dt} + C^{ik}_j\frac{\delta p_k}{dt}, \tag{7.6'}$$

the covariant differential along the curve γ is given by

$$\begin{aligned}\frac{DX}{dt} &= \left(\frac{d\overset{0}{X}{}^i}{dt} + \overset{0}{X}{}^s\frac{\omega^i{}_s}{dt}\right)\delta_i + \\ &+ \left(\frac{d\overset{1}{X}{}^i}{dt} + \overset{1}{X}{}^s\frac{\omega^i{}_s}{dt}\right)\dot\partial_s + \left(\frac{dX_i}{dt} - X_s\frac{\omega^s{}_i}{dt}\right)\dot\partial^i.\end{aligned} \tag{7.7}$$

The theory of the parallelism of the vector fields along a curve γ presented in Sect.2 of this chapter can be applied here. We obtain:

Theorem 10.7.1. *The vector field* $X = \overset{0}{X}{}^i\delta_i + \overset{1}{X}{}^i\dot\partial_i + X_i\dot\partial^i$ *is parallel along the parametrized curve* γ, *with respect to* D, *if, and only if, its coordinates* $\overset{0}{X}{}^i, \overset{1}{X}{}^i, X_i$ *are solutions of the differential equations*

$$\frac{d\overset{\alpha}{X}{}^i}{dt} + \overset{\alpha}{X}{}^s\frac{\omega^i{}_s}{dt} = 0,\ \frac{dX_i}{dt} - X_s\frac{\omega^s{}_i}{dt} = 0,\ (\alpha = 0, 1). \tag{7.8}$$

The proof is immediate, by means of the expression (7.7) for $\frac{DX}{dt}$.

A theorem of existence and uniqueness for the parallel vector fields along a given parametrized curve in $T^{*2}M$ can be formulated in a classical manner.

The vector field $X \in \mathcal{X}(T^{*2}M)$ is called *absolute parallel* with respect to the canonical N–linear connection $D\Gamma(N)$, if $DX = 0$ for any curve γ. It is equivalent to the fact that the following system of Pfaff equations

$$d\overset{\alpha}{X}{}^i + \overset{\alpha}{X}{}^s\omega_s{}^i = 0,\ dX_i - X_s\omega^s{}_i = 0,\ (\alpha = 0, 1) \tag{7.9}$$

is integrable.

The system (7.9) is equivalent to the system

$$(7.9)' \qquad \begin{cases} \overset{\alpha\ i}{X}{}_{|j} = \overset{\alpha\ i}{X}\Big|_j = \overset{\alpha\ i}{X}\big|^j = 0, \quad (\alpha = 0,1) \\ \\ X_{i|j} = X_i\Big|_j = X_i\big|^j = 0 \end{cases},$$

which must be integrable.

Using the Ricci identities, the previous system is integrable if and only if the coordinates $(\overset{\alpha\ i}{X}, X_i)$ of the vector X satisfy the following equations

$$(7.10) \qquad \begin{cases} \overset{\alpha\ h}{X} R_h{}^i{}_{jk} = 0, \ \overset{\alpha\ h}{X} P_h{}^i{}_{jk} = 0, \ \overset{\alpha\ h}{X} P_h{}^i{}_j{}^k = 0, \\ \overset{\alpha\ h}{X} S_h{}^i{}_{jk} = 0, \ \overset{\alpha\ h}{X} S_h{}^i{}_j{}^k = 0, \ \overset{\alpha\ h}{X} S_h{}^{ijk} = 0 \end{cases} (\alpha = 0,1)$$

and

$$(7.11) \qquad \begin{array}{l} X_h R_i{}^h{}_{jk} = 0, \quad X_h P_i{}^h{}_{jk} = 0, \quad X_h P_i{}^h{}_j{}^k = 0 \\ \\ X_h S_i{}^h{}_{jk} = 0, \quad X_h S_i{}^h{}_j{}^k = 0, \quad X_h S_i{}^{hjk} = 0. \end{array}$$

The manifold $T^{*2}M$ is called *with absolute parallelism* of vectors with respect to D, if any vector field on $T^{*2}M$ is absolute parallel.

In this case the systems (7.10), (7.11) are verified for any vector field X. It follows:

Theorem 10.7.2. *The manifold $T^{*2}M$ is with absolute parallelism of vectors, with respect to the N–linear connection D if, and only if, all d–curvature tensors of D vanish.*

The curve γ is autoparallel with respect to D if $D_{\dot\gamma}\dot\gamma = 0$.

By means of (7.2) and (7.7) we deduce

$$(7.12) \qquad \begin{aligned} \frac{D\dot\gamma}{dt} &= \left(\frac{d^2x^i}{dt^2} + \frac{dx^s}{dt}\frac{\omega^i_s}{dt}\right)\delta_i + \left(\frac{d}{dt}\frac{\delta y^i}{dt} + \frac{\delta y^s}{dt}\frac{\omega^i_s}{dt}\right)\dot\partial_i + \\ &+ \left(\frac{d}{dt}\frac{\delta p_i}{dt} - \frac{\delta p_s}{dt}\frac{\omega^s_i}{dt}\right)\dot\partial^i. \end{aligned}$$

Theorem 10.7.3. *A smooth parametrized curve (7.1) is an autoparallel curve with respect to the N–linear connection D if and only if the functions $x^i(t), y^i(t), p_i(t)$,*

$t\in I$, verify the following system of differential equations

$$\frac{d^2x^i}{dt^2}+\frac{dx^s}{dt}\frac{\omega^i_s}{dt}=0,$$

$$\frac{d}{dt}\frac{\delta y^i}{dt}+\frac{\delta y^s}{dt}\frac{\omega^i_s}{dt}=0, \tag{7.13}$$

$$\frac{d}{dt}\frac{\delta p_i}{dt}-\frac{\delta p_s}{dt}\frac{\omega^s_i}{dt}=0.$$

Of course, the theorem of existence and uniqueness for the autoparallel curve can be easily formulated.

We recall that γ is an horizontal curve if $\dot{\gamma}=\dot{\gamma}^H$. The horizontal curves are characterized by

$$x^i=x^i(t),\ \frac{\delta y^i}{dt}=0,\ \frac{\delta p_i}{dt}=0. \tag{7.14}$$

Definition 10.7.1. An horizontal path of an N–linear connection D is a horizontal autoparallel curve with respect to D.

Theorem 10.7.4. *The horizontal paths of an N–linear connection D are characterized by the system of differential equations:*

$$\frac{d^2x^i}{dt^2}+H^i_{jk}(x,y,p)\frac{dx^j}{dt}\frac{dx^k}{dt}=0,\ \frac{\delta y^i}{dt}=0,\ \frac{\delta p_i}{dt}=0. \tag{7.15}$$

Indeed, the equations (7.14), (7.6)′ and (7.13) imply (7.15).

A parametrized curve $\gamma: I\rightarrow T^{*2}M$ is w_α–vertical in the point $x_0\in M$ if its tangent vector field $\dot{\gamma}$ belongs to the distribution W_α, $(\alpha=1,2)$.

Evidently, a w_1–vertical curve γ in the point $x_0\in M$ is represented by the equations of the form

$$x^i=x^i_0,\ y^i=y^i(t),\ p_i=0,\ t\in I \tag{7.16}$$

and a w_2–vertical curve γ in the point $x_0\in M$ is analytically represented by the equations of the form

$$x^i=x^i_0,\ y^i=0,\ p_i=p_i(t),\ t\in I. \tag{7.16$'$}$$

We define a w_α–path $(\alpha=1,2)$ in the point $x_0\in M$ with respect to D to be a w_α–vertical curve γ in the mentioned point, which is an autoparallel curve with respect to D.

By means of (7.16), (7.16)′ and (7.12) we can prove:

Theorem 10.7.5.

1°. *The* w_1*–vertical paths in the point* $x_0 \in M$ *are characterized by the system of differential equations*

$$x^i = x^i_0, \; p_i = 0; \; \frac{d^2y^i}{dt^2} + C^i_{jk}(x_0, y, 0)\frac{dy^j}{dt}\frac{dy^k}{dt} = 0.$$

2°. *The* w_2*-vertical paths in the point* $x_0 \in M$ *are characterized by the system of differential equations*

$$x^i = x^i_0, \; y^i = 0; \; \frac{d^2p_i}{dt^2} - C_i^{jk}(x_0, 0, p)\frac{dp_j}{dt}\frac{dp_k}{dt} = 0.$$

Remark. We assume that there exists the coefficients ${C^i}_{jk}(x_0, y, 0)$.

In the case of the Berwald connection $B\Gamma(N)$, (5.8), the previous characterizations of w_α-paths appear in a very simple form.

10.8 Structure equations of an N–linear connection

For an N–linear connection D, with the coefficients $D\Gamma(N) = (H^i_{jk}, C^i_{jk}, C_i^{jk})$ in the adapted basis $(\delta_i, \dot{\partial}_i, \dot{\partial}^i)$ we can prove:

Lemma 10.8.1.

1°. *Each of the following geometrical object fields*

$$d(dx^i) - dx^m \wedge {\omega^i}_m, \; d(\delta y^i) - \delta y^m \wedge {\omega^i}_m, \; d(\delta p_i) + \delta p_m \wedge {\omega^m}_i,$$

are d-vector fields. However, the last one is a d-covector field, with respect to the index i.

2°. *The geometrical object field*

$$d{\omega^i}_j - {\omega^m}_j \wedge {\omega^i}_m$$

is a d-tensor field, with respect to indices i *and* j.

Using the previous Lemma we can prove, by a straightforward calculus, a fundamental result in the geometry of the Hamilton spaces of order 2.

Theorem 10.8.1. *For any N–linear connection D, with the coefficients* $D\Gamma(N) = (H^i_{jk}, C^i_{jk}, C^{jk}_i)$, *the following structure equations hold good:*

$$
\begin{aligned}
& d(dx^i) - dx^m \wedge \omega^i{}_m = - \overset{(0)}{\Omega}{}^i, \\
(8.1) \qquad & d(\delta y^i) - \delta y^m \wedge \omega^i{}_m = - \overset{(1)}{\Omega}{}^i, \\
& d(\delta p_i) + \delta p_m \wedge \omega^m{}_i = -\Omega_i,
\end{aligned}
$$

and

$$
(8.2) \qquad d\omega^i{}_j - \omega^m{}_j \wedge \omega^i{}_m = -\Omega^i{}_j,
$$

where $\overset{(0)}{\Omega}{}^i, \overset{(1)}{\Omega}{}^i$ *and* Ω_i *are the 2-forms of torsion:*

$$
\begin{aligned}
\overset{(0)}{\Omega}{}^i &= \frac{1}{2} T^i_{jk} dx^j \wedge dx^k + C^i_{jk} dx^j \wedge \delta y^k + C^{ik}_j \delta y^j \wedge \delta p_k, \\
\overset{(1)}{\Omega}{}^i &= \frac{1}{2} \underset{(1)}{R}{}^i{}_{jk} dx^j \wedge dx^k + \underset{(1)}{P}{}^i{}_{jk} dx^j \wedge \delta y^k + \underset{(2)}{B}{}_j{}^{ik} dx^j \wedge \delta p_k + \\
(8.3) \qquad &+ C^{ik}_j \delta y^j \wedge \delta p_k + \frac{1}{2} S^i_{jk} \delta y^j \wedge \delta y^k, \\
\Omega_i &= \frac{1}{2} \underset{(2)}{R}{}_{ijk} dx^j \wedge dx^k + \underset{(2)}{B}{}_{ijk} dx^j \wedge \delta y^k + \underset{(2)}{P}{}^k{}_{ij} dx^j \wedge \delta p_k + \\
&+ C^k{}_{ij} \delta y^j \wedge \delta p_k + \frac{1}{2} S_i^{kj} \delta p_j \wedge \delta p_k,
\end{aligned}
$$

and where $\Omega^i{}_j$ *is the 2-form of curvature:*

$$
\begin{aligned}
\Omega^i{}_j &= \frac{1}{2} R_j{}^i{}_{km} dx^k \wedge dx^m + P_j{}^i{}_{km} dx^k \wedge \delta y^m + P_j{}^i{}_k{}^m dx^k \wedge \delta p_m + \\
(8.4) \qquad &+ \frac{1}{2} S_j{}^i{}_{km} \delta y^k \wedge \delta y^m + S_j{}^i{}_k{}^m \delta y^k \wedge \delta p_m + \frac{1}{2} S_j{}^{ikm} \delta p_k \wedge \delta p_m.
\end{aligned}
$$

In the particular case of the Berwald connection we have $C^i{}_{jk} = C_i{}^{kj} = 0$, $\underset{(1)}{P}{}^i{}_{jk} = 0$, $S^i{}_{jk} = S_i^{jk} = 0$, $S_j{}^i{}_{km} = 0$, $S_j{}^i{}_k{}^m = 0$ and $S_i^{jkm} = 0$.

Remark. The previous theorem is extremely important in a theory of submanifolds embedding in the total space $T^{*2}M$ of the dual bundle $T^{*2}M = TM \times_M T^*M$, endowed with a regular Hamiltonian of order 2.

Chapter 11

Generalized Hamilton spaces of order 2

One of the most important structures on the total space of the dual bundle $T^{*2}M$ is the notion of generalized Hamilton metric of order two, $g^{ij}(x, y, p)$, [110]. It is suggested by the generalized Hamilton metric, described in the section 1 of Ch. 5, which has notable applications in Relativistic Optics of order two. We define the concept of generalized Hamilton space as the pair $GH^{(2)n} = (M, g^{ij}(x, y, p))$ and study a criteria of reducibility, the most general metrical connections, lift of a GH–metric, the almost contact geometrical model. We end this section with some example of remarkable $GH^{(2)n}$–spaces.

11.1 The spaces $GH^{(2)n}$

Definition 11.1.1. A generalized Hamilton space of order two is a pair $GH^{(2)n} = (M, g^{ij}(x, y, p))$, where

1° g^{ij} is a d–tensor field of type $(2, 0)$, symmetric and nondegenerate on the manifold $T^{*2}M$.

2° The quadratic form $g^{ij}X_iX_j$ has a constant signature on $T^{*2}M$.

As usually g^{ij} is called the *fundamental tensor* or metric tensor of the space $GH^{(2)n}$.

In the case when $T^{*2}M$ is a paracompact manifold then on $T^{*2}M$ there exist the metric tensors $g^{ij}(x, y, p)$ positively defined such that (M, g^{ij}) is a generalized Hamilton space.

Definition 11.1.2. A generalized Hamilton metric $g^{ij}(x, y, p)$ of order two (on short GH–metric) is called reducible to an Hamilton metric (H–metric) of order two if there exists a function $H(x,y,p)$ on $T^{*2}M$ such that

$$g^{ij}=\frac{1}{2}\dot{\partial}^i\dot{\partial}^j H. \tag{1.1}$$

Let us consider the d-tensor field

$$C^{ijk}=\frac{1}{2}\dot{\partial}^k g^{ij}. \tag{1.2}$$

We can prove:

Proposition 11.1.1. *A necessary condition for a generalized Hamilton metric g^{ij} of order two to be reducible to a Hamilton metric of order two is that the d–tensors C^{ijk} is totally symmetric.*

Theorem 11.1.1. *Let $g^{ij}(x,y,p)$ be a 0–homogeneous GH–metric with respect to p_i. Then a necessary and sufficient condition that it to be reducible to an H–metric is that the d–tensor field C^{ijk} is totally symmetric.*

Proof. If there exists a GH–metric g^{ij} reducible to a H–metric, i.e. (1.1) holds, then $C^{ijk}=\frac{1}{2}\dot{\partial}^k\dot{\partial}^i\dot{\partial}^j H$ is totally symmetric (Proposition 11.1.1.).

Conversely, assuming that $g^{ij}(x,y,p)$ is 0–homogeneous with respect to p_i, taking into account the formula $H(x,y,p)=g^{ij}(x,y,p)p_ip_j$, and the fact that $C^{ijk}=\dot{\partial}^k g^{ij}$ is totally symmetric, it follows $g^{ij}=\frac{1}{2}\dot{\partial}^i\dot{\partial}^j H$. **q.e.d.**

Remark. Let $\gamma_{ij}(x)$ be a Riemannian metric. Then it is not difficult to prove that the d–tensor field

$$g^{ij}(x,y,p)=e^{-2\sigma(x,y,p)}\gamma^{ij}(x),\ \sigma\in\mathcal{F}(T^{*2}M),$$

is a GH–metric which is not reducible to an H–metric of order two, provided $\dot{\partial}^i\sigma$ does not vanishes .

The covariant tensor field g_{ij} is obtained from the equations

$$g_{ij}g^{jk}=\delta_i^k. \tag{1.3.}$$

Of course, g_{ij} is a symmetric, nondegenerate and covariant of order two, d-tensor field.

Theorem 11.1.2. *The following d–tensor fields*

$$\begin{aligned} C^i{}_{jk} &= \frac{1}{2}g^{is}\left(\dot{\partial}_j g_{sk}+\dot{\partial}_k g_{js}-\dot{\partial}_s g_{jk}\right),\\ C_i^{jk} &= -\frac{1}{2}g_{is}\left(\dot{\partial}^j g^{sk}+\dot{\partial}^k g^{js}-\dot{\partial}^s g^{jk}\right), \end{aligned} \tag{1.4}$$

have the properties:

$$g^{ij}|_k = 0, \;\; g^{ij}|^k = 0, \tag{1.5}$$

and

$$C^i_{jk} = C^i_{kj}, \;\; C^{jk}_i = C^{kj}_i. \tag{1.6}$$

Indeed, the d-tensors C^i_{jk}, C^{jk}_i have the properties (1.6). By a direct calculus we can prove (1.5) taking into account of (1.3). **q.e.d.**

Remarks.

1° The tensors $g^{ij}|_k$ and $g^{ij}|^k$ are the w_1- w_2–covariant derivatives of the fundamental tensor field g^{ij}, *respectively.*

2° The tensors C^i_{jk} and C^{jk}_i are the w_1 and w_2–coefficients of a canonical metrical N–linear connection D, respectively.

Some particular cases

1. Let $g_{ij}(x,y)$ be the fundamental tensor field of a Finsler space $F^n = (M,F)$ and let $g^{ij}(x,y)$ be its contravariant tensor field. Let us consider $\bar{g}^{ij}(x,y,p)$ defined on $T^{*2}M$ by $\bar{g}^{ij}(x,y,p) = g^{ij}(x,y)$.

The tensors C^i_{jk} are given by the first formula (1.4) and by $C^{jk}_i = 0$.

It follows $C^{ijk} = 0$. The GH–metric $g^{ij}(x,y)$ has the covariant metric $g_{ij}(x,y)$ reducible to a particular metric: $g_{ij} = \dfrac{1}{2}\dot{\partial}_i\dot{\partial}_j F^2$.

We have:

Theorem 11.1.3. *The nonlinear connection N of the space $GH^{(2)n} = (M, g^{ij}(x,y))$ has the coefficients:*

$$\begin{cases} N^i{}_j = \dot{\partial}_j(\gamma^i{}_{rs}(x,y)y^r y^s) - (\textit{Cartan nonlinear connection of } F^n), \\ \\ N_{ij} = (\dot{\partial}_i N^h_j)p_h, \end{cases} \tag{1.7}$$

They are determined only on the fundamental function of the Finsler space F^n.

Proof. The tensors N^i_j is exactly the Cartan non–linear connection and $\dot{\partial}_i N^h_j$ is its Berwald connection. A straightforward calculus shows that the rule of transformation of N_{ij} is exactly (3.10), Ch. 6. **q.e.d.**

2. Let $g^{ij}(x,p)$ be the fundamental tensor field of a Cartan space $\mathcal{C}^n = (M, H(x,p))$, [97]. It follows $g^{ij} = \dfrac{1}{2}\dot{\partial}^i\dot{\partial}^j H^2$ and therefore we obtain that H is 1–homogeneous with respect to p_i.

We consider the extension $\bar{g}^{ij}(x,y,p) = g^{ij}(x,p)$ of the tensor g^{ij} to $T^{*2}M$.

The tensor C^i_{jk} vanishes and C^{jk}_i is given by the second formula (1.4), where $g_{ij}g^{jk} = \delta^k_i$.

In this case, we can determine a nonlinear connection N depending only by the fundamental tensor g^{ij} [see Ch.6]. Indeed, let $\gamma^i_{jk}(x,p)$ the Christoffel symbols of $g_{ij}(x,p)$ and let us put $\gamma^0_{jh} = \gamma^i_{jh}p_i$, $\gamma^0_{j0} = \gamma^0_{jh}p^h$, $p^h = g^{hj}p_j = \frac{1}{2}\dot{\partial}^h H^2$.

Theorem 11.1.4. *The space $GH^{(2)n} = (M, g^{ij}(x,p))$, determined by the Cartan space $\mathcal{C}^n$ has a nonlinear connection N with the following coefficients deduced only from g^{ij} :*

$$(1.8)\qquad \begin{cases} N^i_j = (\dot{\partial}^i N_{jh})y^h \\ \\ N_{ij} = \gamma^0_{ij} - \frac{1}{2}\gamma^0_{h0}\dot{\partial}^h g_{ij} \end{cases}$$

To a generalized Hamilton space of order two $GH^{(2)n} = (M, g^{ij})$ we associate the Hamilton absolute energy

$$(1.9)\qquad \mathcal{E}(x,y,p) = g^{ij}(x,y,p)p_ip_j$$

and consider the d-tensor field

$$(1.9)'\qquad g^{*ij} = \frac{1}{2}\dot{\partial}^i\dot{\partial}^j\mathcal{E}.$$

The space $GH^{(2)n}$ is called *weakly regular* if:

$$(1.9)''\qquad \text{rank}\left\|g^{*ij}\right\| = n.$$

We can prove the following fact:

The weakly regular $GH^{(2)n}$–spaces have a nonlinear connection N depending only on the fundamental tensor field g^{ij}.

11.2 Metrical connections in $GH^{(2)n}$–spaces

If a nonlinear connection N, with the coefficients (N^i_j, N_{ij}), is a priori given, let us consider the direct decomposition (see (5.2), Ch.7):

$$(2.1)\qquad T_uT^{*2}M = N(u) \oplus W_1(u) \oplus W_2(u),\ \forall\, u \in T^{*2}M,$$

and the adapted basis to it, $(\delta_i, \dot{\partial}_i, \dot{\partial}^i)$, where

$$(2.2)\qquad \delta_i = \partial_i - N^j_i\dot{\partial}_j + N_{ij}\dot{\partial}^j.$$

The dual adapted basis is $(dx^i, \delta y^i, \delta p_i)$ where

$$\delta y^i = dy^i + N^i_j dx^j,\ \ \delta p_i = dp_i - N_{ji} dx^j \tag{2.3}$$

An N–linear connection $D\Gamma(N) = (H^i_{jk}, C^i_{jk}, C^{jk}_i)$ determines the h–, w_1–, w_2–covariant derivatives in the tensor algebra of d–tensor fields.

Definition 11.2.1. An N–linear connection $D\Gamma(N)$ is called *metrical* with respect to GH–metric g^{ij} if

$$g^{ij}{}_{|h} = 0,\ \ g^{ij}|_h = 0,\ \ g^{ij}|^h = 0. \tag{2.4}$$

In the case when g^{ij} is positively defined we can introduce the lengths of a d-covector field X_i by

$$\|X\| = \{g^{ij}(x,y,p)X_i X_j\}^{1/2}. \tag{2.5}$$

The following property is not difficult to prove:

Theorem 11.2.1. *An N–linear connection $D\Gamma(N)$ is metrical with respect to GH–metric g^{ij} if and only if along to any smooth curve $\gamma : I \to T^{*2}M$, and for any parallel d-covector field X, $\dfrac{DX}{dt} = 0$, we have $\dfrac{d\|X\|}{dt} = 0$.*

The tensorial equations (2.4) imply:

$$g_{ij|h} = 0,\ \ g_{ij}|_h = 0,\ \ g_{ij}|^h = 0. \tag{2.6}$$

Now, using the same technique as in the case of Ch. 5, we can prove the following important result:

Theorem 11.2.2.

1. *There exists a unique N–linear connection $D\Gamma(N) = (H^i_{jk}, C^i_{jk}, C^{jk}_i)$ having the properties:*

   1°. *The nonlinear connection N is a priori given.*

   2°. *$D\Gamma(N)$ is metrical with respect to GH–metric g^{ij} i.e. (2.6) are verified.*

   3°. *The torsion tensors $T^i{}_{jk}, S^i{}_{jk}$ and $S_i{}^{jk}$ vanish.*

2. *The previous connection has the coefficients $C^i{}_{jk}$ and C^{jk}_i given by (1.4) and $H^i{}_{jk}$ are the generalized Christoffel symbols:*

(2.7) $$H^i_{jk} = \frac{1}{2}g^{is}(\delta_j g_{sk} + \delta_k g_{js} - \delta_s g_{jk}).$$

The known Obata's operators, are given by

(2.8) $$\Omega^{ij}_{hk} = \frac{1}{2}(\delta^i_h\delta^j_k - g_{hk}g^{ij}),\ \Omega^{*ij}_{hk} = \frac{1}{2}(\delta^i_h\delta^j_k + g_{hk}g^{ij}).$$

They are the supplementary projectors on the module of d-tensor fields $\tau^1_2(T^{*2}M)$. Also, they are covariant constants with respect to any metrical connection $D\Gamma(N)$.

Exactly as in Ch.5, we can prove:

Theorem 11.2.3. *The set of all N–linear connections $D\overline{\Gamma}(N) = (\overline{H}^i_{jk}, \overline{C}^i_{jk}, \overline{C}^{jk}_i)$, which are metrical with respect to g^{ij}, is given by*

(2.9) $$\begin{cases} \overline{H}^i_{jk} = H^i_{jk} + \Omega^{is}_{rj}X^r_{sk}, \\ \overline{C}^i_{jk} = C^i_{jk} + \Omega^{is}_{rj}Y^r_{sk}, \\ \overline{C}^{jk}_i = C^{jk}_i + \Omega^{is}_{rj}Z^{rk}_s. \end{cases}$$

where $D\Gamma(N) = (H^i_{jk}, C^i_{jk}, C^{jk}_i)$ is given by (1.4), (2.7) and $X^i_{jk}, Y^i_{jk}, Z^{jk}_i$ are arbitrary d-tensor fields.

We obtain:

Corollary 11.2.1. *The mapping $D\Gamma(N) \longrightarrow D\overline{\Gamma}(N)$ determined by (2.9) and the composition of these mappings is an Abelian group.*

Remark. It is important to determine the geometrical object fields invariant to the previous group of transformations of metrical connections [105].

From Theorem 11.2.3 we can deduce:

Theorem 11.2.4. *There exists a unique metrical connection $D\overline{\Gamma}(N) = (\overline{H}^i_{jk}, \overline{C}^i_{jk}, \overline{C}^{jk}_i)$ with respect to GH–metric g^{ij}, having the torsion d–tensor fields $T^i{}_{jk}, S^i{}_{jk}, S_i{}^{jk}$ a priori given. The coefficients of $D\overline{\Gamma}(N)$ are given by the following formulas*

(2.10) $$\begin{aligned} \overline{H}^i_{jk} &= \frac{1}{2}g^{is}(\delta_j g_{sk} + \delta_k g_{js} - \delta_s g_{jk}) + \frac{1}{2}g^{is}(g_{sh}T^h{}_{jk} - g_{jh}T^h{}_{sk} + g_{kh}T^h{}_{js}), \\ \overline{C}^i_{jk} &= \frac{1}{2}g^{is}(\dot\partial_j g_{sk} + \dot\partial_k g_{js} - \dot\partial_s g_{jk}) + \frac{1}{2}g^{is}(g_{sh}S^h{}_{jk} - g_{jh}S^h{}_{sk} + g_{kh}S^h{}_{js}), \\ \overline{C}^{jk}_i &= -\frac{1}{2}g_{is}(\dot\partial^j g^{sk} + \dot\partial^k g^{js} - \dot\partial^s g^{jk}) - \frac{1}{2}g_{is}(g^{sh}S_h{}^{jk} - g^{jh}S_h{}^{sk} + g^{kh}S_h{}^{js}). \end{aligned}$$

We can introduce the notions of Rund connection, Berwald connection and Hashiguchi connection as in Chapter 2, and prove the existence of a commutative diagram from the mentioned chapter.

Finally, if we denote $R^{hi}{}_{jk} = g^{hs}R_s{}^i{}_{jk}$ etc., then applying the Ricci identitiesto g^{ij}, and taking into account the equations (1.6), we get

Theorem 11.2.5. *The curvature tensor fields $g^{hs}R_s{}^i{}_{jk}$, $g^{hs}P_s{}^i{}_{jk}$, etc. are skew symmetric in the indices h, i.*

11.3 The lift of a GH–metric

Let the nonlinear connection N be given, then the adapted basis $(\delta_i, \dot{\partial}_i, \dot{\partial}^i)$ and its dual basis $(dx^i, \delta y^i, \delta p_i)$ can be determined.

Therefore, a generalized Hamilton space of order two $GH^{(2)n} = (M, g^{ij})$ allows to introduce the N–lift:

$$\mathbb{G} = g_{ij}dx^i \otimes dx^j + g_{ij}\delta y^i \otimes \delta y^j + g^{ij}\delta p_i \otimes \delta p_j \tag{3.1}$$

defined in every point $u \in T^{*2}M$.

Theorem 11.3.1.

1°. *The N–lift $\mathbb{G}$ is a nonsingular tensor field on the manifolds $T^{*2}M$, symmetric, of type $(0,2)$ depending only by the GH–metric g^{ij} and by the nonlinear connection N.*

2°. *The pair $(T^{*2}M, \mathbb{G})$ is a (pseudo)–Riemannian space.*

3°. *The distributions N, W_1, W_2 are orthogonal with respect to $\mathbb{G}$, respectively.*

Indeed, every term from (3.1) is defined on $T^{*2}M$, because g_{ij} is a d-tensor field, and $dx^i, \delta y^i, \delta p_i$ have the rule of transformations (5.6)″, Ch. 9. The determinant of $\mathbb{G}$ is equal to the determinant of matrix $\|g_{ij}\|$. Hence $\det\|\mathbb{G}\| \neq 0$. Now it is clear that $\mathbb{G}$ is a (pseudo)–Riemannian metric. And it is evident that the distributions N, W_1, W_2 are orthogonal with respect to $\mathbb{G}$, *respectively.*

The tensor $\mathbb{G}$ is of the form

(3.2)
$$\mathbb{G} = \mathbb{G}^H + \mathbb{G}^{W_1} + \mathbb{G}^{W_2},\ \mathbb{G}^H = g_{ij}dx^i \otimes dx^j,\ \mathbb{G}^{W_1} = g_{ij}\delta y^i \otimes \delta y^j,\ \mathbb{G}^{W_2} = g^{ij}\delta p_i \otimes \delta p_j.$$

Here $\mathbb{G}^H$ is the restriction of the metric $\mathbb{G}$ to the distribution H, $\mathbb{G}^{W_1}$ is its restriction to W_1 and $\mathbb{G}^{W_2}$ is its restriction to the distribution W_2. Moreover $\mathbb{G}^H$, $\mathbb{G}^{W_1}$ $\mathbb{G}^{W_2}$ are d-tensor fields.

It is not difficult to prove:

Theorem 11.3.2. *The tensors* $\mathbb{G}$, $\mathbb{G}^H$, $\mathbb{G}^{W_1}$ $\mathbb{G}^{W_2}$ *are covariant constant with respect to any metrical* N*–linear connection* $D\Gamma(N)$.

Therefore, the equation $D\mathbb{G} = 0$, $\forall\, X \in \mathcal{X}(T^{*2}M)$ is equivalent to $g_{ij|h} = g_{ij}\big|_h = g^{ij}|^h = 0$. The same property holds for the d-tensors $\mathbb{G}^H$, $\mathbb{G}^{W_1}$ $\mathbb{G}^{W_2}$.

The geometry of the (pseudo)–Riemannian space $(T^{*2}M, \mathbb{G})$ can be studied by means of a metrical N-linear connection.

Let $\mathbb{F}$ be the natural almost contact structure determined by N and is given in the section 8, Ch.9.

Theorem 11.3.3. *The pair* $(\mathbb{G}, \mathbb{F})$ *is a Riemannian almost contact structure determined only by* GH*–metric* g^{ij} *and by the nonlinear connection* N.

Proof. In the adapted basis $(\delta_i, \dot{\partial}_i, \dot{\partial}^i)$ it follows that the equation

$$\mathbb{G}(\mathbb{F}X, Y) = -\mathbb{G}(\mathbb{F}Y, X),\ \forall\, X, Y \in \mathcal{X}(T^{*2}M) \tag{3.3}$$

is verified.

The 2-form associated to the structure $(\mathbb{G}, \mathbb{F})$ is given by

$$\breve{\theta}(X, Y) = \mathbb{G}(\mathbb{F}X, Y). \tag{3.4}$$

Since $D_X\mathbb{G} = 0$, $D_X\mathbb{F} = 0$, we get that $D_X\breve{\theta} = 0$.

In the local adapted basis $\breve{\theta}$ has the expression:

$$\breve{\theta} = g_{ij}\delta y^i \wedge dx^j. \tag{3.4$'$}$$

Theorem 11.3.4. *The 2-form* $\breve{\theta}$ *determines an almost presymplectic structure on the manifold* $T^{*2}M$. *It is not an integrable structure if the metric* g^{ij} *depends on the moments* p_i.

Indeed, $\breve{\theta}$ is a 2-form of rank $2n < 3n$ and for $\dot{\partial}^i g^{jk} \neq 0$ the exterior diferential of $\breve{\theta}$ does not vanish.

The last theorem suggests to consider another almost contact Riemannian structure on $T^{*2}M$.

In order to do this, let us consider some new geometrical object fields on $T^{*2}M$:

$$p^i = g^{ij}p_j, \tag{3.5}$$

$$\breve{\partial}_i = g_{ij}\dot{\partial}^i. \tag{3.6}$$

Let us define the $\mathcal{F}(T^{*2}M)$–linear mapping

$$\check{\mathbb{F}} : \mathcal{X}(T^{*2}M) \longrightarrow \mathcal{X}(T^{*2}M)$$

given in the local adapted basis by

$$\check{\mathbb{F}}(\delta_i) = -\check{\partial}_i,\ \check{\mathbb{F}}(\dot{\partial}_i) = 0,\ \check{\mathbb{F}}(\check{\partial}_i) = \delta_i,\ (i = 1, ..., n). \tag{3.7}$$

Theorem 11.3.5. *The mapping* $\check{\mathbb{F}}$ *has the following properties:*

1°. $\check{\mathbb{F}}$ *is globally defined on* $T^{*2}M$.

2°. $\check{\mathbb{F}}$ *is a tensor field of type* $(1,1)$ *on* $T^{*2}M$.

3°. $\mathrm{Ker}\,\check{\mathbb{F}} = W_1,\ \mathrm{Im}\,\check{\mathbb{F}} = N \oplus W_1$.

4°. $\mathrm{rank}\left\|\check{\mathbb{F}}\right\| = 2n$.

5°. $\check{\mathbb{F}}^3 + \check{\mathbb{F}} = 0$.

The proof is completely similar with the one of exactly like Theorem 9.8.2. The mapping $\check{\mathbb{F}}$ will be called the (p)–almost contact structure determined by g^{ij} and by N. The Nijenhuis tensor of the (p)–almost contact structure is

$$\mathcal{N}_{\check{\mathbb{F}}}(X,Y) = \check{\mathbb{F}}^2[X,Y] + [\check{\mathbb{F}}X, \check{\mathbb{F}}Y] - \check{\mathbb{F}}[\check{\mathbb{F}}X, Y] - \check{\mathbb{F}}[X, \check{\mathbb{F}}Y],$$

and the condition of normality of $\check{\mathbb{F}}$ is as follows

$$\mathcal{N}_{\check{\mathbb{F}}}(X,Y) + \sum_{i=1}^{n} d(\delta y^i)(X,Y), = 0,\ \forall\, X, Y \in \mathcal{X}(T^{*2}M). \tag{3.8}$$

The relation (8.3) can be explicitely written in adapted basis.

Theorem 11.3.6. *The pair* $(\mathbb{G}, \check{\mathbb{F}})$ *is a Riemannian almost contact structure determined by* g^{ij} *and by* N.

Indeed, we have verified the property:

$$\mathbb{G}(\check{\mathbb{F}}X, Y) = -\mathbb{G}(\check{\mathbb{F}}Y, X).$$

The 2-form associated to $(\mathbb{G}, \check{\mathbb{F}})$ is

$$\hat{\theta}(X,Y) = \mathbb{G}(\check{\mathbb{F}}X, Y). \tag{3.9}$$

In the adapted basis $\hat{\theta}$ is given by

$$\hat{\theta} = \delta p_i \wedge dx^i. \tag{3.9$'$}$$

As we know, if the torsion $\tau_{ij} = N_{ij} - N_{ji}$ of the nonlinear connection vanishes (Ch.9), then we have:

$$\hat{\theta} = \theta = d\omega = dp_i \wedge dx^i.$$

Hence for $\tau_{ij} = 0$, the 2-form $\hat{\theta}$ is canonical presymplectic structure. It does not depend on the nonlinear connection N (see, Theorem 9.1.3).

Theorem 11.3.7. *The associated 2-form $\hat{\theta}$ of the almost contact structure $(\mathbb{G}, \check{\mathbb{F}})$ has the properties:*

1°. *$\hat{\theta}$ is globally defined on $T^{*2}M$.*

2°. $\mathrm{rank}\left\|\hat{\theta}\right\| = 2n$.

3°. *$\hat{\theta}$ depends by g^{ij} and by N.*

4°. *$\hat{\theta}$ defines an almost presymplectic structure on $T^{*2}M$.*

5°. *If the torsion τ_{ij} of the nonlinear connection N vanishes, then $\hat{\theta}$ is canonical presymplectic structure:*

$$\hat{\theta} = d\omega = dp_i \wedge dx^i. \tag{3.9$''$}$$

6°. *$\hat{\theta}$ is covariant constant to any N–linear connection $D\Gamma(N)$.*

The Riemannian almost contact space $(T^{*2}M, \mathbb{G}, \check{\mathbb{F}})$ will be called the *geometrical model* of the generalized Hamilton space $GH^{(2)n} = (M, g^{ij})$.

11.4 Examples of spaces $GH^{(2)n}$

We shall consider a generalized Hamilton space of order two, $GH^{(2)n} = (M, g^{ij})$, whose fundamental tensor is as follows:

$$g^{ij}(x, y, p) = e^{-2\sigma(x,y,p)}\gamma^{ij}(x, y), \tag{4.1}$$

where $\gamma_{ij}(x, y)$ is the fundamental tensor of a Finsler space $F^n = (M, F)$, γ^{ij} is its contraviant tensor field and $\sigma : T^{*2}M \to R$ is a smooth function.

In the particular case where $\dot{\partial}^i\sigma = 0$, and $\gamma_{ij}(x, y) = \gamma_{ij}(x)$ is a Lorentz metric, this structure was used for a constructive axiomatic theory of General Relativity by R. Miron and R. Tavakol [121].

If $\gamma_{ij}(x, y)$ is a locally Minkowski Finsler space and $\dot{\partial}^i\sigma = 0$, then (4.1) gives a class of a generalized P.L. Antonelli and H. Shimada metric [19].

In order to study the $GH^{(2)n}$ spaces with the metric (4.1) in the general case when the d–vector field $\dot{\partial}^i\sigma$ does not vanish, we prove at the beginning that this metric is not reducible to an H–metric.

Theorem 11.4.1. *The generalized Hamilton space of order two with the metric* (4.1) *is not reducible to an Hamilton space of order two.*

Proof. Taking into account Proposition 11.1.1 is sufficiently to prove that if $\dot{\partial}^i\sigma \neq 0$, then the tensor field $C^{ijk} = \dfrac{1}{2}\dot{\partial}^k g^{ij}$ is not totally symmetric. **q.e.d.**

From the formula (4.1), we deduce

$$C^{ijk} = -g^{ij}\dot{\partial}^k\sigma.$$

Consequently, C^{ijk} is totally symmetric if and only if $\dot{\partial}^k\sigma = 0$.

Let us consider the Cartan nonlinear connection of the Finsler space F^n, with the coefficients$N^i_j(x, y)$. Thus, using Theorem 11.1.3, we can a priori take the nonlinear connection N with the coefficients (1.7) as the nonlinear connection of the considered space $GH^{(2)n}$.

Proposition 11.4.1. *The nonlinear connection N, with the coefficients* $(N^i{}_j, N_{ij})$ *from the formulas* (1.7) *depend only on the GH–metric* (4.1).

Now we can determine the metrical connection $D\Gamma(N)$ of the space $GH^{(2)n}$ using Theorem 11.2.3. This metrical N–linear connection will be called canonical.

It is not difficult to prove:

Theorem 11.4.2. *The canonical metrical connection $D\Gamma(N)$ of the space $GH^{(2)n}$, with fundamental tensor field* (4.1), *has the coefficients:*

$$\begin{aligned} H^i_{jk} &= \overset{0}{F^i}{}_{jk} + \delta^i_j\delta_k\sigma + \delta^i_k\delta_j\sigma - \gamma_{jk}\gamma^{is}\delta_s\sigma, \\ C^i_{jk} &= \overset{0}{C^i}{}_{jk} + \delta^i_j\dot{\partial}_k\sigma + \delta^i_k\dot{\partial}_j\sigma - \gamma_{jk}\gamma^{is}\dot{\partial}_s\sigma, \\ C^{jk}_i &= -(\delta^j_i\dot{\partial}^k\sigma + \delta^k_i\dot{\partial}^j\sigma - \gamma_{is}\gamma^{jk}\dot{\partial}^s\sigma), \end{aligned} \tag{4.2}$$

where $(\overset{0}{F^i}{}_{jk}, \overset{0}{C^i}{}_{jk})$ *is the Cartan metrical connection of the Finsler space F^n.*

Now, applying the theory from the previous chapters and using Theorem 11.4.2, we can develop the geometry of the spaces $GH^{(2)n}$, with the metric (4.1). For instant, we can write the structure equations of the canonical connections (4.2), etc.

Other important example suggested by the Relativistic Optics is given by the following GH–metric of order two:

$$g^{ij}(x,y,p) = \gamma^{ij}(x,y) + \frac{1}{1-n^2(x,y)}p^i p^j, \ p^i = \gamma^{ij}p_j, \tag{4.3}$$

where $n^2(x,y) > 1$, on the manifold $T^{*2}M$, and $\gamma^{ij}(x,y)$ is contravariant tensor of a fundamental tensor field $\gamma_{ij}(x,y)$ of a Finsler space $F^n = (M, F(x,y))$.

In this case we can prove that $GH^{(2)n}$, with the metric (4.3), is not reducible to a Hamilton space of order two.

Taking into account the nonlinear connection N with coefficients (1.7) we can determine, by means of Theorem 11.2.3, a canonical metrical connection $D\Gamma(N)$ depending only by the considered space $GH^{(2)n}$.

As a final example, we can study by the previous methods "the Antonelli–Shimada metric" defined in the preferential charts of an atlas on the manifold $T^{*2}M$ by

$$g^{ij}(x,y,p) = e^{-2\sigma(x,y,p)}\gamma^{ij}(p), \tag{4.4}$$

where

$$\begin{aligned} &\gamma^{ij}(p) = \frac{1}{2}\dot{\partial}^i\dot{\partial}^j H^2(p), \\ &H^2(p) = \{(p_1)^m + \cdots + (p_n)^m\}^{\frac{2}{m}}, \ m \geq 3. \end{aligned} \tag{4.5}$$

Finally, we remark that the theory exposed in this chapter will be useful in the next chapters for study the geometry of Hamilton spaces of order two.

Chapter 12

Hamilton spaces of order 2

The theory of dual bundle $(T^{*2}M, \pi^{*2}, M)$ mentioned in the last three Chapters allows to study a natural extension to order two of the notion of Hamilton spaces studied in the Chapters 4,5,6. A Hamilton space of order two is a pair $H^{(2)n} = (M, H(x,y,p))$ formed by a real, n-dimensional smooth manifold M and a regular Hamiltonian function $H : (x,y,p) \in T^{*2}M \to H(x,y,p) \in R$. The geometry of the spaces $H^{(2)n}$ can be constructed step by step following the same ideas as in the classical case of the spaces $H^{(1)n} = (M, H(x,p))$, by using the geometry of manifold $T^{*2}M$ endowed with an regular Hamiltonian $H(x,y,p)$.

12.1 The spaces $H^{(2)n}$

Let us consider again a differentiable manifold M, real and of dimension n and the dual bundle $T^{*2}M$ of the 2-osculator bundle T^2M.

Definition 12.1.1. A regular Hamilton of order two is a function $H : T^{*2}M \to R$, differentiable on $\widetilde{T^{*2}M}$ and continuous on the zero section of the projection $\pi^{*2} : T^{*2}M \to M$, whose Hessian, with entries

$$(1.1) \qquad g^{ij}(x,y,p) = \frac{1}{2}\, \dot{\partial}^i \dot{\partial}^j H$$

is nondegenerate.

In other words, the following condition holds

$$(1.1)' \qquad \text{rank}\left\| g^{ij}(x,y,p) \right\| = n, \quad \text{on } \widetilde{T^{*2}M}.$$

Moreover, since $g^{ij}(x,y,p)$ being a d-tensor field, of type $(2,0)$, the condition $(1.1)'$ has geometrical meaning.

Of course, if the base manifold M is paracompact, then on $T^{*2}M$ there exist the regular Hamiltonians.

The d–tensor g^{ij} is symmetric and contravariant. Its covariant d-tensor field will be denoted by $g_{ij}(x,y,p)$ and it is given by the elements of the matrix $\|g^{ij}\|^{-1}$. Hence we have:

$$g_{ij}g^{jk} = \delta^k_i. \tag{1.2}$$

Definition 12.1.2. A Hamilton space of order two is a pair $H^{(2)n} = (M, H(x,y,p))$, where H is a regular Hamiltonian having the property that the tensor field $g^{ij}(x,y,p)$ has a constant signature on the manifold $T^{*2}M$.

As usually, H is called the *fundamental function* and g^{ij} *fundamental tensor* field of the Hamilton space or order two, $H^{(2)n}$.

In the case when the fundamental tensor field g^{ij} is positively defined, then the condition $(1.1)'$ is verified.

Theorem 12.1.1. *If the manifold M is paracompact then always exists a regular Hamiltonian H such that the pair (M,H) gives rise to a Hamilton space of order two.*

Proof. Let $F^n = (M, F(x,y))$ be a Finsler space having $\gamma_{ij}(x,y)$ as fundamental tensor. Then, the function defined on the manifold $T^{*2}M$ by

$$H(x,y,p) = \alpha\gamma^{ij}(x,y)p_ip_j,\ \alpha \in R^+$$

is a regular Hamiltonian or order two, and the pair $H^{(2)n} = (M, H(x,y,p))$ is an Hamilton space of order two. Its fundamental tensor is $g^{ij}(x,y,p) = \alpha\gamma^{ij}(x,y)$. Obviously, M being paracompact, a Finsler space $F^n = (M, F(x,y))$ exists and therefore $H(x,y,p)$ exists. **q.e.d.**

One of the important d-tensor field derived from the fundamental function H of the space $H^{(2)n}$ is:

$$C^{ijh} = -\frac{1}{2}\,\dot{\partial}^h g^{ij} = -\frac{1}{4}\,\dot{\partial}^i\dot{\partial}^j\dot{\partial}^h H. \tag{1.3}$$

Proposition 12.1.1. *We have:*

1° *C^{ijh} is a totally symmetric d-tensor field.*

2° *C^{ijh} vanishes, if and only if the fundamental tensor field g^{ij} does not depend on the momenta p_i.*

Other geometrical object fields which are entirely determined by means of the Hamiltonian of order two, $H(x,y,p)$, are the w_1- and w_2- coefficients of a metrical connection, respectively.

Theorem 12.1.2. *The d-tensor fields*

$$\begin{aligned} C^i{}_{jk} &= \frac{1}{2}\, g^{is}\left(\dot{\partial}_j g_{sk} + \dot{\partial}_k g_{js} - \dot{\partial}_s g_{jk}\right), \\ C_i^{jk} &= -\frac{1}{2}\, g_{is}\left(\dot{\partial}^j g^{sk} + \dot{\partial}^k g^{js} - \dot{\partial}^s g^{jk}\right) \end{aligned} \tag{1.4}$$

have the following properties:

1° *They depend only on the fundamental function H.*

2° *They are symmetric in the indices jk.*

3° *The formula*

$$C_i^{jk} = g_{is} C^{sjk} \tag{1.5}$$

holds.

4° *They are the coefficients of the w_1- and w_2- metrical connection. So we get:*

$$g^{is}|_k = 0, \;\; g^{ij}|^k = 0. \tag{1.6}$$

The proof is not difficult.

The curvature d-tensor fields $S_h{}^i{}_{jk}$, $S_h{}^i{}_j{}^k$ and $S_h{}^{ijk}$ expressed in formulae (6.4)″, Ch.10, depend only on the fundamental function H.

The w_1 and w_2–vertical paths of the Hamilton space of order two are given by Theorem 10.7.5, respectively. Namely

Theorem 12.1.3.

1° *The w_1–vertical paths of the space $H^{(2)n}$ in the point $(x_0^i) \in M$ are characterized by the system of differential equations*

$$x^i = x_0^i,\; p_i = 0,\; \frac{d^2y^i}{dt^2} + C^i{}_{jk}(x_0, y, 0)\frac{dy^j}{dt}\frac{dy^k}{dt} = 0.$$

2° *The w_2–paths of the space $H^{(2)n}$ in the point $(x_0^i) \in M$ are characterized by the system of differential equations*

$$x^i = x_0^i,\; y^i = 0,\; \frac{d^2p_i}{dt^2} - C_i^{jk}(x_0, 0, p)\frac{dp_j}{dt}\frac{dp_k}{dt} = 0.$$

The horizontal paths of the Hamilton space of order two will be studied after a canonical nonlinear connection will be introduced.

12.2 Canonical presymplectic structures and canonical Poisson structures

As we know on the total space $T^{*2}M$ of the dual of 2-osculator bundle there exist different remarkable canonical structures and object fields. Namely:

$$\varphi = p_i y^i, \tag{2.1}$$

$$\omega = p_i dx^i, \tag{2.2}$$

$$\theta = d\omega = dp_i \wedge dx^i, \tag{2.3}$$

where θ is a presymplectic structure on $T^{*2}M$ of rank $2n$.

There exist, also, the canonical Poisson structures $\{\ \}_0$ and $\{\ \}_1$ defined for any $f, g \in \mathcal{F}(T^{*2}M)$ by

$$\begin{aligned}\{f,g\}_0 &= \frac{\partial f}{\partial x^i}\frac{\partial g}{\partial p_i} - \frac{\partial g}{\partial x^i}\frac{\partial f}{\partial p_i},\\ \{f,g\}_1 &= \frac{\partial f}{\partial y^i}\frac{\partial g}{\partial p_i} - \frac{\partial g}{\partial y^i}\frac{\partial f}{\partial p_i}.\end{aligned} \tag{2.4}$$

Each of these Poisson brackets are invariant with respect to changes of coordinates on the manifold $T^{*2}M$ they are $\mathbb{R}$–linear with respect to each argument, skewsymmetric, satisfies the Jacobi identities and the mapping

$$\{f, \cdot\}_\alpha : \mathcal{F}(T^{*2}M) \to \mathcal{F}(T^{*2}M)$$

is a derivation in algebra of the functions $\mathcal{F}(T^{*2}M)$.

Proposition 12.2.1. *The following identities hold:*

1°

$$\begin{aligned}&\{x^i, x^j\}_\alpha = \{y^i, y^j\}_\alpha = \{p_i, p_j\}_\alpha = 0, \quad (\alpha = 0, 1),\\ &\{x^i, y^j\}_\alpha = \{y^i, p_j\}_\alpha = 0, \qquad (\alpha = 0, 1),\\ &\{x^i, p_j\}_0 = \delta^i_j,\ \{y^i, p_j\}_1 = \delta^i_j.\end{aligned} \tag{2.5}$$

2o *For any $H \in \mathcal{F}(T^{*2}M)$ we have:*

$$\begin{aligned}&\{x^i, H\}_0 = \frac{\partial H}{\partial p_i};\ \{x^i, H\}_1 = 0,\\ &\{y^i, H\}_0 = 0,\ \{y^i, H\}_1 = \frac{\partial H}{\partial p_i}\\ &\{p_i, H\}_0 = -\frac{\partial H}{\partial x^i},\ \{p_i, H\}_1 = -\frac{\partial H}{\partial y^i}.\end{aligned} \tag{2.6}$$

Assuming that the manifold $T^{*2}M$ is endowed with a regular Hamiltonian H such that $H^{(2)n} = (M, H)$ is an Hamilton space of order two, we get an *Hamiltonian system of order* 1 given by the triple $(T^{*2}M, H(x, y, p), \theta)$ and we can treat it by the classical methods [cf. M. de Leon and Gotay [85]]. In this case, evidently only the Poisson structure $\{\ \}_0$ will be considered.

Therefore we will study the induced canonical symplectic structures and the induced Poisson structures on the submanifolds Σ_0 and Σ_1 of the manifold $T^{*2}M$, where Σ_0 and Σ_1 will be described below.

Let us consider the bundle $(T^{*2}M, \overline{\pi}^*, T^*M)$ and its canonical section, σ_0 : $(x, p) \in T^*M \rightarrow (x, 0, p) \in T^{*2}M$. Let us denote by $\Sigma_0 = \operatorname{Im}\sigma_0$. It follows that Σ_0 is a submanifold of the manifold $T^{*2}M$. Let us denote the restriction of θ to the submanifold Σ_0 by θ_0, and let us remark that Σ_0 has the equation $y^i = 0$, where (x^i, p_i) are the coordinates of the points $(x, p) \in \Sigma_0$.

Theorem 12.2.1. *The pair* (Σ_0, θ_0) *is a symplectic manifold.*

Proof. Indeed,

$$\theta_0 = dp_i \wedge dx^i \tag{2.7}$$

is a closed 2–form and $\operatorname{rank}\|\theta_0\| = 2n = \dim\Sigma_0$. **q.e.d.**

In a point $u = (x, p) \in \Sigma_0$ the tangent space $T_u\Sigma_0$ has the natural basis $\left(\dfrac{\partial}{\partial x^i}, \dfrac{\partial}{\partial p_i}\right)_u$, $(i = 1, ..., n)$, and natural cobasis $(dx^i, dp_i)_u$.

Let us consider $\mathcal{F}(\Sigma_0)$–module $\mathcal{X}(\Sigma_0)$ and $\mathcal{F}(\Sigma_0)$–module $\mathcal{X}^*(\Sigma_0)$ of tangent vector fields to Σ_0 and of cotangent vector fields to Σ_0, respectively.

Then, the following $\mathcal{F}(\Sigma_0)$–linear mapping

$$S_{\theta_0} : \mathcal{X}(\Sigma_0) \longrightarrow \mathcal{X}^*(\Sigma_0),$$

defined by

$$S_{\theta_0}(X) = i_X\theta_0, \ \forall X \in \mathcal{X}(\Sigma_0), \tag{2.8}$$

has the property:

$$S_{\theta_0}\left(\frac{\partial}{\partial x^i}\right) = -dp_i, \ S_{\theta_0}\left(\frac{\partial}{\partial p_i}\right) = dx^i. \tag{2.8$'$}$$

A glance at the formula (2.8)′, gives:

Proposition 12.2.2. *The mapping* S_{θ_0} *is an isomorphism.*

Let us consider the space $H^{(2)n} = (M, H(x,y,p))$ and denote $H_0 = H_{|\Sigma_0}$. Then the pair $(M, H_0(x,p))$ is a classical Hamilton space (see Ch. 5) having the fundamental tensor field $g^{ij}(x,0,p) = \dfrac{1}{2}\dfrac{\partial H_0}{\partial p_i \partial p_j}$.

The Proposition 12.2.2 shows that there exists a unique vector field $X_{H_0} \in \in \mathcal{X}(\Sigma_0)$, such that

$$S_{\theta_0}(X_{H_0}) = i_{X_{H_0}}\theta = -dH_0. \tag{2.9}$$

In local basis, we get

$$X_{H_0} = \frac{\partial H_0}{\partial p_i}\frac{\partial}{\partial x^i} - \frac{\partial H_0}{\partial x^i}\frac{\partial}{\partial p_i}. \tag{2.10}$$

Theorem 12.2.2. *The integral curves of the vector field X_{H_0} are given by the "Σ_0-canonical equations":*

$$\frac{dx^i}{dt} = \frac{\partial H_0}{\partial p_i},\ \frac{dp_i}{dt} = -\frac{\partial H_0}{\partial x^i},\ y^i = 0. \tag{2.11}$$

For two functions $f, g \in \mathcal{F}(\Sigma_0)$, let X_f, X_g be the corresponding vector fields given by $i_{X_f}\theta_0 = -df$, $i_{X_g}\theta_0 = -dg$.

Theorem 12.2.3. *The following formula holds*

$$\{f, g\}_0 = \theta_0(X_f, X_g). \tag{2.12}$$

Proof. Indeed, we have

$$\begin{aligned}\theta_0(X_f, X_g) &= (i_{X_f}\theta_0)(X_g) = S_{\theta_0}(X_f)(X_g) = -df(X_g) = -X_g f = \\ &= \left(\frac{\partial f}{\partial x^i}\frac{\partial g}{\partial p_i} - \frac{\partial g}{\partial x^i}\frac{\partial f}{\partial p_i}\right) = \{f, g\}_0.\end{aligned} \tag{2.13}$$

Now, taking a canonical 2-form θ_1 on the fibres of the bundle $(T^{*2}M, \pi^{*2}, M)$ we can obtain a similar relation for the Poisson structure $\{\ \}_1$.

Let Σ_1 be the fibre $(\pi^{*2})^{-1}(x_0) \subset T^{*2}M$ in the point $x_0 \in M$. Then Σ_1 is an immersed submanifold given by $\Sigma_1 = \{(x,y,p) \in T^{*2}M \mid x = x_0\}$. In a point $u = (x_0, y, z) \in \Sigma_1$, the natural basis of the tangent space $T_u\Sigma_1$ is given by $\left(\dfrac{\partial}{\partial y^i}, \dfrac{\partial}{\partial p_i}\right)_u$ and natural cobasis by (dy^i, dp_i).

We can prove, without difficulties, that the following expressions give rise to geometrical object fields on the manifold Σ_1:

(2.14)
$$\omega_1 = p_i dy^i,$$
$$\theta_1 = dp_i \wedge dy^i,$$
$$\{f, g\}_1 = \frac{\partial f}{\partial y^i}\frac{\partial g}{\partial p_i} - \frac{\partial g}{\partial y^i}\frac{\partial f}{\partial p_i}.$$

Consequently, we get:

Theorem 12.2.4. *The following properties hold:*

1° θ_1 *is canonical symplectic structure on the manifold* Σ_1.

2° $\{\ \}_1$ *is a canonical Poisson structure on* Σ_1.

The relationship between these two canonical structures can be deduced by the same techniques in the case of the pair of structures $(\theta_0, \{\ \}_0)$.

Indeed, the mapping
$$S_{\theta_1} : \mathcal{X}(\Sigma_1) \longrightarrow \mathcal{X}^*(\Sigma_1)$$
defined by

(2.15)
$$S_{\theta_1}(X) = i_X\theta_1,\ \forall X \in \mathcal{X}(\Sigma_1)$$

has the properties

(2.15)′
$$S_{\theta_1}\left(\frac{\partial}{\partial y^i}\right) = -dp_i,\ S_{\theta_1}\left(\frac{\partial}{\partial p_i}\right) = dy^i.$$

Proposition 12.2.3. S_{θ_1} *is an isomorphism.*

That means that there exists an unique vector field X_{H_1} such that

(2.16)
$$i_{X_{H_1}}\theta_1 = -dH_1,$$

where $H_1 : \Sigma_1 \to R$ is the regular Hamiltonian, $H_1(y,p) = H(x_0, y, p)$.

Locally, X_{H_1} is given by

(2.17)
$$X_{H_1} = \frac{\partial H_1}{\partial p_i}\frac{\partial}{\partial y^i} - \frac{\partial H_1}{\partial y^i}\frac{\partial}{\partial p_i},$$

and its integral curve are as follows

(2.18)
$$x^i = x^i_0;\ \frac{dy^i}{dt} = \frac{\partial H_1}{\partial p_i};\ \frac{dp_i}{dt} = -\frac{\partial H_1}{\partial y^i}.$$

These are called "the Σ_1–canonical equations" of the space $H^{(2)n}$. Therefore, we can state:

Theorem 12.2.5. *The integral curve of the vector field X_{H_1} are given by the Σ_1 canonical equations* (2.18).

Finally, we can prove:

Theorem 12.2.6. *The following formula holds*

$$\{f, g\}_1 = \theta_1(X_f, X_g),\ \forall\, f, g \in \mathcal{F}(\Sigma_1).$$

The previous theory shows the intimate relations between symplectic structures θ_α and the Poisson structure $\{\ \}_\alpha$ on the manifolds Σ_α, $(\alpha = 0, 1)$.

12.3 Lagrange spaces of order two

We shall prove the existence of a natural diffeomorphism between the Hamilton space of order two, $H^{(2)n} = (M, H(x, y, p))$ and the Lagrange spaces of order two $L^{(2)n} = (M, L(x, y^{(1)}, y^{(2)}))$. To this purpose, we shall briefly sketch the general theory of the space $L^{(2)n}$ (see, §1,2, Ch.6, of the book [106]). The fundamental function of the space $L^{(2)n}$ is a Lagrangian of order two, $L : (x, y^{(1)}, y^{(2)}) \in T^2M \to L(x, y^{(1)}, y^{(2)}) \in R$, which is regular and the fundamental tensor field

$$a_{ij}(x, y^{(1)}, y^{(2)}) = \frac{1}{2}\frac{\partial^2 L}{\partial y^{(2)i}\partial y^{(2)j}} \tag{3.1}$$

has a constant signature.

On the manifold T^2M there exist two distribution V_1 and V_2. The distribution V_1 is the vertical distribution of dimension $2n$ and $V_2 \subset V_1$, $\dim V_2 = n$. Clearly, $\dim M = n$, $T^2M = 3n$.

A transformation of local coordinates on $T^2M : (x^i, y^{(1)i}, y^{(2)i}) \to (\tilde{x}^i, \tilde{y}^{(1)i}, \tilde{y}^{(2)i})$, $(i, j, h, k, ... = 1, 2, ..., n)$ is given by the formula (1.1), Ch.6, for $k = 2$. Namely,

$$\begin{cases} \tilde{x}^i = \tilde{x}^i(x^1, ..., x^n),\ \det\left\|\dfrac{\partial \tilde{x}^i}{\partial x^j}\right\| \neq 0, \\ \tilde{y}^{(1)i} = \dfrac{\partial \tilde{x}^i}{\partial x^j}\, y^{(1)j}, \\ 2\tilde{y}^{(2)i} = \dfrac{\partial \tilde{y}^{(1)i}}{\partial x^j}\, y^{(1)j} + 2\dfrac{\partial \tilde{y}^{(1)i}}{\partial y^{(1)j}}\, y^{(2)j}, \end{cases} \tag{3.2}$$

where

$$\frac{\partial \tilde{x}^i}{\partial x^j} = \frac{\partial \tilde{y}^{(1)i}}{\partial y^{(1)j}} = \frac{\partial \tilde{y}^{(2)i}}{\partial y^{(2)j}}; \quad \frac{\partial \tilde{y}^{(1)i}}{\partial x^j} = \frac{\partial \tilde{y}^{(2)i}}{\partial y^{(1)j}}. \tag{3.2$'$}$$

Of course, for every point $u = (x, y^{(1)}, y^{(2)}) \in T^2M$ the natural basis of the tangent space $T_u(T^2M)$ transforms as:

$$\begin{aligned} \frac{\partial}{\partial x^i} &= \frac{\partial \tilde{x}^j}{\partial x^i}\frac{\partial}{\partial \tilde{x}^j} + \frac{\partial \tilde{y}^{(1)j}}{\partial x^i}\frac{\partial}{\partial \tilde{y}^{(1)j}} + \frac{\partial \tilde{y}^{(2)j}}{\partial x^i}\frac{\partial}{\partial \tilde{y}^{(2)j}} \\ \frac{\partial}{\partial y^{(1)i}} &= \frac{\partial \tilde{y}^{(1)j}}{\partial y^{(1)i}}\frac{\partial}{\partial \tilde{y}^{(1)j}} + \frac{\partial \tilde{y}^{(2)j}}{\partial y^{(1)j}}\frac{\partial}{\partial \tilde{y}^{(2)j}} \\ \frac{\partial}{\partial y^{(2)i}} &= \frac{\partial \tilde{y}^{(2)j}}{\partial y^{(2)j}}\frac{\partial}{\partial \tilde{y}^{(2)j}}. \end{aligned} \tag{3.3}$$

By means of these formulae one can prove that the vector fields $\left(\frac{\partial}{\partial y^{(2)1}}, \dots, \frac{\partial}{\partial y^{(2)n}}\right)$ determine a local basis of the distribution V_2, and

$$\left\{\frac{\partial}{\partial y^{(1)1}}, \dots, \frac{\partial}{\partial y^{(1)n}}, \frac{\partial}{\partial y^{(2)1}}, \dots, \frac{\partial}{\partial y^{(2)n}}\right\}$$

is a local basis of the vertical distribution V_1. These distributions are integrable and $V_2 \subset V_1$, $\dim V_1 = 2n$, $\dim V_2 = n$.

Let us remark that there are two Liouville vector fields:

$$\overset{1}{\Gamma} = y^{(1)i}\frac{\partial}{\partial y^{(2)i}}, \quad \overset{2}{\Gamma} = y^{(1)i}\frac{\partial}{\partial y^{(1)i}} + 2y^{(2)i}\frac{\partial}{\partial y^{(2)i}} \tag{3.4}$$

with the properties that $\overset{1}{\Gamma}$ belongs to the distribution V_2 and $\overset{2}{\Gamma}$ belongs to the distribution V_1. The vector fields $\overset{1}{\Gamma}$ and $\overset{2}{\Gamma}$ are linear independent.

There exists a 2-tangent structure J, on T^2M, defined by

$$J\left(\frac{\partial}{\partial x^i}\right) = \frac{\partial}{\partial y^{(1)i}}, \ J\left(\frac{\partial}{\partial y^{(1)i}}\right) = \frac{\partial}{\partial y^{(2)i}}, \ J\left(\frac{\partial}{\partial y^{(2)i}}\right) = 0, \ (i = 1, ..., n). \tag{3.5}$$

The following properties of the 2-tangent structure J hold:

1° J is a tensor field on T^2M of type $(1,1)$.

2° J is an integrable structure.

3° $\operatorname{Im} J = V_1$, $\operatorname{Ker} J = V_2$, $J(V_1) = V_2$.

4° $\operatorname{rank}\|J\| = 2n$.

5° $J\overset{2}{\Gamma} = \overset{1}{\Gamma}, \ J\overset{1}{\Gamma} = 0.$

6° $J^3 = 0.$

A 2-semispray on T^2M is a vector field S on T^2M with the property

$$JS = \overset{2}{\Gamma}. \tag{3.6}$$

Locally S is given by

$$S = y^{(1)i}\frac{\partial}{\partial x^i} + 2y^{(2)i}\frac{\partial}{\partial y^{(1)i}} - 3G^i(x, y^{(1)}, y^{(2)})\frac{\partial}{\partial y^{(2)i}}, \tag{3.7}$$

where G^i are *the coefficients* of S and they characterize the vector field S.

A nonlinear connection N on the manifold $T^2M = \mathrm{Osc}^2M$ is a vector subbundle $N(T^2M)$ of the tangent bundle $T(T^2M)$ which, together with the vertical subbundle $V(T^2M)$, give the Whitney sum:

$$TT^2M = NT^2M \oplus VT^2M.$$

Noticing that $J(N_0) = N_1$, $N_0 = N$, and that N_1 is a subdistribution of the vertical distribution V_1, we obtain the direct decomposition of linear spaces

$$T_uT^2M = N_0(u) \oplus N_1(u) \oplus V_2(u), \ \forall u \in T^2M. \tag{3.8}$$

An adapted basis to this direct decomposition is given by

$$\left\{\frac{\delta}{\delta x^i}, \frac{\delta}{\delta y^{(1)i}}, \frac{\partial}{\partial y^{(2)i}}\right\}, \tag{3.9}$$

where

$$\frac{\delta}{\delta x^i} = \frac{\partial}{\partial x^i} - \underset{(1)}{N}{}^j_i\frac{\partial}{\partial y^{(1)j}} - \underset{(2)}{N}{}^j_i\frac{\partial}{\partial y^{(2)i}}, \ \frac{\delta}{\delta y^{(1)i}} = \frac{\partial}{\partial y^{(1)i}} - \underset{(1)}{N}{}^j_i\frac{\partial}{\partial y^{(2)i}}. \tag{3.9)$'$}$$

The systems of functions $\left\{\underset{(1)}{N}{}^j_i, \underset{(2)}{N}{}^j_i\right\}$ give *the coefficients* of the nonlinear connection N.

The adapted cobasis, which is the dual basis of (3.9),

$$\{dx^i, \delta y^{(1)i}, \delta y^{(2)i}\}, \tag{3.10}$$

where

$$\delta y^{(1)i} = dy^{(1)i} + \underset{(1)}{M}{}^i_j dx^j, \ \delta y^{(2)i} = dy^{(2)i} + \underset{(1)}{M}{}^i_j dy^{(1)j} + \underset{(2)}{M}{}^i_j dx^j \tag{3.10)$'$}$$

The new coefficients $\underset{(1)}{M}{}^i_j, \underset{(2)}{M}{}^i_j$ of the nonlinear connection N that appear here are called the *dual coefficients* of N. They are related with the primary coefficients $\underset{(1)}{N}{}^i_j, \underset{(2)}{N}{}^i_j$ by the formula:

$$\underset{(1)}{M}{}^i_j = \underset{(1)}{N}{}^i_j, \; \underset{(2)}{M}{}^i_j = \underset{(2)}{N}{}^i_j + \underset{(1)}{N}{}^i_m \underset{(1)}{N}{}^m_j \tag{3.11}$$

Conversely, the previous formulae, uniquely determine $\underset{(1)}{N}, \underset{(2)}{N}$ as functions of $\underset{(1)}{M}, \underset{(2)}{M}$.

R. Miron [106] and I. Bucătaru [36], [37] showed that a 2-semispray with the coefficients G^i, uniquely determines a nonlinear connection. The dual coefficients given by Bucătaru are very simple:

$$\underset{(1)}{M}{}^i_j = \frac{\partial G^i}{\partial y^{(2)j}}, \; \underset{(2)}{M}{}^i_j = \frac{\partial G^i}{\partial y^{(1)j}}. \tag{3.12}$$

Studying the variational problem for the regular Lagrangian of order 2, $L(x, y^{(1)}, y^{(2)})$ we can determine a canonical nonlinear connection of the Lagrange space of order two, $L^{(2)n} = (M, L(x, y^{(1)}, y^{(2)}))$.

12.4 Variational problem in the spaces $L^{(2)n}$

Let us consider a Lagrange space of order two, $L^{(2)n} = (M, L(x, y^{(1)}, y^{(2)}))$. If $X \in \in \mathcal{X}(E)$, let us denote the operator of Lie derivation with respect to X by $\mathcal{L}_X$. Applying this operator with respect to the Liouville vector fields $\overset{1}{\Gamma}, \overset{2}{\Gamma}$, (2.4) we get two important scalar fields determined by the Lagrangian L:

$$\overset{1}{I}(L) = \mathcal{L}_{\overset{1}{\Gamma}} L, \; \overset{2}{I}(L) = \mathcal{L}_{\overset{2}{\Gamma}} L. \tag{4.1}$$

They are called *the main invariants* of the space $L^{(2)n}$.

Let $c : [0, 1] \to M$ be a smooth parameterized curve and assume $\operatorname{Im} c \subset U \subset M$, where U is a domain of a local chart on the manifold M. The curve c is represented by the equations $x^i = x^i(t)$, $t \in [0, 1]$. The extension $\widetilde{c}$ of c to the manifold T^2M is:

$$x^i = x^i(t), \; y^{(1)i} = \frac{dx^i}{dt}(t), \; y^{(2)i} = \frac{1}{2}\frac{d^2x^i}{dt^2}(t), \; t \in [0, 1]. \tag{4.2}$$

The integral of action of the Lagrangian $L(x, y^{(1)}, y^{(2)})$ along the curve c is defined by

$$I(c) = \int_0^1 L\left(x(t), \frac{dx}{dt}, \frac{1}{2}\frac{d^2x}{dt^2}\right) dt. \tag{4.3}$$

It is known that if $I(c)$ does not depend by the parameterization of the curve c then $\overset{1}{I}(L) = 0$, $\overset{2}{I}(L) = L$ (Zermelo conditions). In this case rank$\|a_{ij}\| < n$. So, the fundamental tensor of the space $L^{(2)n}$ is singular. Consequently, the functional $I(c)$ depends on the parametrization of the curve c.

Along the curve c, the following operators can be introduced:

$$\begin{aligned} \overset{0}{E}_i &= \frac{\partial}{\partial x^i} - \frac{d}{dt}\frac{\partial}{\partial y^{(1)i}} + \frac{1}{2}\frac{d^2}{dt^2}\frac{\partial}{\partial y^{(2)i}}, \\ \overset{1}{E}_i &= -\frac{\partial}{\partial y^{(1)i}} + \frac{d}{dt}\frac{\partial}{\partial y^{(2)i}}, \\ \overset{2}{E}_i &= \frac{1}{2}\frac{\partial}{\partial y^{(2)i}}. \end{aligned} \tag{4.4}$$

In the monograph [106], the following theorems are proved:

Theorem 12.4.1.

1° *For any differentiable function* $\phi(t)$, $t \in [0,1]$ *we have*

$$\overset{0}{E}_i\,(\phi L) = \phi\,\overset{0}{E}_i\,(L) + \frac{d\phi}{dt}\,\overset{1}{E}_i(L) + \frac{d^2\phi}{dt^2}\,\overset{2}{E}_i(L). \tag{4.5}$$

2° $\overset{\alpha}{E}_i(L)$, $(\alpha = 0,1,2)$ *are d-covector fields.*

3° $\dfrac{dL}{dt} = \dfrac{dx^i}{dt}\,\overset{0}{E}_i\,(L) + \dfrac{d}{dt}\,\overset{2}{I}(L) - \dfrac{1}{2}\dfrac{d^2}{dt^2}\,\overset{1}{I}(L).$

Theorem 12.4.2. *The variational problem on the integral of action $I(c)$ leads to the Euler–Lagrange equations*

$$\overset{0}{E}_i\,(L) := \frac{\partial L}{\partial x^i} - \frac{d}{dt}\frac{\partial L}{\partial y^{(1)i}} + \frac{1}{2}\frac{d^2}{dt^2}\frac{\partial L}{\partial y^{(2)i}} = 0,\ y^{(1)i} = \frac{dx^i}{dt},\ y^{(2)i} = \frac{1}{2}\frac{d^2x^i}{dt^2}. \tag{4.6}$$

Taking into account of the Theorem 12.4.1 it follows that $\dfrac{dL}{dt}$ do not vanishes along with the integral curve of the Euler–Lagrange equations. Therefore, it introduce the notion of (Hamiltonian) energy, [106]. However we point out that in the case of the space $L^{(2)n}$ it will depend on the curve $c : [0,1] \to M$.

Definition 12.4.1. Along the smooth curve $c : [0,1] \to M$ the following functions

$$\overset{2}{\mathcal{E}}_c(L) = \overset{2}{I}(L) - \frac{1}{2}\frac{d}{dt}\,\overset{1}{I}(L) - L,\quad \overset{1}{\mathcal{E}}_c(L) = -\frac{1}{2}\overset{1}{I}(L) \tag{4.7}$$

are called the energy of order two and of order 1 of the Lagrange space $L^{(2)n} = (M, L)$, *respectively*.

In the monograph [106] it is proved:

Theorem 12.4.3. *For any differentiable Lagrangian $L(x, y^{(1)}, y^{(2)})$ the energy of order two $\overset{2}{\mathcal{E}}_c(L)$ is conserved along every solution curve c of the Euler–Lagrange equations $\overset{0}{E}_i(L) = 0$.*

Along the smooth curve $c : [0,1] \to M$, the energies $\overset{2}{\mathcal{E}}_c(L)$, and $\overset{1}{\mathcal{E}}_c(L)$ can be written in the form

$$\overset{2}{\mathcal{E}}_c(L) = p_{(1)i}\frac{dx^i}{dt} + p_i\frac{d^2x^i}{dt^2} - L,\ \overset{1}{\mathcal{E}}_c(L) = p_i\ \frac{dx^i}{dt} \tag{4.8}$$

where

$$p_{(1)i} = \frac{\partial L}{\partial y^{(1)i}} - \frac{1}{2}\ \frac{d}{dt}\ \frac{\partial L}{\partial y^{(2)i}},\ p_i = \frac{1}{2}\ \frac{\partial L}{\partial y^{(2)i}} \tag{4.9}$$

are the *Jacobi–Ostrogradski momenta.*

Theorem 12.4.4. *Along a smooth curve c we have*

$$\begin{aligned} &\frac{dp_{(1)i}}{dt} - \frac{\partial L}{\partial x^i} = -\overset{0}{E}_i(L) \\ &\frac{dp_i}{dt} - p_{(1)i} = \overset{1}{E}_i(L). \end{aligned} \tag{4.10}$$

This property is useful in order to prove:

Theorem 12.4.5. *Along each solution curve c of the Euler–Lagrange equations $\overset{0}{E}_i(L) = 0$, the following Hamilton equations hold:*

$$\begin{aligned} &\frac{\partial \overset{2}{\mathcal{E}}_c(L)}{\partial p_{(1)i}} = \frac{dx^i}{dt},\ \frac{\partial \overset{2}{\mathcal{E}}_c(L)}{\partial x^i} = -\frac{dp_{(1)i}}{dt} \\ &\frac{\partial \overset{2}{\mathcal{E}}_c(L)}{\partial p_i} = \frac{d^2x^i}{dt^2},\ \frac{\partial \overset{2}{\mathcal{E}}_c(L)}{\partial y^{(1)i}} = -\frac{dp_i}{dt}. \end{aligned} \tag{4.11}$$

From this reason $\overset{2}{\mathcal{E}}_c(L)$ is called the Hamiltonian energy of the space $L^{(2)n}$.

Some of the previous results hold even in the case when L is not a regular Lagrangian. If L is the fundamental function of a Lagrange space of order 2, $L^{(2)n} = (M, L)$ we can determine a canonical nonlinear connection N depending only on L.

Indeed, we consider *the Synge equation* [106]:

$$a^{ij}\overset{1}{E}_i(L) = \frac{d^3x^i}{dt^2} + 3!G^i\left(x, \frac{dx}{dt}, \frac{1}{2}\frac{d^2x}{dt^2}\right) = 0, \tag{4.12}$$

where

$$3G^i = \frac{1}{2}\, a^{ij}\left\{y^{(1)m}\frac{\partial}{\partial x^m}\left(\frac{\partial L}{\partial y^{(2)j}}\right) + 2y^{(2)m}\frac{\partial}{\partial y^{(1)m}}\left(\frac{\partial L}{\partial y^{(2)j}}\right) - \frac{\partial L}{\partial y^{(1)j}}\right\}. \tag{4.13}$$

Thus, the canonical semispray of $L^{(2)n}$ is given as follows:

Theorem 12.4.6. *Any Lagrange space $L^{(2)n} = (M, L)$ has a canonical semispray, determined only by the fundamental function L. It is given by:*

$$S = y^{(1)i}\frac{\partial}{\partial x^i} + 2y^{(2)i}\frac{\partial}{\partial y^{(1)i}} - 3G^i(x, y^{(1)}, y^{(2)})\frac{\partial}{\partial y^{(2)i}}, \tag{4.14}$$

where the coefficients G^i are expressed in the formula (4.13).

Consequently, we obtain [106]:

Theorem 12.4.7. *For any Lagrange space $L^{(2)n} = (M, L)$ there exist the nonlinear connections determined only by the fundamental function L. One of them is given by the dual coefficients $(\underset{(1)}{M}{}^i_j, \underset{(2)}{M}{}^i_j)$ from the formulae* (3.12), (4.13).

The nonlinear connection mentioned in the previous theorem will be called canonical for the space $L^{(2))n}$.

12.5 Legendre mapping determined by a space $L^{(2)n}$

If a Lagrange space of order two, $L^{(2)n} = (M, L(x, y^{(1)}, y^{(2)}))$ is given, then it determines a local diffeomorphism $\varphi : \widetilde{T^2M} \longrightarrow \widetilde{T^{*2}M}$, which preserves the fibres. The mapping φ transforms the canonical semispray S of $L^{(2)n}$ in the dual semispray S_ξ, where $\xi = \varphi^{-1}$, and determines a nonlinear connection N^* on $T^{*2}M$. Still, like in the classical case, φ does not transform the regular Lagrangian $L(x, y^{(1)}, y^{(2)})$ in a regular Hamiltonian $H(x, y, p)$. However, a formula of type (1.6), $Ch.8$ can be introduced. We investigate these problems in the following.

If we denote $x = y^{(0)}$, the fundamental function L in $L^{(2)n}$ will be written as $L(y^{(0)}, y^{(1)}, y^{(2)})$ and its fundamental tensor will be given by

$$a_{ij}(y^{(0)}, y^{(1)}, y^{(2)}) = \frac{1}{2}\,\frac{\partial^2 L}{\partial y^{(2)i}\partial y^{(2)j}}.$$

Proposition 12.5.1. *If L is the fundamental function of a Lagrange space of order two, $L^{(2)n}$, then the following mapping $\varphi : (y^{(0)}, y^{(1)}, y^{(2)}) \in \widetilde{T^2M} \longrightarrow (x, y, p) \in \widetilde{T^{*2}M}$, given by*

$$x^i = y^{(0)i},\quad y^i = y^{(1)i},\quad p_i = \frac{1}{2}\frac{\partial L}{\partial y^{(2)i}}, \tag{5.1}$$

is a local diffeomorphism which preserve the fibres.

Proof. The mapping φ is differentiable and its Jacobian has the determinant equal to $\det \|a_{ij}\|$, which do not vanish on $\widetilde{T^2M}$.

Of course, we have $\pi^2(y^{(0)}, y^{(1)}, y^{(2)}) = \pi^{*2} \circ \varphi(y^{(0)}, y^{(1)}, y^{(2)})$.

This diffeomorphism is called the *Legendre mapping* (or transformation).

We denote

$$p_i = \frac{1}{2}\frac{\partial L}{\partial y^{(2)i}} = \varphi_i(y^{(0)}, y^{(1)}, y^{(2)}). \tag{5.2}$$

Clearly, φ_i is a d-covector field on $L^{(2)n}$.

The local inverse diffeomorphism $\xi = \varphi^{-1}$ is given by

$$\begin{cases} y^{(0)i} = x^i, \\ y^{(1)i} = y^i, \\ y^{(2)i} = \xi^i(x, y, p). \end{cases} \tag{5.3}$$

The mapping ξ^i has the same rule of transformation as the variables $y^{(2)i}$ from (3.2), with respect to a changing of local coordinates on $T^{*2}M$.

The mappings φ and ξ satisfy the conditions:

$$\xi \circ \varphi = 1_{\check{U}},\ \varphi \circ \xi = 1_{\hat{U}},\ \check{U} = (\pi^2)^{-1}(U),\ \hat{U} = (\pi^{*2})^{-1}(U),\ U \subset M.$$

We have the following identities

$$a_{ij}(y^{(0)}, y^{(1)}, y^{(2)}) = \frac{\partial \varphi_i}{\partial y^{(2)j}},\ a^{ij}(x, y, \xi(x, y, p)) = \frac{\partial \xi^i}{\partial p_j} \tag{5.4}$$

and

$$\begin{cases} \dfrac{\partial \varphi_i}{\partial x^j} = -a_{is}\dfrac{\partial \xi^s}{\partial x^j};\ \dfrac{\partial \varphi_i}{\partial y^j} = -a_{is}\dfrac{\partial \xi^s}{\partial y^j};\ \dfrac{\partial \varphi_i}{\partial y^{(2)j}} = a_{ij} \\ \dfrac{\partial \xi^i}{\partial x^j} = -a^{is}\dfrac{\partial \varphi_s}{\partial x^j};\ \dfrac{\partial \xi^i}{\partial y^j} = -a^{is}\dfrac{\partial \varphi_s}{\partial y^j};\ \dfrac{\partial \xi^i}{\partial p_j} = a^{ij} \end{cases} \tag{5.5}$$

The differential $\varphi_* : T_u(T^2M) \to T_{\varphi(u)}(T^{*2}M)$, of the diffeomorphism φ is expressed in the natural basis as follows

$$\begin{aligned}\frac{\partial}{\partial y^{(0)i}} &= \frac{\partial}{\partial x^i} + \frac{\partial \varphi_m}{\partial x^i}\frac{\partial}{\partial p_m},\\ \frac{\partial}{\partial y^{(1)i}} &= \frac{\partial}{\partial y^i} + \frac{\partial \varphi_m}{\partial y^i}\frac{\partial}{\partial p_m},\\ \frac{\partial}{\partial y^{(2)i}} &= a_{im}\frac{\partial}{\partial p_m}.\end{aligned} \tag{5.6}$$

Theorem 12.5.1. *The mapping $\varphi : \widetilde{T^2M} \to \widetilde{T^{*2}M}$, (5.1), transforms the semispray*

$$S = y^{(1)i}\frac{\partial}{\partial x^i} + 2y^{(2)i}\frac{\partial}{\partial y^{(1)i}} - 3G^i(x, y^{(1)}, y^{(2)})\frac{\partial}{\partial y^{(2)i}}, \tag{5.7}$$

in the dual semispray S_ξ on $\widetilde{T^{*2}M}$:

$$S_\xi = y^i\frac{\partial}{\partial x^i} + 2\xi^i\frac{\partial}{\partial y^i} + f_i\frac{\partial}{\partial p_i}, \tag{5.7$'$}$$

which has the following coefficients

$$f_i = -a_{is}\left(\frac{\partial \xi^s}{\partial x^r}y^r + 2\frac{\partial \xi^s}{\partial y^r}\xi^r + 3G^s(x, y, \xi(x, y, p))\right). \tag{5.7$''$}$$

Proof. If $X \in \mathcal{X}(\widetilde{T^2M})$ have the local expression

$$X = X^{(0)i}\frac{\partial}{\partial y^{(0)i}} + X^{(1)i}\frac{\partial}{\partial y^{(1)i}} + X^{(2)i}\frac{\partial}{\partial y^{(2)i}},$$

then

$$\begin{aligned}\varphi_* X = X^{(0)i}(x, y, \xi)\frac{\partial}{\partial x^i} + X^{(1)i}(x, y, \xi)\frac{\partial}{\partial y^i} +\\ + \left(X^{(0)m}\frac{\partial \varphi_i}{\partial x^m} + X^{(1)m}\frac{\partial \varphi_i}{\partial y^m} + X^{(2)m}a_{mi}\right)\frac{\partial}{\partial p_i}.\end{aligned}$$

Consequently $\varphi_* S = S_\xi$ holds.

Corollary 12.5.1. *The dual semispray S_ξ (5.7)$'$, determines on $\widetilde{T^{*2}M}$ a nonlinear connection with the coefficients:*

$$N^i{}_j = \dot{\partial}^i f_j, \;\; N_{ij} = \dot{\partial}_i f_j. \tag{5.8}$$

We have to ask ourselves if by means of the mapping φ we can transform a regular Lagrangian $L(y^{(0)}, y^{(1)}, y^{(2)})$ in a regular Hamiltonian. Notice that $y^{(2)i} = \xi^i(x, y, p)$ is not a vector field. Therefore the product $p_i\xi^i(x, y, p)$ is not a scalar field as in the classical case of the Hamilton spaces $H^{(1)n} = (M, H(x, p))$.

Let us fix a nonlinear connection $\overset{0}{N}$ with the coefficients $\overset{0}{N}{}^i_j(x, y)$ on $T^1M \subset$ $\subset T^2M$. Then on $\widetilde{T^2M}$ we get the d-vector field

$$z^{(2)i} = y^{(2)i} + \frac{1}{2}\overset{0}{N}{}^i_j(y^{(0)}, y^{(1)})y^{(1)j}. \tag{5.9}$$

This d-vector field is transformed by φ in the following d-vector field on $T^{*2}M$:

$$\check{z}^{(2)i} = \xi^i(x, y, p) + \frac{1}{2}\overset{0}{N}{}^i_j(x, y)y^j. \tag{5.9$'$}$$

Let us consider the following Hamiltonian:

$$H(x, y, p) = 2p_i\check{z}^{(2)i} - L(x, y, \xi(x, y, p)). \tag{5.10}$$

Then we have:

Theorem 12.5.2. *The Hamiltonian function H, (5.10), is the fundamental function of a Hamilton space $H^{(2)n}$ and its fundamental tensor field $a^{is}(x, y, \xi(x, y, p))$ is the contravariant of the fundamental tensor field a_{ij} of the space $L^{(2)n} = (M, L)$.*

Proof. From the formula (5.10) we deduce

$$\frac{1}{2}\dot{\partial}^j H = \check{z}^{(2)j}. \tag{5.10$'$}$$

Therefore, we get

$$g^{ij}(x, y, p) = \frac{1}{2}\dot{\partial}^i\dot{\partial}^j H = \frac{\partial\check{z}^{(2)j}}{\partial p_i} = \frac{\partial\xi^j}{\partial p_i} = a^{ij}(x, y, \xi(x, y, p)). \tag{5.10$''$}$$

Consequently, the pair $H^{(2)n} = (M, H)$, (5.10), is an Hamilton space or order two. **q.e.d.**

The space $H^{(2)n} = (M, H)$, (5.10), is called the *dual* of the space $L^{(2)n} = (M, L)$. Of course, this dual depends on the choice of the nonlinear connection of $\overset{0}{N}$.

12.6 Legendre mapping determined by $H^{(2)n}$

Now let us pay attention to the inverse problem: Being given a Hamilton space of order two, $H^{(2)n} = (M, H(x, y, p))$, let us determine its dual, i.e., a Lagrange space of order two.

In this case, we will start from $H^{(2)n}$ and try to determine a local diffeomorphism of form (5.3) by means of the fundamental function $H(x,y,p)$ of $H^{(2)n}$. But $y^{(2)i}$ is not a vector field. Therefore we cannot define it only by $\frac{1}{2}\dot{\partial}^i H$, which is a d-vector field. As in the previous section we assume that the nonlinear connection $\overset{\circ}{N}$, with coefficients $\overset{\circ}{N}{}^i{}_j(x,y)$, which does not depend on the momenta p_i, is apriori given.

Consequently,

$$z^i = y^{(2)i} + \frac{1}{2}\overset{\circ}{N}{}^i{}_j(x,y^{(1)})y^{(1)j}, \tag{6.1}$$

is a d-vector field on $\widetilde{T^2M}$.

The mapping $\xi_1 : \widetilde{T^{*2}M} \longrightarrow \widetilde{T^2M}$ defined by

$$y^{(0)i} = x^i,\ y^{(1)i} = y^i,\ y^{(2)i} = \xi_1^i(x,y,p), \tag{6.2}$$

where

$$\xi_1^i(x,y,p) = \frac{1}{2}\{\dot{\partial}^i H(x,y,p) - \overset{\circ}{N}{}^i{}_j(x,y)y^j\}, \tag{6.2$'$}$$

is the *Legendre transformation* determined by the pair $(H(x,y,p), \overset{\circ}{N}{}^i_j)$.

Theorem 12.6.1. *The mapping given by* (6.2), (6.2)$'$, *is a local diffeomorphism, which preserves the fibres of* $T^{*2}M$ *and* T^2M.

Proof. The determinant of the Jacobian of ξ_1 is equal to $\det\|g^{ij}(x,y,p)\|$ and $\pi^* = \pi \circ \xi_1$.

The formula (6.1), (6.2), (6.2)$'$ imply:

$$\dot{\partial}^i \xi_1^j = g^{ij}(x,y,p), \tag{6.3}$$

$$z^i(x,y,\xi_1^i) = \frac{1}{2}\dot{\partial}^i H(x,y,p). \tag{6.4}$$

Let us consider the inverse mapping φ_1 of the Legendre transformation ξ_1 :

$$x^i = y^{(0)i},\ y^i = y^{(1)i},\ p_i = \varphi_{1i}(y^{(0)}, y^{(1)}, y^{(2)}). \tag{6.5}$$

It follows

$$\frac{\partial \varphi_{1i}}{\partial y^{(2)j}} = g_{ij}(x,y,\varphi_1). \tag{6.6}$$

In a regular Lagrangian it is interesting to remark that the Hamiltonian $H(x,y,p)$ is transformed by ξ_1 exactly as in the classical case:

$$L(x,y^{(1)},y^{(2)}) = 2p_i z^i - H(x,y,p),\ p_i = \varphi_{1i}(x,y^{(1)},y^{(2)}). \tag{6.7}$$

Theorem 12.6.2. *The Lagrangian L from* (6.7) *is a regular one. Its fundamental tensor field is given by* $g_{ij}(x, y^{(1)}, \varphi_1)$.

Proof. Because $\dfrac{\partial z^i}{\partial y^{(2)j}} = \delta^i_j$, it results $\dfrac{1}{2}\dfrac{\partial L}{\partial y^{(2)i}} = \varphi_{1i}(x, y^{(1)}, y^{(2)})$ and from (6.6) we get $\dfrac{1}{2}\dfrac{\partial^2 L}{\partial y^{(2)i}\partial y^{(2)j}} = g_{ij}(x, y^1, \varphi_1)$. **q.e.d.**

The space $L^{(2)n} = (M, L)$ with L given in (6.7) is called the *dual* of the space $H^{(2)n}$.

12.7 Canonical nonlinear connection of the space $H^{(2)n}$

The Lagrange space of order two, $L^{(2)n} = (M, L(x, y^{(1)}, y^{(2)})$, with the fundamental function

(7.1) $$L(x, y^1, y^2) = 2p_i z^i - H(x, y, p),\ p_i = \varphi_{1i}(x, y^{(1)}, y^{(2)})$$

is the dual of the Hamilton space of order two, $H^{(2)n} = (M, H(x, y, p))$. Its canonical semispray

$$S = y^{(1)i}\frac{\partial}{\partial x^i} + 2y^{(2)i}\frac{\partial}{\partial y^{(1)i}} - 3G^i(x, y^{(1)}, y^{(2)})\frac{\partial}{\partial y^{(2)i}}$$

is transformed by the Legendre transformation ξ_1 in the *canonical* dual semispray S_{ξ_1}:

$$S_{\xi_1} = y^i\frac{\partial}{\partial x^i} + 2\xi^i_1\frac{\partial}{\partial y^i} + f_i(x, y, p)\frac{\partial}{\partial p_i}.$$

The relation between the coefficients G^i and f_i is as follows

(7.2) $$f_i = y^m\frac{\partial \varphi_{1i}}{\partial x^m} + 2\xi^m\frac{\partial \varphi_{1i}}{\partial y^m} - 3G^m(x, y, \xi_1)g_{mi},\ y^{(2)i} = \xi^i_1(x, y, p).$$

And since ξ_1 and φ_1 are the inverse mappings, respectively, we have

$$\frac{\partial \varphi_{1i}}{\partial x^m} = -g_{is}\frac{\partial \xi^s_1}{\partial x^m},\ \frac{\partial \varphi_{1i}}{\partial y^m} = -g_{is}\frac{\partial \xi^s_1}{\partial y^m}.$$

Substituting in (7.2) we get

(7.2)′ $$-g^{ij}f_j = y^m\frac{\partial \xi^i_1}{\partial x^m} + 2\xi^m_1\frac{\partial \xi^i_1}{\partial y^m} + 3G^i(x, y, \xi_1).$$

The expression (4.13) of $G^i(x, y^{(1)}, y^{(2)})$ give

$$-3G^i(x,y,\xi_1) = y^m \frac{\partial \xi_1^i}{\partial x^m} + 2\xi_1^m \frac{\partial \xi_1^i}{\partial y^m} + \frac{1}{2} g^{im} \left\{ \frac{\partial}{\partial y^m}(p_s \dot{\partial}^s H - H) - p_s \frac{\partial \xi_1^s}{\partial y^m} \right\}.$$

The last formula and (7.2)′ leads to

$$f_m = \frac{1}{2}\left[\frac{\partial}{\partial y^m}(p_s \dot{\partial}^s H) - p_s \frac{\partial \xi_1^s}{\partial y^m}\right]. \tag{7.3}$$

Because ξ_1^i is in (6.2)′, we have:

Theorem 12.7.1. *The coefficients f_i of the canonical semispray S_{ξ_1} of the Hamilton spaces $H^{(2)n}$ are given by:*

$$f_j(x,y,p) = \frac{1}{2}\frac{\partial}{\partial y^m}\left[-H + p_s \overset{\circ}{N}{}^s_m(x,y) y^m\right] \tag{7.4}$$

Finally, applying Theorem 9.5.2., we can formulate:

Theorem 12.7.2. *The coefficients of the canonical nonlinear connection N of the Hamilton space $H^{(2)n}$ are as follows:*

(7.5)

$$N^i{}_j = \frac{1}{2}\frac{\partial}{\partial y^j}\left[-\dot{\partial}^i H + \overset{\circ}{N}{}^i_m(x,y) y^m\right]; N_{ij} = \frac{1}{2}\frac{\partial^2}{\partial y^i \partial y^j}\left[-H + p_s \overset{\circ}{N}{}^s_m(x,y) y^m\right].$$

Remarks.

1° If $\frac{\partial H}{\partial y^i} = 0$, $\overset{\circ}{N}{}^i_j(x,y) = \overset{\circ}{N}{}^i{}_{jk}(x) y^k$, then N coincides with $\overset{\circ}{N}$.

2° The torsion $\tau_{ij} = N_{ij} - N_{ji}$ vanishes.

The Theorem 12.7.2 is important in applications.

12.8 Canonical metrical N connection of space $H^{(2)n}$

For a Hamilton space of order two, $H^{(2)n} = (M, H(x, y, p))$ let us consider the canonical nonlinear connection N determined in the previous section. We are going to investigate the N-linear connections which are metrical with respect to fundamental tensor field g^{ij} of $H^{(2)n}$, i.e.:

$$g^{ij}{}_{|h} = 0,\ g^{ij}|_h = 0,\ g^{ij}|^h = 0 \tag{8.1}$$

Considering the space $H^{(2)n}$ as a generalized space $GH^{(2)n}$ with the fundamental tensor g^{ij} we can apply Theorems 11.2.1 and 11.2.2, one obtain:

Theorem 12.8.1. *For a Hamilton space of order two, $H^{(2)n} = (M, H(x,y,p))$ the following properties hold:*

1° *There exists a unique N-connection $C\Gamma(N) = (H^i{}_{jk}, C^i{}_{jk}, C_i^{jk})$ which satisfies (8.1), as well as the conditions*

$$H^i{}_{jk} = H^i{}_{kj},\ C^i{}_{jk} = C^i{}_{kj},\ C_i^{jk} = C_i^{kj} \tag{8.2}$$

2° *The coefficients of $D\Gamma(N)$ are given by the generalized Christoffel symbols:*

$$\begin{aligned} H^i{}_{jk} &= \frac{1}{2}g^{is}(\delta_j g_{sk} + \delta_k g_{js} - \delta_s g_{jk}),\\ C^i{}_{jk} &= \frac{1}{2}g^{is}(\dot{\partial}_j g_{sk} + \dot{\partial}_k g_{js} - \dot{\partial}_s g_{jk}),\\ C_I^{jk} &= -\frac{1}{2}g_{is}(\dot{\partial}^j g^{sk} + \dot{\partial}^k g^{js} - \dot{\partial}^s g^{jk}), \end{aligned} \tag{8.3}$$

where the operators $\delta_j = \partial_j - N^k{}_j\dot{\partial}_k + N_{jk}\dot{\partial}^k$ are constructed using the canonical nonlinear connection N.

The connection $C\Gamma(N)$ is called the *metrical N–connection* of $H^{(2)n}$.

Now, applying the theory from Ch.9, we can write the structure equations of the metrical N-connection $C\Gamma(N)$, Ricci identities and Bianchi identities. The parallelism theory as well as, the theory of the special curves, horizontal paths etc. can be obtained.

We can conclude that the geometry of the second order Hamilton spaces, can be constructed from the canonical connections N and $C\Gamma(N)$.

The geometrical model of the space $H^{(2)n}$ is determined the N-lift $\mathbb{G}$, ((3.1), Ch.11) of the fundamental tensor g^{ij} and by the (p)-almost contact structure $\check{\mathbb{F}}$, ((3.7), Ch.11).

We obtain, without difficulties:

Theorem 12.8.2. *The pair $(\mathbb{G}, \check{\mathbb{F}})$ is a Riemannian almost contact structure determined only by canonical nonlinear connection N and by fundamental function H of the space $H^{(2)n}$. If N is torsion free (i.e. $N_{ij} = N_{ji}$), then its associated 2-form $\check{\theta}$ is canonical presymplectic structure $\check{\theta} = dp_i \wedge dx^i$.*

The geometrical space $\mathcal{H}^{3n} = (\widetilde{T^{*2}M}, \mathbb{G}, \check{\mathbb{F}})$ is called the *geometrical model* of the space $H^{(2)n} = (M, H(x,y,p))$. It can be used to study the main geometrical features of the space $H^{(2)n}$.

12.9 The Hamilton spaces $H^{(2)n}$ of electrodynamics

Let us consider the Hamilton spaces of order two, $H^{(2)n} = (M, H(x,y,p))$, with the fundamental function

$$(9.1)\quad H(x,y,p) = \frac{1}{mc} g^{ij}(x,y)p_ip_j - \frac{2e}{m} p_iA^i(x,y) + \frac{e^2}{mc^2} A^i(x,y)A^j(x,y)g_{ij}(x,y),$$

where $g_{ij}(x,y)$ is the fundamental tensor of a Finsler space $F^n = (M, F(x,y))$, $A^i(x,y)$ is a vector field, on T^1M and m, c, e are the physical constants. The functions $g_{ij}(x,y)$ can be considered as gravitational potential and $A^i(x,y)$ the electromagnetic potentials.

In the classical theory of the electrodynamics $H(x,p)$ is obtained from the known Lagrangian of electrodynamics via Legendre transformation, [97], [105].

The fundamental tensor field of the space $H^{(2)n}$ is given by:

$$(9.2)\quad \gamma^{ij}(x,y,p) = \frac{1}{mc} g^{ij}(x,y)$$

It is homothetic to the fundamental tensor $g^{ij}(x,y)$ of Finsler space F^n.

This remark leads to the fact that $H^{(2)n} = (M,H)$, (9.1), is the Hamilton space of order two of the Electrodynamics.

The covariant tensor $\gamma_{ij}(x,y,p)$ is given by

$$(9.2)'\quad \gamma_{ij}(x,y,p) = mcg_{ij}(x,y).$$

The tensor field C^{ijk}, (1.3), vanishes. It follows:

$$(9.3)\quad C^{ijk} = 0,\ C_i^{jk} = 0.$$

The tensor $C^i{}_{jk}$ from (1.4) reads:

$$(9.4)\quad C^i{}_{jk}(x,y) = \frac{1}{2} g^{is}(\dot{\partial}_j g_{sk} + \dot{\partial}_k g_{sj} - \dot{\partial}_s g_{jk}).$$

Then, Theorem 12.1.3 leads to:

Theorem 12.9.1.

1° *The w_1-paths of space $H^{(2)n}$ in the point $x_0 \in M$ are characterized by the system of differential equations*

$$x^i = x^i_0,\ p_i = 0,\ \frac{d^2y^i}{dt^2} + C^i{}_{jk}(x_0,y)\frac{dy^j}{dt}\frac{dy^k}{dt} = 0.$$

2° *The w_2-path of the space $H^{(2)n}$ in the point $x_0 \in M$ are characterized by the system of differential equations*

$$x^i = x^i_0,\ y^i = 0,\ \frac{d^2 p_i}{dt^2} = 0.$$

Let $\overset{\circ}{N}{}^i_j(x, y)$ be the Cartan nonlinear connection of the Finsler space F^n and z^i the d-vector field on T^2M:

$$z^i = y^{(2)i} + \frac{1}{2}\overset{\circ}{N}{}^i_j y^k.$$

Remember that the Christoffel symbols $\gamma^i{}_{jk}(x, y)$ of F^n and $\gamma^i{}_{jk}y^j y^k = \gamma^i_{00}$ gives

$$\overset{\circ}{N}{}^i_j = \frac{1}{2}\frac{\partial \gamma^i_{00}}{\partial y^j},\ y^j\overset{\circ}{N}{}^i_j = \gamma^i_{00}. \tag{9.5}$$

Let us consider the functions

$$\xi^i_1 = \frac{1}{2}\{\dot{\partial}^i H - \gamma^i_{00}\}. \tag{9.6}$$

We obtain the Legendre transformation (6.2) determined by the Hamilton space $H^{(2)n}$. Then, it follows:

$$z^i(x, y, \xi) = \frac{1}{2}\dot{\partial}^i H = \gamma^{ij}p_j - \frac{e}{m}A^i. \tag{9.7}$$

The dual space $L^{(2)n} = (M, L(x, y^{(1)}, y^{(2)}))$ of the space $H^{(2)n}$ has the property

$$L(x, y, \xi_1) = p_i\dot{\partial}^i H - H = \gamma^{ij}p_i p_j - \frac{e^2}{c}A^i A^j \gamma_{ij}, \tag{9.8}$$

and the canonical dual semispray S_{ξ_1} has the coefficients (7.4) given in our case by

$$f_j(x, y, p) = \frac{1}{2}\frac{\partial}{\partial y^j}\{-H + \gamma^i_{00}p_i\}. \tag{9.9}$$

Taking into account Theorem 12.7.2 we obtain the coefficients $N^i{}_j$ and N_{ij} of the canonical nonlinear connection N of the space $H^{(2)n}$:

$$N^i{}_j = \frac{1}{2}\frac{\partial}{\partial y^j}\{-\dot{\partial}^i H + \gamma^i_{00}\},\ N_{ij} = \frac{1}{2}\frac{\partial^2}{\partial y^i \partial y^j}\left[-H + \gamma^s_{00}p_s\right]. \tag{9.10}$$

Therefore the coefficients $N^i{}_j$ can be written:

$$N^i{}_j = \overset{\circ}{N}{}^i_j - A^i_j,\ A^i_j = -\frac{1}{2}\dot{\partial}_j\dot{\partial}^i H. \tag{9.10'}$$

Clearly A^i_j is a d-tensor field of type $(1,1)$.

The adapted basis $\left(\dfrac{\delta}{\delta x^i}, \dfrac{\partial}{\partial y^i}, \dfrac{\partial}{\partial p_i}\right)$ to the distributions N, W_1, W_2 has the first vector fields $\dfrac{\delta}{\delta x^i}$ of the form

$$\frac{\delta}{\delta x^i} = \frac{\partial}{\partial x^i} - (\overset{\circ}{N}{}^j_i - A^j_i)\frac{\partial}{\partial y^j} + N_{ij}\frac{\partial}{\partial p_j}$$

In other words:

$$(*) \qquad \frac{\delta \gamma_{ij}}{\delta x^k} = \frac{mc}{e}\left[\frac{\partial g_{ij}}{\partial x^k} - \overset{\circ}{N}{}^m_k \frac{\partial g_{ij}}{\partial y^m} + A^m_k \frac{\partial g_{ij}}{\partial y^m}\right].$$

Therefore, the coefficients $H^i{}_{jk}$ of the canonical connection $C\Gamma(N)$ of the space $H^{(2)n}$ are given in the following

Theorem 12.9.2. *The canonical metrical connection $C\Gamma(N)$ of the Hamilton space $H^{(2)n}$ of electrodynamics has the following coefficients*

$$(9.11) \qquad H^i{}_{jk} = F^i{}_{jk} + A^i{}_{jk}, \;\; C^i{}_{jk}, \;\; C_i^{jk} = 0$$

where $(F^i{}_{jk}, C^i{}_{jk})$ is Cartan metrical connection of the Finsler space F^n and $A^i{}_{jk}$ is a d-tensor field expressed by

$$(9.12) \qquad A^i{}_{jk} = \frac{1}{2}g^{is}\left(A^m_j \frac{\partial g_{sk}}{\partial y^m} + A^m_k \frac{\partial g_{js}}{\partial y^k} - A^m_s \frac{\partial g_{ij}}{\partial y^m}\right).$$

Indeed, the last theorem follows from a straightforward calculus, using the formula $(*)$ and the expression of $H^i{}_{jk}$ from (8.3).

We remark that the geometry of the Hamilton spaces of electrodynamics $H^{(2)n}$, (9.1) can be developed by means of the canonical nonlinear connection N, (9.10) and on the canonical metrical connection $C\Gamma(N)$, (9.11).

Chapter 13

Cartan spaces of order 2

The Hamilton spaces $H^{(2)n} = (M, H(x,y,p))$ for which the fundamental function $H(x,y,p)$ is 2-homogeneous with respect to momenta p_i form an interesting class of Hamilton spaces of order two, called *Cartan spaces or order two.*

For these spaces it is important to determine the fundamental geometrical object fields, as canonical nonlinear connections and canonical metrical N-connections.

13.1 $C^{(2)n}$–spaces

Definition 13.1.1. A Cartan space of order two is a pair $C^{(2)n} = (M, K(x,y,p))$, for which the following axioms hold:

1° K is a real function on $T^{*2}M$, differentiable on $\widetilde{T^{*2}M}$ and continuous on zero section of the projection π^{*2}.

2° $K > 0$ on $T^{*2}M$.

3° K is positively 1-homogeneous with respect to momenta p_i.

4° The Hessian of K^2, with elements:

$$g^{ij}(x,y,p) = \frac{1}{2}\dot{\partial}^i\dot{\partial}^j K^2 \tag{1.1}$$

is positively defined on $\widetilde{T^{*2}M}$.

It follows that g^{ij} from (1.1) is contravariant of order two, symmetric and non-degenerate d-tensor field. It is called *fundamental* (or metric) *tensor* of space $C^{(2)n}$. $K(x,y,p)$ is called *fundamental function* of $C^{(2)n}$.

Let us start noticing:

Theorem 13.1.1. *If the base manifold M is paracompact, then there exist on $T^{*2}M$ functions K such that the pair (M,K) is a Cartan space of order two.*

Proof. The manifold M being paracompact, it follows that $T^{*2}M$ is paracompact, too, and therefore there exists a real function F on TM, which is fundamental for the Finsler space $F^n = (M, F(x,y))$. Let $a_{ij}(x,y)$ be the fundamental tensor of F^n and $a^{ij}(x,y)$ be its contravariant tensor. Obviously a^{ij} is positively defined. If we consider the function

$$K(x,y,p) = \{a^{ij}(x,y)p_ip_j\}^{1/2}, \tag{1.2}$$

then we obtain the fundamental function of a Cartan space of order two. **q.e.d.**

The Cartan spaces $C^{(2)n}$ with fundamental function (1.2) are special. They can be characterized by the vanishing of the d-tensor field

$$C^{ijk} = -\frac{1}{4}\dot{\partial}^i\dot{\partial}^j\dot{\partial}^h K^2 \tag{1.3}$$

Proposition 13.1.2. *For any Cartan spaces of order 2, we obtain:*

1° *The components $g^{ij}(x,y,p)$ of the fundamental tensor are 0-homogeneous with respect to p_i.*

2° $\dfrac{1}{2}\dfrac{\partial K^2}{\partial p_i} = g^{ij}p_j.$

3° $g^{ij}(x,y,p)p_ip_j = K^2(x,y,p).$

4° $p_iC^{ijk} = 0.$

Let $g_{ij}(x,y,p)$ be the covariant tensor of $g^{ij}(x,y,p)$.
A similar theorem with that given in Ch.12, can be formulated:

Theorem 13.1.2. *For any Cartan space $C^{(2)n} = (M,K)$ the following d-tensors*

$$\begin{aligned} C^i{}_{jk} &= \frac{1}{2}g^{is}(\dot{\partial}_j g_{sk} + \dot{\partial}_k g_{js} - \dot{\partial}^s g_{jk}), \\ C_i^{jk} &= -\frac{1}{2}g_{is}(\dot{\partial}^j g^{sk} + \dot{\partial}^k g^{js} - \dot{\partial}^s g^{jk}), \end{aligned} \tag{1.4}$$

have the properties:

1° *They are the w_1- and w_2-coefficients of a canonical metrical connection, i.e.*

$$g^{ij}|_k = 0,\ \ g^{ij}|^k = 0. \tag{1.5}$$

2° *The following identities hold:*

(1.6) $$C_i^{\ jk} = g_{is}C^{sjk}$$

(1.7) $$p_i|^k = \delta_i^{\ k} = \delta_i^k$$

(1.8) $$S^i_{\ jk} = C^i_{\ jk} - C^i_{\ kj} = 0; \ S_i^{\ jk} = C_i^{\ jk} - C_i^{\ kj} = 0.$$

The proof is similar with that given in §, Ch.5.

Let us consider a w_α–vertical curve $\gamma : I \longrightarrow \widetilde{T^{*2}M}$ in the point $x_0 \in M$ (Ch.10). Applying Theorem 10.7.5., we have:

Theorem 13.1.3.

1) *The w_1–vertical paths in the point $x_0 \in M$ of a Cartan space of order two, $C^{(2)n} = (M, K)$ are characterized by the system of differential equations*

$$x^i = x^i_0, \ p_i = 0, \ \frac{d^2y^i}{dt^2} + C^i_{\ jk}(x_0, y, 0)\frac{dy^j}{dt}\frac{dy^k}{dt} = 0.$$

2) *In the space $C^{(2)n} = (M, K)$ the w_2–paths in the point $x_0 \in M$ are characterized by the system:*

$$x^i = x^i_0, \ y^i = 0, \ \frac{d^2p_i}{dt^2} - C_i^{\ jk}(x_0, 0, p)\frac{dp_j}{dt}\frac{dp_k}{dt} = 0.$$

13.2 Canonical presymplectic structure of space $C^{(2)n}$

The natural geometrical object fields on the manifolds $\widetilde{T^{*2}M}$, in the case of Cartan space of order two, together with the fundamental function $K(x, y, p)$ of space $C^{(2)n}$ give rise to some important properties, especially in the case of canonical equations (§2, ch.12).

We have $\omega = p_i dx^i$ and

(2.1) $$\theta = d\omega = dp_i \wedge dx^i$$

is the canonical presymplectic structure.

The canonical Poisson structure $\{,\}_0$ and $\{,\}_1$ from (2.4), ch.12, can be also considered.

Remarking that

(2.2) $$H(x, y, p) = K^2(x, y, p)$$

is a regular Hamiltonian, by means of Proposition 12.2.1, we get:

Proposition 13.2.1. *In a Cartan space $C^{(2)n} = (M, K)$ the following equations hold:*

$$\tag{2.3} \begin{aligned} &\{x^i, K^2\}_0 = \frac{\partial K^2}{\partial p_i},\ \{y^i, K^2\}_0 = 0,\ \{p_i, K^2\}_0 = -\frac{\partial K^2}{\partial x^i}, \\ &\{x^i, K^2\}_1 = 0,\ \{y^i, K^2\}_1 = \frac{\partial K^2}{\partial p_i},\ \{p_i, K^2\}_1 = -\frac{\partial K^2}{\partial y^i}. \end{aligned}$$

Notice that the triple $(T^{*2}M, K^2(x,y,p), \theta)$ is a Hamiltonian system.

Let us consider the canonical section of π^*, given by $\sigma_0 : (x,p) \in T^*M \longrightarrow (x,0,p) \in T^{*2}M$ and $\Sigma_0 = \text{Im}\sigma_0$. Then Σ_0 is an (immersed) submanifold of the manifold $\widetilde{T^{*2}M}$ and let us denote the restriction of θ to Σ_0 by θ_0. We remark that Σ_0 has the equation $y^i = 0$.

Theorem 12.2.1. affirmes that the pair (Σ_0, θ) is a symplectic manifold.

Therefore the mapping $S_{\theta_0} : \mathcal{X}(\Sigma_0) \longrightarrow \mathcal{X}^*(\Sigma_0)$ defined by

$$\tag{2.4} S_{\theta_0}(X) = i_X\theta_0,\ \forall X \in \mathcal{X}(\Sigma_0)$$

is an isomorphism.

We denote $K_0 = K_{|\Sigma_0}$. Then we have

Proposition 13.2.2. *The pair (M, K_0) is a classical Cartan space.*

Indeed, in this case, the definition from Ch.6 is satisfied. Its fundamental tensor field is

$$\tag{2.5} g^{ij}(x,0,p) = \frac{1}{2}\frac{\partial^2 K_0^2}{\partial p_i \partial p_j}.$$

We obtain, also:

Theorem 13.2.1. *There exists a unique vector field $X_{K_0^2} \in \mathcal{X}(\Sigma_0)$ with the property*

$$\tag{2.6} S_0(X_{K_0^2}) = i_{X_{K_0^2}}\theta_0 = -dK_0^2.$$

In the local basis, the vector field $X_{K_0^2}$ has the expression

$$\tag{2.7} X_{H_{K_0^2}} = \frac{\partial K_0^2}{\partial p_i}\frac{\partial}{\partial x^i} - \frac{\partial K_0^2}{\partial x^i}\frac{\partial}{\partial p_i}.$$

The integral curves of the vector field $X_{K_0^2}$ determines the Σ_0–canonical equations of the space $C^{(2)k}$.

Theorem 13.2.2. *The Σ_0–canonical equations of the space $C^{(2)n}$ are as follows:*

$$\frac{dx^i}{dt} = \frac{\partial K_0^2}{\partial p_i},\ \frac{dp_i}{dt} = -\frac{\partial K_0^2}{\partial x^i},\ y^i = 0. \tag{2.8}$$

Corollary 13.2.1. *The equations* (2.8) *can be written in the form*

$$\frac{dx^i}{dt} = \{K_0^2, x^i\}_0,\ \frac{dp_i}{dt} = \{K_0^2, p_i\},\ y^i = 0. \tag{2.9}$$

Remark. It is clear that the Jacobi method, described in Section 2, ch.4, for integration of the equations (2.8), can be used in this case.

Now, let Σ_1 be the fibre, in the point $x_0 \in M$, of the bundle $(T^{*2}M, \pi^*, M)$. Then Σ_1 is an immersed submanifold in $\widetilde{T^{*2}M}$.

Let us consider (cf. §2, ch.12) the following differential forms on Σ_1:

$$\omega_1 = p_i dy^i,\ \theta_1 = dp_i \wedge dy^i, \tag{2.10}$$

and the Poisson bracket:

$$\{f, g\}_1 = \frac{\partial f}{\partial y^i}\frac{\partial g}{\partial p_i} - \frac{\partial g}{\partial y^i}\frac{\partial f}{\partial p_i}. \tag{2.11}$$

The relations between these canonical structures on the manifold Σ_1 can be studied applying the same method as in the case of structures $(\Sigma_0, \theta_0, K_0^2)$.

If we denote $K_1 = K_{|\Sigma_1}$, we get the following Σ_1–canonical equations.

Theorem 13.2.3.
The Σ_1–canonical equations of the Cartan space $C^{(2)n} = (M, K(x,y,p))$ are the following

$$x^i = x_0^i,\ \frac{dy^i}{dt} = \frac{\partial K_1^2}{\partial p_i};\ \frac{dp_i}{dt} = -\frac{\partial K_1^2}{\partial y^i}. \tag{2.12}$$

Taking into account the formula (2.3) we obtain

Corollary 13.2.2. *The equations* (2.12) *are equivalent to the system of equations*

$$x^i = x_0^i,\ \frac{dy^i}{dt} = \{K_1^2, y^i\}_1,\ \frac{dp_i}{dt} = \{K_1^2, p_i\}.$$

For the integration of the Σ_1–canonical equations we can use the Jacobi method. Let us try to determine a solution curve $\sigma(t)$ in the fibre Σ_1 at point $x_0 \in M$ of the form

$$x^i = x^i_0,\ y^i = y^i(t),\ p_i = \frac{\partial S_1}{\partial y^i}(y(t)), \tag{2.13}$$

where $S_1 \in \mathcal{F}(\Sigma_1)$.

Substituting in (2.12) we get:

$$\begin{aligned} x^i = x^i_0,\ \ \frac{dy^i}{dt} &= \frac{\partial K^2}{\partial p_i}\left(x_0, y(t), \frac{\partial S_1}{\partial y}(y(t))\right),\\ \frac{dp_i}{dt} &= \frac{\partial^2 S_1}{\partial y^i \partial y^j}\frac{\partial K^2}{\partial p_i} = -\frac{\partial K^2}{\partial y^i}. \end{aligned} \tag{2.14}$$

It follows from (2.14)

$$dK^2\left(x_0, y, \frac{\partial S_1}{\partial y}\right) = \{K^2, K^2\}_1 = 0.$$

and therefore

$$K^2\left(x_0, y, \frac{\partial S_1}{\partial y}\right) = \text{const.} \tag{2.15}$$

By integration of (2.15), we can determine S_1 and from (2.14) we obtain the curve $\sigma(t)$.

13.3 Canonical nonlinear connection of $C^{(2)n}$

We can associate the Hamilton space of order two, $H^{(2)n} = (M, K^2(x,y,p))$ to a Cartan space of order two, $C^{(2)n} = (M, K(x,y,p))$. Then the canonical nonlinear connection N of the space $H^{(2)n}$ will be called the canonical nonlinear connection of the Cartan space $C^{(2)n}$. Therefore, we can apply the theory from the section 7 of the previous chapter.

Let $\overset{\circ}{N}$ be a fixed connection with the coefficients $\overset{\circ}{N}{}^i{}_j(x,y)$. Thus, on T^2M, we have the vector field

$$z^i = y^{(2)i} + \overset{\circ}{N}{}^i{}_j(x,y)y^i, \tag{3.1}$$

and we obtain the Legendre ξ_1 mapping determined by $C^{(2)n}$ and by $\overset{\circ}{N}$:

$$y^{(0)i} = x^i,\ y^{(1)i} = y^i,\ y^{(2)i} = \xi^i_1(x,y,p), \tag{3.2}$$

where

$$\xi_1^i(x,y,p) = \frac{1}{2}\{\dot{\partial}^i K^2(x,y,p) - \overset{\circ}{N}{}^i{}_j(x,y)y^i\}. \tag{3.2)'$$

It follows that (3.2), (3.2)' is a local diffeomorphism which preserves the fibres of $T^{*2}M$ and T^2M.

From (3.1), (3.2) we obtain:

$$\dot{\partial}^i \xi_1^j = g^{ij}(x,y,p), \tag{3.3}$$

$$\check{z}^i(x,y,p) = \frac{1}{2}\dot{\partial}^i K^2(x,y,p). \tag{3.4}$$

Let φ_1 be the inverse mapping of ξ_1. Then it is of the form

$$x^i = y^{(0)i},\ y^i = y^{(1)i},\ p_i = \varphi_{1i}(y^{(0)}, y^{(1)}, y^{(2)}). \tag{3.5}$$

From (3.5) we get

$$\frac{\partial \varphi_{1i}}{\partial y^{(2)j}} = g_{ij}(x,y,\varphi_1). \tag{3.6}$$

The Lagrangian (6.7), determined from K^2 by means of Legendre transformation ξ_1, is given by

$$L(x, y^{(1)}, y^{(2)}) = 2p_i z^i - K^2(x,y,p),\ p_i = \varphi_{1i}(x, y^{(1)}, y^{(2)}). \tag{3.7}$$

The formula (3.7) and the property of homogeneity of K^2, with respect to momenta p_i, lead to the equation:

$$L(x, y^{(1)}, \xi_1(x,y,p)) = K^2(x,y,p). \tag{3.8}$$

Indeed, from (3.4) and (3.7) it follows

$$L(x,y,\xi_1) = p_i \dot{\partial}^i K^2 - K^2 = K^2.$$

The canonical dual semispray of $C^{(2)n}$ is

$$S_{\xi_1} = y^i \frac{\partial}{\partial x^i} + 2\xi_1^i \frac{\partial}{\partial y^i} + f_i(x,y,p)\frac{\partial}{\partial p_i}. \tag{3.9}$$

Theorems 12.7.1 and 12.7.2 imply:

Theorem 13.3.1. *The canonical semispray S_ξ has the coefficients*

$$f_j(x,y,p) = \frac{1}{2}\frac{\partial}{\partial y^j}[-K^2(x,y,p) + p_s \overset{\circ}{N}{}^s{}_m(x,y)y^m]. \tag{3.10}$$

Theorem 13.3.2. *The coefficients of the canonical nonlinear connection N of the Cartan space of order two, $C^{(2)n}$ are given by:*

(3.11)

$$N^i{}_j = \frac{1}{2}\frac{\partial}{\partial y^j}\left[-\dot{\partial}^i K^2 + \overset{\circ}{N}{}^i{}_m(x,y)y^m\right], \quad N_{ij} = \frac{1}{2}\frac{\partial^2}{\partial y^i \partial y^j}\left[-K^2 + p_s \overset{\circ}{N}{}^s{}_m y^m\right].$$

Remarks.

1° The coefficients of canonical nonlinear connection N depend on an a priori given nonlinear connection $\overset{\circ}{N}$ $(\overset{\circ}{N}{}^i{}_j(x,y))$. The previous theory is more simple if $\overset{\circ}{N}$ is the Cartan nonlinear connection of a Finsler space $F^n = (M, F(x,y))$.

2° The torsion τ_{ij} of the canonical nonlinear connection vanishes. In this case the canonical presymplectic structure θ, (2.1), can be written in the form:

$$\theta = \delta p_i \wedge dx^i, \tag{3.12}$$

where $\delta p_i = dp_i - N_{ji}dx^j$.

3° Since $\delta y^i = dy^i + N^i{}_j dx^j$, Prop. 6.3, ch.9, we have:

Theorem 13.3.3. *A horizontal curve on $C^{(2)n}$ is characterized by the system of differential equations:*

$$x^i = x^i(t), \ \delta y^i = 0, \ \frac{\delta p_i}{dt} = 0, \tag{3.13}$$

where $x^i = x^i(t)$, $t \in I$, are a priori given.

13.4 Canonical metrical connection of space $C^{(2)n}$

Let N be the canonical nonlinear connection N of the Cartan space of order two, $C^{(2)n} = (M, K(x,y,p))$. The adapted basis to the distributions N, W_1, W_2 is $\left(\frac{\delta}{\delta x^i}, \frac{\partial}{\partial y^i}, \frac{\partial}{\partial p_i}\right)$ and its dual basis is $(dx^i, \delta y^i, \delta p_i)$, where

$$\delta_j = \frac{\delta}{\delta x^i} = \frac{\partial}{\partial x^i} - N^j{}_i \frac{\partial}{\partial y^j} + N_{ij}\frac{\partial}{\partial p_j}, \tag{4.1}$$

and

$$\delta y^i = dy^i + N^i{}_j dx^j,\ \delta p_i = dp_i - N_{ji}dx^j. \tag{4.1'}$$

Now, using an already very known method, we can prove:

Theorem 13.4.1.

1) *In a Cartan space of order two, $C^{(2)n} = (M, K)$ there exists a unique N-linear connection $C\Gamma(N) = (H^i{}_{jk}, C^i{}_{jk}, C_i^{jk})$ which verifies the following axioms:*

   1° *N is the canonical nonlinear connection with coefficients* (3.11).

   2° *$C\Gamma(N)$ is metrical with respect to fundamental tensor g^{ij} of space $C^{(2)n}$, i.e.*

$$g^{ij}{}_{|k} = 0,\ g^{ij}\big|_k = 0,\ g^{ij}\big|^k = 0 \tag{4.2}$$

   3° *The d-tensor of torsion $T^i{}_{jk}$, $S^i{}_{jk}$, S_i^{jk} vanish.*

2) *This connection has the coefficients:*

$$\begin{aligned} H^i{}_{jk} &= \frac{1}{2} g^{is}(\delta_j g_{sk} + \delta_k g_{js} - \delta_s g_{jk}),\\ C^i{}_{jk} &= \frac{1}{2} g^{is}(\dot\partial_j g_{sk} + \dot\partial_k g_{js} - \dot\partial_s g_{jk}),\\ C_i^{jk} &= -\frac{1}{2} g_{is}(\dot\partial^j g^{sk} + \dot\partial^k g^{js} - \dot\partial^s g^{jk}). \end{aligned} \tag{4.3}$$

The N-connection $C\Gamma(N)$, (4.3), is called the canonical metrical connection of the space $C^{(2)n}$.

Let us consider the covariant curvature tensors (6.5)′, Ch.10 of $C\Gamma(N)$. Then, applying the Ricci identities (see 9.7) to the covariant of fundamental tensor field, g_{ij} and taking into account the equations

$$g_{ij|k} = 0,\ g_{ij}\big|_k = 0,\ g_{ij}\big|^k = 0 \tag{4.2'}$$

we obtain:

Theorem 13.4.2. *The tensors of curvature $R_{ijkh}, P_{ijkh}, P_{ijk}{}^h$ and $S_{ijkh}, S_{ijk}{}^h, S_{ij}{}^{kh}$ are skew–symmetric in the first two indices.*

The (y)–deflection tensors of $C\Gamma(N)$ are given by (5.5)′, ch.10, and (p)–deflection of the same connection are

$$\Delta_{ij} = N_{ij} - p_m H^m_{ij},\ \delta_{ij} = -p_m C^m{}_{ij},\ \delta_i^j = \delta_i^j. \tag{4.4}$$

Now, let us solve the problem of determination of a metrical N-linear connection for which the p-deflection tensor Δ_{ij} vanishes.

Theorem 13.4.3.

1) *In a Cartan space* $C^{(2)n} = (M, K)$ *there exists a unique* N*-linear connection* $D\Gamma(N) = (H^i{}_{jk}, C^i{}_{jk}, C_i{}^{jk})$ *for which the following axioms hold:*

   1° $\Delta_{ij} = 0$ *and the coefficients* $\overset{\circ}{N}{}^i{}_j(x,y)$ *of the nonlinear connection* $N(\overset{\circ}{N}{}^i{}_j, N_{ij})$ *are given a priori.*

   2° $D\Gamma(N)$ *is metrical with respect to* g^{ij}, *i.e.* (4.2) *holds.*

2) 1°. *The coefficients* $(H^i{}_{jk}, C^i{}_{jk}, C_i{}^{jk})$ *of* $D\Gamma(N)$ *are given by* (4.3).

   2°. *The coefficients* N_{ij} *of the nonlinear connection* N *are expressed by:*

$$\begin{aligned} N_{ij} &= \gamma^0_{ij} - \frac{1}{2}\dot{\partial}^m g_{ij}\left(\gamma^0_{m0} + \frac{1}{2}\dot{\partial}_s K^2 \cdot \overset{\circ}{N}{}^s{}_m\right) + \\ &+\frac{1}{2}\left[(\overset{\circ}{N}{}^m{}_i g_{sj} + \overset{\circ}{N}{}^m{}_j g_{is})\frac{\partial p^s}{\partial y^m} + \overset{\circ}{N}{}^m{}_0 \frac{\partial g_{ij}}{\partial y^m}\right], \end{aligned} \tag{4.5}$$

where $\gamma^i{}_{jk}(x,y,p)$ *are the Christoffel symbols of* $g_{ij}(x,y,p)$ *and the index "0" means the contraction by* p_i *or by* $p^i = g^{ij}p_j$.

Proof. If the nonlinear connection $N\ (\overset{\circ}{N}{}^i{}_j(x,y), N_{ij}(x,y,p))$ is known, then Theorem 13.4.1 can be applied and it follows the existence and uniqueness of the coefficients $(H^i{}_{jk}, C^i{}_{jk}, C_i{}^{jk})$ from (4.3) which satisfy the axioms 2° and 3°. Let us determine the coefficients N_{ij} of N in the condition of axiom 1°, i.e. $\Delta_{ij} = 0$.

Taking into account (4.4), $\Delta_{ij} = 0$ is equivalent to $N_{ij} = H^m{}_{ij}p_m$. Since the operators $\dfrac{\delta}{\delta x^j}$ have the expressions $\dfrac{\delta}{\delta x^j} = \dfrac{\partial}{\partial x^j} - \overset{\circ}{N}{}^i{}_j \dfrac{\partial}{\partial y^i} + N_{ij}\dfrac{\partial}{\partial p_i}$, the equations $g_{ij|h} = 0$, $T^i{}_{jh} = 0$ leads to

$$\begin{aligned} H^h{}_{ij} &= \gamma^h{}_{ij} + \frac{1}{2}\left[(\overset{\circ}{N}{}^m{}_i g_{sj} + \overset{\circ}{N}{}^m{}_j g_{is})\dot{\partial}_m g^{hs} + \overset{\circ}{N}{}^m{}_s g^{sh}\dot{\partial}_m g_{ij}\right] - \\ &-\frac{1}{2}[(N_{im}g_{sj} + N_{jm}g_{is})\dot{\partial}^m g^{hs} + N_{sm}g^{sh}\dot{\partial}^m g_{ij}]. \end{aligned} \tag{4.6}$$

Thus $N_{ij} = p_h H^h{}_{ij}$ allows to write:

$$N_{ij} = \gamma^0{}_{ij} + \frac{1}{2}[(\overset{\circ}{N}{}^m{}_i g_{sj} + \overset{\circ}{N}{}^m{}_j g_{is})\dot{\partial}_m p^s + \overset{\circ}{N}{}^m{}_0 \dot{\partial}_m g_{ij}] - \frac{1}{2}N_{0m}\dot{\partial}^m g_{ij}. \tag{4.7}$$

A new contraction by $p^j = g^{js}p_s$ leads to

$$N_{0i} = \gamma^0_{i0} + \frac{1}{2}\partial_m K^2 \overset{\circ}{N}{}^m_i.$$

Substituting in the formula (4.7) we have the solution (4.5) of the equation $\Delta_{ij} = 0$, and the theorem is proved. **q.e.d.**

For the canonical metrical connection $C\Gamma(N)$, (4.3) the Ricci identities are given in Theorem 10.6.1, taking into account the axiom 3°, i.e.:

$$T^i{}_{jk} = 0,\ S^i{}_{jk} = 0,\ S_i^{jk} = 0. \tag{4.8}$$

The torsions are given by (6.3), (6.3)″ and the curvature tensors are expressed by (6.4) and (6.4)′.

Theorem 10.6.3 gives some important identities for $C\Gamma(N)$, in which we take into account (4.6) and (4.4).

13.5 Parallelism of vector fields. Structure equations of $C\Gamma(N)$

Let us consider the canonical metrical connection $C\Gamma(N)$, (4.3), and let $\gamma : I \longrightarrow \widetilde{T^{*2}M}$ be a smooth parameterized curve as in section 7, Ch.10.

For a vector field $X \in \mathcal{X}(T^{*2}M)$ given in the adapted basis $(\delta_i, \partial_i, \dot{\partial}_i)$ by

$$X = \overset{\circ}{X}{}^i\delta_i + \overset{1}{X}{}^i\dot{\partial}_i + X_i\dot{\partial}^i \tag{5.1}$$

the covariant differential DX is expressed by:

$$DX = (d\overset{\circ}{X}{}^i + X^s\omega^i{}_s)\delta_i + (d\overset{1}{X}{}^i + X^s\omega^i{}_s)\dot{\partial}_i + (dX_i - X_s\omega^s{}_i)\dot{\partial}^i, \tag{5.2}$$

where the 1-*forms of connection* $\omega^i{}_s$ of the Cartan space of order two are:

$$\omega^i{}_j = H^i{}_{jk}dx^k + C^i{}_{jk}\delta y^k + C_j^{ik}\delta p_k. \tag{5.3}$$

Therefore, Theorem 10.7.1 gives the necessary and sufficient condition for the parallelism of vector field X, (5.1) with respect to $C\Gamma(N)$ along the parametrized curve γ.

Theorem 10.7.2 states that the Cartan space $C^{(2)n}$ is with the absolute parallelism of vectors if, and only if, all d–tensors of curvature vanish:

$$R_h{}^i{}_{jk} = 0,\ P_h{}^i{}_{jk} = 0,\ P_h{}^i{}_j{}^k = 0,$$
$$S_h{}^i{}_{jk} = 0,\ S_h{}^i{}_j{}^k = 0,\ S_h{}^{ijk} = 0$$

We recall Theorem 10.7.3, also:

Theorem 13.5.1. *A smooth parametrized curve* $\gamma : I \longrightarrow \widetilde{T^{*2}M}$ *is autoparallel with respect to the canonical metrical connection* $C\Gamma(N)$ *of Cartan space* $C^{(2)n}$ *if and only if the system of differential equations* (7.13), *Ch.7, is verified.*

Of course, a theorem of existence and uniqueness for the autoparallel curve one can formulate without difficulties.

Taking into account Theorem 10.3.3, and (7.15), Ch.10, we get:

Theorem 13.5.2. *The horizontal paths of the canonical metrical connection* $C\Gamma(N)$ *of the space* $C^{(2)n}$ *are characterized by the system of differential equations:*

$$\frac{d^2x^i}{dt^2} + H^i{}_{jk}\frac{dx^j}{dt}\frac{dx^k}{dt} = 0, \quad \frac{\delta y^i}{dt} = 0, \quad \frac{\delta p_i}{dt} = 0.$$

Finally, Theorem 10.7.5 has as consequence:

Theorem 13.5.3.

1° *The* w_1*–vertical paths of* $C\Gamma(N)$ *in the point* $x_0 \in M$ *is characterized by the system of differential equations*

$$x^i = x^i_0, \ p_i = 0, \ \frac{d^2y^i}{dt^2} + C^i{}_{jk}(x_0, y, 0)\frac{dy^j}{dt}\frac{dy^k}{dt} = 0.$$

2° *The* w_2*–vertical paths of* $C\Gamma(N)$ *in the point* $x_0 \in M$ *is characterized by*

$$x^i = x^i_0, \ y^i = 0, \ \frac{d^2p_i}{dt^2} - C_i{}^{jk}(x_0, 0, p)\frac{dp_j}{dt}\frac{dp_k}{dt} = 0.$$

Remark. We assume that the restrictions of the coefficients $C^i{}_{jk}$ to the zero section of $\pi^* : T^{*2}M \longrightarrow M$ exist.

The structure equations of the canonical metrical connection $C\Gamma(N)$ of the Cartan space of order two, $C^{(2)n} = (M, K(x, y, p))$ are given in the section 7 of Ch.7, taking into account the particular properties of this connection.

So, we obtain:

Theorem 13.5.4. *The canonical metrical connection* $C\Gamma(N)$ *of the Cartan space of order two,* $C^{(2)n} = (M, K)$*, has the following structure equations:*

$$(5.4) \qquad \begin{aligned} d(dx^i) - dx^m \wedge \omega^i{}_m &= -\overset{(0)}{\Omega}{}^i, \\ d(\delta y^i) - \delta y^m \wedge \omega^i{}_m &= -\overset{(1)}{\Omega}{}^i, \\ d(\delta p_i) + \delta p_m \wedge \omega^m{}_i &= -\Omega_i \end{aligned}$$

and

$$(5.5) \qquad d\omega^i{}_j - \omega^m{}_j \wedge \omega^i{}_m = -\Omega^i{}_j$$

where the 2-forms of torsion $\overset{(o)}{\Omega}{}^i, \overset{(1)}{\Omega}{}^i, \Omega_i$ *and 2-forms of curvature* $\Omega^i{}_j$ *are given by the formulae* (8.6), *ch.7, and* (4.8).

The Bianchi identities of the connection $C\Gamma(N)$ can be obtained taking the exterior differential of the system of equations (5.4), (5.5) modulo the same system.

Remark. Using the structure equations, we can study, without major difficulties, the theory of Cartan subspaces of order two in a Cartan space $C^{(2)n}$.

13.6 Riemannian almost contact structure of a space $C^{(2)n}$

Consider a Cartan space of order two, $C^{(2)n} = (M, K(x,y,p))$ and its canonical nonlinear connection N, with coefficients $(N^i{}_j, N_{ij})$ from (3.11). The adapted basis $(\delta_i, \dot{\partial}_i, \dot{\partial}^i)$ and its dual $(dx^i, \delta y^i, \delta p_i)$ are determined by N.

The associated $GH^{(2)n} = (M, g^{ij}(x,y,p))$ space of $C^{(2)n}$ is uniquely determined. Therefore, using the theory from section 3, ch.11, we can investigate the notion of Riemannian almost contact structure of the space $C^{(2)n}$.

We introduce the N-lift to $\widetilde{T^{*2}M}$ of the fundamental tensor $\mathbb{G}^{ij}(x,y,p)$ by:

$$(6.1) \qquad \mathbb{G} = g_{ij}dx^i \otimes dx^j + g_{ij}\delta y^i \otimes \delta y^j + g^{ij}\delta p_i \otimes \delta p_j,$$

which is defined in every point $u^* \in T^{*2}M$.

Theorem 13.6.1.

1° $\mathbb{G}$ *is a tensor field on the manifold* $\widetilde{T^{*2}M}$, *of type* $(0,2)$, *nonsingular depending only on the fundamental function* K *of* $C^{(2)n}$ *and by the nonlinear connection* $\overset{\circ}{N}{}^i{}_j(x,y)$.

2° *The pair* $(\widetilde{T^{*2}M}, \mathbb{G})$ *is a Riemannian space.*

3° *The distributions* N, W_1, W_2 *are respectively orthogonal with respect to* $\mathbb{G}$.

Proof. Since g^{ij} and g_{ij} are d-tensor fields, symmetric and positively defined, it follows that $\mathbb{G}$ has the properties 1°, 2° and 3°.

Applying Theorem 11.3.2, we deduce:

Theorem 13.6.2. *The tensor* $\mathbb{G}$ *is covariant constant with respect to canonical metrical connection* $C\Gamma(N)$ *of the space* $C^{(2)n}$, *i.e.*

$$D\mathbb{G} = 0.$$

In other words, the canonical metrical connection $C\Gamma(N)$ is an N-linear metrical connection with respect to the Riemannian structure $\mathbb{G}$.

Let us consider the $\mathcal{F}(\widetilde{T^{*2}M})$-linear mapping

$$\check{\mathbb{F}} : \mathcal{X}(\widetilde{E_2^*M}) \longrightarrow \mathcal{X}(\widetilde{E_2^*M})$$

defined by

$$\check{\mathbb{F}}(\delta_i) = -g_{ij}\dot{\partial}^j, \ \check{\mathbb{F}}(\dot{\partial}_i) = 0, \ \check{\mathbb{F}}(\dot{\partial}^i) = g^{ij}\delta_j. \tag{6.2}$$

Theorem 13.6.3. *The mapping* $\check{\mathbb{F}}$ *has the following properties:*

1° $\check{\mathbb{F}}$ *is globally defined on* $\widetilde{T^{*2}M}$.

2° $\check{\mathbb{F}}$ *is a tensor field of type* $(1,1)$ *on* $\widetilde{T^{*2}M}$, *i.e.*

$$\check{\mathbb{F}} = -g_{ij}\dot{\partial}^i \otimes dx^j + g^{ij}\delta_j \otimes \delta p_i. \tag{6.3}$$

3° $\ker\check{\mathbb{F}} = W_1, \ \operatorname{Im}\check{\mathbb{F}} = N \oplus W_2.$

4° $\operatorname{rank}\left\|\check{\mathbb{F}}\right\| = 2n.$

5° $\check{\mathbb{F}}^3 + \check{\mathbb{F}} = 0.$

The proof is exactly like the case of Theorem 9.8.2.

Hence, $\check{\mathbb{F}}$ is called the (p)-almost contact structure determined by g^{ij} and by $\overset{\circ}{N}$.

The Nijenhuis tensor $\mathcal{N}_{\check{\mathbb{F}}}$ and the condition of normality of $\check{\mathbb{F}}$ can be explicitely written in adapted basis (see ch.11).

Now if we remark the tensor (6.3) and the fact that g_{ij} and g^{ij} are covariant constant with respect to the canonical metrical connection, it follows:

Theorem 13.6.4. *The tensor field* $\check{\mathbb{F}}$ *is covariant constant with respect to the canonical metrical connection* $C\Gamma(N)$, *i.e.*

$$D\check{\mathbb{F}} = 0.$$

Finally, let us notice that the pair $(\mathbb{G}, \check{\mathbb{F}})$ has some remarkable properties.

Theorem 13.6.5.

1° *The pair* $(\mathbb{G}, \check{\mathbb{F}})$ *is a Riemannian almost contact structure on the manifold* $\widetilde{T^{*2}M}$ *determined only by the space* $C^{(2)n}$ *and by the nonlinear connection* $\overset{\circ}{N}{}^i{}_j(x, y)$.

2° *The tensor fields* $\mathbb{G}$ *and* $\check{\mathbb{F}}$ *are covariant constant with respect to canonical metrical connection* $C\Gamma(N)$ *of the space* $C^{(2)n}$.

3° *The associated 2-form of the structure* $(\mathbb{G}, \check{\mathbb{F}})$ *is the canonical presymplectic structure on* $\widetilde{T^{*2}M}$:

$$\theta = dp_i \wedge dx^i.$$

Indeed, we have $\mathbb{G}(\check{\mathbb{F}}X, Y) = -\mathbb{G}(\check{\mathbb{F}}Y, X)$, $\forall X, Y \in \mathcal{X}(T^{*2}M)$ and $\theta(X, Y) = \mathbb{G}(\check{\mathbb{F}}X, Y)$. In adapted basis we get $\theta = \delta p_i \wedge dx^i$. But the torsion $\tau_{ij} = N_{ij} - N_{ji}$ vanishes, hence as consequence $\theta = dp_i \wedge dx^i$, and therefore Theorems 13.6.2 and 13.6.4 implies 2°, etc.

The Riemannian almost contact space $(\widetilde{T^{*2}M}, \mathbb{G}, \mathbb{F})$ is called the *Riemannian almost contact model* of the Cartan space of order 2, $C^{(2)n}$.

It is extremely useful in applications.

Bibliography

[1] ABATE, M., PATRIZIO, G., *Finsler Metrics. A Global Approach with Applications to Geometric Function Theory*, Springer–Berlag, L.N. in Mathematics 1591(1994).

[2] ABED, S.H., *On differential Forms in Finsler bundles,* Colloquia Mathematica Societatis Janos Bolyai Debrecen, 1984, 23–56.

[3] ABRAHAM, R., MARSDEN, J.E., *Foundations of Mechanics* (second edition), Reading Mass: The Benjamin–Cummings Publ. Co. (1978).

[4] AIKOU, T., *Some remarks on the conformal equivalence of complex Finsler structures*, in the vol. Finslerian Geometry, Kluwer Acad. Publ. FTPH vol. 109, 2000, 35–53.

[5] AIKOU, T., *A note on infinitesimal deformations of complex Finsler structures*, Analele Stiinţifice ale Univ. Al.I. Cuza, Iaşi, s. Mat. (N.S.), 43(1997), no.2, 1998, 295–305.

[6] AKHBAR–ZADEH, H., *Sur les espaces de Finsler isotropes*, C.R. Acad. Sci. de Paris 252(1961), 2061–2063.

[7] ANASTASIEI, M., *A historical remark on the connections of Chern and Rund,* Contemporary mathematics, Finsler Geometry 176(1996), 171–177.

[8] ANASTASIEI, M., SHIMADA, H., SABĂU, V.S., *On the nonlinear connection of a Second Order Finsler Space*, Algebras, Groups and Geometries, Hadronic Press vol.16, no.1, march 1999, 1–10.

[9] ANASTASIEI, M., SHIMADA, M., *Deformations of Finsler metrics*, In the vol.: Finslerian Geometries, Kluwer Acad. Publ. FTPH, vol. 109, 2000, 53–67.

[10] ANASTASIEI, M. and ANTONELLI, P.L., *The Differential Geometry of Lagrangians which Generate Sprays,* in vol. of Kluwer Acad. Publ. FTPH, no. 76, 1996, 15–34.

[11] ANTONELLI, P.L., INGARDEN, R.S., MATSUMOTO, M., *The Theory of Sprays and Finsler Spaces with Applications in Physics and Biology,* Kluwer Acad. Publ. FTPH no. 58, 1993.

[12] ANTONELLI, P.L., HRIMIUC, D., *A New Class of Spray-Generating Lagrangians,* in vol. Lagrange and Finsler Geometry. Applications to Physics and Biology, Kluwer Acad. Publ. FTPH, no. 76, 1996, 81–92.

[13] ANTONELLI, P.L.,, HRIMIUC, D., *On the theory of N-Lagrange manifolds with applications in Biology and Physics*, Nonlinear World 3(1996), 299–333.

[14] ANTONELLI, P.L.,, HRIMIUC, D., *Symplectic transformation of the differential geometry of* T^*M, Nonlinear Analysis 36(1999), 529–557.

[15] ANTONELLI, P.L.,, HRIMIUC, D., *On the geometry of a homogeneous contact transformation*, Revue Roumaine de Mathématiques Pure et Appliquées, 9–10 (1997), 697–705.

[16] ANTONELLI, P.L.,, BUCĂTARU, I., *Finsler connections in an homologic geometry of a Kropina space*, preprint.

[17] ANTONELLI, P.L., LACKEY, B.C. (eds.), *The Theory of Finslerian Laplacians and Applications*, Kluwer Acad. Publ. 459 (1998).

[18] ANTONELLI, P.L., MIRON, R. (eds.), *Lagrange and Finsler Geometry. Applications to Physics and Biology,* Kluwer Acad. Publ. FTPH no. 76, 1996.

[19] ANTONELLI, P.L., SHIMADA, H., *On 1-form connections with constant coefficients*, Tensor N.S. 50(1991), 263–275.

[20] ANTONELLI, P.L., ZASTAWNIAK, T.J., *Fundamentals of Finslerian Diffusion with Aplications*, Kluwer Acad. Publ. 101 (1999).

[21] ANTONELLI, P.L. (ed.), *Finslerian Geometries. A Meeting of Minds*, Kluwer Acad. Publ. (FTPH), vol. 1909(2000).

[22] ASANOV, G.S., *Finsler Geometry, Relativity and Gauge Theories,* D. Reidel Publ. Co., Dordrecht, 1985.

[23] BACSO, S., MATSUMOTO, M., *On Finsler Spaces of Douglas Type*, Publ. Math. Debrecen 51/3-4 (1997), 385–406.

[24] BAO, D., CHERN, S.S., ZHEN, Z., *An introduction to Riemannian–Finsler Geometry* (Graduate Texts in Mathematics; 200), Springer–Verlag, 2000.

[25] BAO, D., CHERN, S.S., *On a Notable Connection in Finsler Geometry*, Houston J. Math. 19, no. 1, 1993, 135–180.

[26] BAO, D., CHERN, S.S., SHEN, Z. (eds.), *Finsler Geometry*, Contemporary Mathematics, AMS, no.196, 1995.

[27] BAO, D., CHERN, S.S., SHEN, Z, *Rigidity issues on Finsler surfaces*, Rev. Roumaine de Mathématiques Pures et Appliqués, no. 7,8, 1997.

[28] BATES, S., WEINSTEIN, A., *Lecture on the Geometry of Quantization*, Berkeley Mathematics. Lecture Notes, vol. 8, 1997, 135 p.

[29] BEIL, R., *Finsler geometry and a unified field theory,* p.265–272.

[30] BEIL, R.G., *Equations of Motion from Finsler Geometric Methods*, In the vol.: Finslerian Geometries, Kluwer Acad. Publ. FTPH, vol. 109, 2000, 95–111.

[31] BEJANCU, A., *Finsler Geometry and Applications*, Ellis Harwood Ltd. (1990).

[32] BEJANCU, A., *On the theory of Finsler submanifolds*, In the vol.: Finslerian Geometries, Kluwer Acad. Publ. FTPH, vol. 109, 2000, 111—131.

[33] BEJANCU, A., FARRAN, H.R., *Geometry of Pseudo–FinslerSubmanifolds,* Kluwer Acad.Publ., 2000.

[34] BRANDT, H.E., *Finslerian fields*, in the vol. Finslerian Geometry, Kluwer Acad. Publ. FTPH vol. 109, 2000, 131–139.

[35] BRYANT, R., *Finsler structures on the 2–sphere satisfying $K = 1$.* In the volume: *Finsler Geometry*, edited by D. Bao, S.S. Chern, Z. Shen, Contemporary Mathematics, 196, p.27–42.

[36] BUCĂTARU, I., *Connection Map in the Higher Order Geometry*, An.Şt. Univ. "Al.I.Cuza" Iaşi, T.XLII, Supl., s.I-a, Mat., (1996), 253-258.

[37] BUCĂTARU, I., *Horizontal lifts in the higher order geometry*, Publ. Math. Debrecen, 56/1-2, 2000, 21–32.

[38] CHEN, B.Y., *Total mean curvature and submanifolds of finite type*, World Scientific (1984).

[39] CHEN, B.Y., VERSTRAELEN, L., *Laplace transformations of submanifolds,* PADGE, KU Leuven, KU Brussel (1995).

[40] CHERN, S.S., *A simple intrinsic proof of the Gauss–Bonnet formula for classed Riemannian manifolds*, Ann. of math., 45, 1944, 747–752.

[41] CHERN, S.S., *On Finsler Geometry,* C.R. Acad. Sci. Paris, t. 314, 1992, 757–761.

[42] CHERN, S.S., *A Mathematician and His Mathematical Work* (Selected papers of S.S. Chern), Edited by S.Y. Cheng, P.Li and G.Titan, World Scientific, 1996.

[43] CRAMPIN, M., *On the inverse problem of the calculus of variations for systems of second–order ordinary differential equations*, in the vol. Finslerian Geometry, Kluwer Acad. Publ. FTPH vol. 109, 2000, 139–151.

[44] CRAMPIN, M., SARLET, W., CANTRIJN, F., *Higher–Order Equations and Higher–Order Lagrangian Mechanics*, Math. Proc. Camb. Phil. Soc. 1986, 99, 565–587.

[45] CRÂŞMĂREANU, M., *First integrals generated by pseudosymmetries in Nambu-Poisson dynamics,* J. of Nonlinear Math. Phys. 7(2000), no.2, 126-135.

[46] CRUCEANU, V., *Sur la structure presque–produit associée à une connection sur un espace fibré*, An. St. Univ. "Al.I. Cuza" Iaşi, s.I, t. XV, f.1 (1969), 159–167.

[47] CRUCEANU, V., *Sur la théorie de sous–fibrés vectoriels*, C.R. Acad. Sci. Paris, 302, 1986, 705–708.

[48] CRUCEANU, V., FURTUNY, P., GADEA, P.M., *A Survey on Paracomplex Geometry*, Rockey Mountain J. of Math. 26, no.1, Winter 1996, 83–115.

[49] CRAIOVEANU, M., PUTA, M., *Geometric Quantization Mechanical Systems*, Proc. of 22-th Conf. on Diff. Geom. and Topology, Buc, (1991), 87–91.

[50] DAVIES, E.T., *On the invariant theory of contact transformations,* Math. Zeits 57(1953), 415-427.

[51] DAZORD, P., *Tores Finslériens sans points conjuguées*, Bull. Soc. Math. France, 99, 1971, 171–192.

[52] DOYLE, T.C., *Tensor decomposition with applications to the contact and couple groups*, Ann. Math. 42(1941), 698–721.

[53] EISENHART, L.P., *Contact transformations*, Ann. Math. 30(2) (1929), 211–249.

[54] EISENHART, L.P., *Finsler spaces derived from Riemannian spaces by contact transformations*, Ann. Math. 37 (1936), 747–765.

[55] EISENHART, L.P., KNEBELMAN, M.S., *Invariant theory of homogeneous contact transformations,* Ann. Math. 37(1936), 747-765.

[56] FARAH, J.J., *Complex Finsler Geometry via the equivalence problem on the tangent bundle,* in the vol. Finslerian Geometry, Kluwer Acad. Publ. FTPH vol. 109, 2000, 151–169.

[57] GRIFONE, J., *Structures Presque–Tangentes et Connexions*, I, An. Inst. Fourier Grenoble, 22, 1972, 287–334; 1972, 287–334.

[58] GRIFONE, J., *Structures Presque–Tangentes et Connexions*, II, An. Inst. Fourier Grenoble, 23, 1972, 291–338.

[59] GROMOV, M., *Filling Riemannian Manifolds*, J. Diff. Geom. 18 (1983), 1–147.

[60] HASHIGUCHI, M., *On Conformal Transformations of Finsler Metrics,* J. Math. Kyoto Univ. 16, 1976, 25–50.

[61] HASSAN, B.T., *The Theory of Geodesics in Finsler Spaces* (Ph.D. Thesis), Southampton Univ. (1967).

[62] HASSAN, B.T., MESBAH, F., *The Gauss–Bonnet theorem in Finsler Geometry* (to appear).

[63] HRIMIUC, D., *Hamilton Geometry,* Pergamon Press, Math. Comput. Modelling 20, no.415 (1994), 57–65.

[64] HRIMIUC, D., *On the geometry of an infinite dimensional Hamilton space,* Tensor N.S., 49(1990), 238–249.

[65] HRIMIUC, D., *The generalized Legendre transformation*, Proc. of Nat. Seminar on Lagrange Space, Braşov (1988), 167–171.

[66] HRIMIUC, D., SHIMADA, H., *On the L-duality between Lagrange and Hamilton manifolds*, Nonlinear World, 3(1996), 613–641.

[67] HRIMIUC, D., SHIMADA, H., *On some special problems concerning the L-duality between Finsler and Cartan spaces*, Tensor N.S., 58(1997), 48–61.

[68] ICHIJYŌ, Y., *Finsler manifolds modeled on a Minkowski space*, J. Math. Kyoto Univ. 16(1976), 639–652.

[69] ICHIJYŌ, Y., *On the Finsler group and an almost symplectic structure on a tangent bundle*, J. Math. Kyoto Univ. 28(1988), 153–163.

[70] IZUMI, H., *On the Geometry of Generalized Metric Spaces*, I. *Connections and identities*, Publicationes Mathematicae, Debrecen 39, 1991, 113–134.

[71] IKEDA, S., *Advanced Studies in Applied Geometry,* Seizansha, Sagamihara, 1995.

[72] INGARDEN, R.S., *On Physical interpretations of Finsler and Kawaguchi spaces,* Tensor N.S. 46(1987), 354–360.

[73] ISHIKAWA, H., *Note on Finslerian relativity,* J. Math. Phys. 22(5) (1981), 995–1004.

[74] KAWAGUCHI, M., *On the vectors of Higher Order and the extended affine connection,* Ann. di Matem. Pura ed Appl. IV(55) (1961), 105–118.

[75] KAWAGUCHI, M., *An Introduction to the Theory of Higher Order Space,* I: *The Theory of Kawaguchi Spaces,* RAAG Memoirs, vol. 3, 1962.

[76] KERN, J., *Lagrange Geometry,* Arch. Math. 25(1974), 438–443.

[77] KIKUCHI, S., *On the condition that a space with (α, β)–metric be locally Minkowskian,* Tensor N.S. 33(1979), 242–246.

[78] KITAYAMA, M., *Hypersurfaces in generalized Lagrange spaces,* In the vol.: Finslerian Geometries, Kluwer Acad. Publ. FTPH, vol. 109, 2000, 179–193.

[79] KLEIN, J., *Espaces Variationnels et Mécaniques,* Ann. Inst. Fourier (Grenoble), 12, 1962, 1–124.

[80] KOBAYASHI, S., NOMIZU, K., *Foundations of Differential Geometry,* vol. I, II, John Wiley and Sons, Inc., New York, (1963), 1969.

[81] KONDO, K., *On the Physical Meanings of the Zermelo Conditions of Kawaguchi Space,* Tensor N.S. 14, 1963, 191–215.

[82] KONDO, K., *Epistemological Foundations of Quasi–Microscopic Phenomena from the Standpoint of Finsler's and Kawaguchi's Higher Order Geometry,* Post RAAG Reports no. 241, 242, 243 (1991).

[83] KRUPKOVA, O., *The Geometry of Ordinary Variational Equations,* Lecture Notes in Math. no.1678, Springer, 1997.

[84] KRUPKA, D., JANYŠKA, J., *Lectures on Differential Invariants,* Univ. Brno, 1990.

[85] DE LEON, M., RODRIGUEZ, P.R., *Generalized Classical Mechanics and Field Theory,* North–Holland (1985), 73–124.

[86] DE LEON, M., RODRIGUEZ, P.R., *Methods of Differential Geometry in Analitical Mechanics*, North–Holland (1989), 73–124.

[87] LIBERMANN, P., MARLE, CH.M., *Symplectic Geometry and Analytical Mechanics,* D. Reidel Publ. Comp. 1987.

[88] MATSUMOTO, M., *Foundations of Finsler Geometry and Special Finsler Spaces,* Kaisheisha Press, Otsu, 1986.

[89] MATSUMOTO, M., *Theory of Finsler spaces with (α, β)–metric*, Rep. Math. Phys. 31(1992), 43–83.

[90] MATSUMOTO, M., *A special class of locally Minkowski space with (α, β)–metric and conformally flat Kropina spaces*, Tensor N.S., 50 (1991), 202–207.

[91] MATSUMOTO, M., SHIMADA, H., *On Finsler spaces with 1-form metric*, Tensor N.S. 32(1978), 161–169.

[92] MAZÉTIS, E., *Geometry in Cartan spaces*, Lithuanian Math. Jour. vol. 38, no.2, 1998, 172–180.

[93] MIKHAIL, F.I., WANAS, M.I., *A generalized field theory.* I. *Field equations*, Proc. R. Soc. London A 356, 1977, 471–481.

[94] MIHAI, I., *Capitole Speciale de Geometria Varietăţilor Complexe*, Ed. Univ. Bucureşti (1995).

[95] MIHAI, I., ROŞCA, R., VERSTAELEN, L., *Some Aspects of the Differential Geometryof Vector Fields*, PADGE, KU Leuven, KU Brussel (1996).

[96] MIRON, R., *A Lagrangian theory of relativity, (I,II)* Analele Şt. Univ. "Al.I.Cuza" Iaşi, XXXII, s.1., Math., f.2,f.3, (1986), 37-62, 7-16.

[97] MIRON, R., *Hamilton Geometry,* An. Şt. "Al.I.Cuza" Univ., Iaşi, s. I-a Mat., 35, 1989, 33–67.

[98] MIRON, R., *Cartan Spaces in a new point of view by considering them as duals of Finsler Spaces*, Tensor N.S. 46(1987), 330–334.

[99] MIRON, R., *The geometry of Cartan spaces*, Prog. of Math., India 22 (1&2) (1988), 1–38.

[100] MIRON, R., *General Randers Spaces*, Lagrange and Finsler Geometry, Ed. by P.L. Antonelli and R. Miron, no. 76, 1996, 123–140.

[101] MIRON, R., *Sur la géométrie des espaces Hamilton*, C.R. Acad. Sci. Paris, Ser.I, 306, No.4, (1988), 195-198.

[102] MIRON, R., *Lagrange Geometry*, Math. Comput. Modelling 20 (4/5) (1994), 25–40.

[103] MIRON, R., *Noether Theorems in Higher–Order Lagrangian Mechanics*, Int. Journal of Theoretical Physics 30 (1995), 1123–1146.

[104] MIRON, R., *Some problems in the geometries of Finsler type*, Contemporary Mathematics, Finsler Geometry 196 (1996), 187–195.

[105] MIRON, R., *Hamilton Geometry*, Univ. Timişoara,Sem. Mecanică, 3(1987), p.54.

[106] MIRON, R., *The Geometry of Higher–Order Lagrange Spaces, Applications to Mechanics and Physics*, Kluwer Acad. Publ. FTPH, no.82(1997).

[107] MIRON, R., *The Geometry of Higher–Order Finsler Spaces*, Hadronic Press., Inc., USA (1998).

[108] MIRON, R., *On the Inverse Problem in the Higher–Order Analytical Mechanics*, New Frontiers in Algebras, Groups and Geometries, Ed. by G. Tsagas, Hadronic Press., Inc., USA (1996), 157–167.

[109] MIRON, R., *The homogeneous lift to tangent bundle of a Finsler metric*, Publicationes Mathematicae, Debrecen, 57, 3-4, (2000), 445-453.

[110] MIRON, R., *Hamilton Spaces of order k greater than or equal to* 1, Int. Journal of Theoretical Phys., vol.39, No.9, (2000), 2327-2336.

[111] MIRON, R., *The notion of higher order Finsler spaces. Theory and applications*, In the vol.: Finslerian Geometries, Kluwer Acad. Publ. FTPH, vol. 109, 2000, 193–200.

[112] MIRON, R., ANASTASIEI, M., *Vector Bundles and Lagrange Spaces with Applications to Relativity*, Geometry Balkan Press, Bucureşti, no.1, 1997.

[113] MIRON, R., ANASTASIEI, M., *The Geometry of Lagrange Spaces: Theory and Applications*, Kluwer Acad. Publ., FTPH, no.59,(1994).

[114] MIRON, R., IANUS, S., ANASTASIEI, M., *The Geometry of the Dual of a Vector Bundle*, Publ. de l'Inst. Math. 46(60), 1989, 145–162.

[115] MIRON, R., KAWAGUCHI, T., *Relativistic Geometrical Optics*, Int. Journ. of Theor. Phys. vol. 30, no. 11, 1991, 1521–1543.

[116] MIRON, R., KIKUCHI, S., *Legendre transformations of the Generalized Lagrange spaces*, Tensor N.S. 50(1991), 177–182.

[117] MIRON, R., SABĂU, S.V., *On the foundations of Higher Order Finsler Spaces*, Algebras, Groups and Geometries, Hadronic Press, vol. 16, no.1, March 1999, 83–90.

[118] MIRON, R., SABĂU, S.V., SHIMADA, H., *Abstract from the Higher Order Finsler Spaces Geometry and examples*, Proceedings of Hokkaido Tokai Univ. 9(1998), 1-12.

[119] MIRON, R., WATANABE, S., IKEDA, S., *Some Connections on Tangent Bundle and Their Applications to the General Relativity,* Tensor N.S. 46, 1987, 8–22.

[120] MIRON, R., ROŞCA, R., ANASTASIEI, M., BUCHNER, K., *New aspects of Lagrangian Relativity,* Found of Physics Letters, 5, no. 2, 1992, 141–171.

[121] MIRON, R., TAVAKOL, R., BALAN, V., ROXBURG, I., *Geometry of Space-time and Generalized Lagrange Gauge Theory,* Publ. Math. Debrecen 42, 3–4, 1993, 215–224.

[122] MUTO, Y., *On the connections in the manifolds admitting homogeneous contact transformations*, Proc. Phys. Math. Soc. of Japan 20(1938), 451–457.

[123] MUTO, Y., YANO, K., *Sur les transformations de contact et les espaces de Finsler*, Tohoku Math J. 45(1939), 293–307.

[124] MODUGNO, M., *Systems of vector valued forms on a fibered manifold and applications to gauge theories*, Proc. Conf. Diff. Geom. in Math. Phys. Salamanca, L.N. in Math. Springer Verlag 1251 (1987).

[125] MORIMOTO, A., *Prolongations of Geometric Structures*, Nagoya Math. Jour. 32 (1968), 67–108.

[126] MUNTEANU, GH., *Generalized complex Lagrange spaces*, In the vol.: Finslerian Geometries, Kluwer Acad. Publ. FTPH, vol. 109, 2000, 209–223.

[127] NICOLESCU, L., *Grupuri Lie*, Ed. Univ. Bucureşti (1994).

[128] NEAGU, AL., BORCEA, T. V., *Lebesgue Measure and Regular Mappings in Finsler Spaces*, In the volume: Lagrange and Finsler Geometry, Kluwer Academic Publishers, FTPH, no.76 , (1996), 195-204.

[129] NUMATA, S., *On Landsberg spaces of scalar curvature*, J. Korean Math. Soc. 12(1975), 97-100.

[130] OKADA, T., *Minkowskian product of Finsler spaces and the Berwald connections*, J. Math. Kyoto Univ. 22(1982), 323–332.

[131] OPRIŞ, D., BUTULESCU, I., *Metode geometrice in studiul Sistemelor de Ecuaţii Diferenţiale.* Ed. Mirton, Timişoara, 1997.

[132] PITIŞ, GH., *Pontriagin Classes in Finsler Geometry*, An. st. "Al.I. Cuza" Univ., Iaşi, mat. XXX, s.I.a (1984), 61–64.

[133] PRIPOAE, G., *Propriétés de rigidité concernant la courbure des métriques indéfinies*, J. Geom. Phys. 7(1990), 13–20.

[134] PUTA, M. *Hamiltonian Mechanical Systems and Geometric Quantization*, Kluwer Acad. Publ. 260, 1993.

[135] RANDERS, G., *On asymmetric metric in the four space of general relativity*, Phys. Rev. 59 (1941), 195-199.

[136] ROS, A., VERSTRAELEN, L., *On a conjecture of K. Ogiue*, Jour. Diff. Geom. 19(1984), 561–566.

[137] ROMAN, M., *On the homogeneous hyperbolic lift of a Finslerian metric*, Algebra, Groups and Geometries, Proceedings of 11th Conference on Finsler, Lagrange and Hamilton Geometries, Bacău, Romania, Hadronic Press, (2000), (to appear).

[138] ROY SISIR, *Statistical Geometry and Applications to Microphysics and Cosmology*, Kluwer Acad. Publ., FTPH, vol. 92, 1998.

[139] RUND, H., *The Differential Geometry of Finsler Spaces*, Grundlehr. Math. Wiss., Springer–Verlag, Berlin, 101, 1959.

[140] SABĂU, V.S., *Higher order ecological metrics*, In the vol.: Finslerian Geometries, Kluwer Acad. Publ. FTPH, vol. 109, 2000, 245–261.

[141] SABĂU, S.V., SHIMADA, H., *Finsler Geometry*, In the vol.: Finslerian Geometries, Kluwer Acad. Publ. FTPH, vol. 109, 2000, 15–24.

[142] SABĂU, S.V., SHIMADA, H., *The Electromagnetic Field in the Second Order Relativistic Optics*, Memoriile Academiei Române (1996) (to appear).

[143] SABĂU, S.V., SHIMADA, H., *Some remarks on the Geometry of Higher Order Osculator Bundle*, Proceedings of Hokkaido Tokai Univ. 9 (1996), 47–58.

[144] SANTILLI, M.R., *Foundations of Theoretical Mechanics*, I: *The inverse problem in Newtonian Mechanics*, Springer–Verlag, 1978.

[145] SANTILLI, M.R., *Foundations of Theoretical Mechanics*, II: *Birkhoffian Generalization of Hamiltonian Mechanics*, Springer–Verlag 1981.

[146] SASAKI, S., *Homogeneous contact transformations*, Tohoku Math. J. 14(1962), 369–397.

[147] SASAKI, S., *On the Differential Geometry of Tangent Bundles of Riemannian Manifolds*, Tôhoku Math. J., 10, 1958, 338–354.

[148] SAUNDERS, D., *The Geometry of Jet Bundles,* Cambridge Univ.Press, New York, London, 1989.

[149] SHEN, Z., *On a Connection in Finsler Geometry* (Houston Jour. Math. 20, 1994, 591–602.

[150] SHIBATA, C., *On Finsler space with Kropina metric*, Rep. on Math. Phys. 13 (1978), 117–128.

[151] SHIBATA, H., SHIMADA, H., AZUMA, M., YASUDA, H. *On Finsler spaces with Randers metric*, Tensor N.S. 33(1977).

[152] SHIMADA, H., *On Finsler spaces with the metric* $L = \sqrt[m]{a_{i_1 \dots i_m}(x) y^{i_1 \dots i_m}}$, Tensor N.S. 33(1979), 365–372.

[153] SHIMADA, H., SABĂU, S.V., *Higher order Miron–Tavakol metric*, Rev.Roumaine de Mathématiques Pures et Appliqués, XLII (9–10), 1998, 813–822.

[154] STOICA, E., *A geometrical characterization of normal Finsler connections,* An. Şt. Univ. "Al.I.Cuza" Iaşi, XXX, Math., 1984-1.

[155] STAVRE, P., *The d-Linear Connections,* Proc. Nat. Sem. on Finsler and Lagrange Spaces, Braşov, Romania, 1998, 375-382.

[156] SYNGE, J.L., *Relativity: General Theory,* North–Holland, 1966.

[157] SZABO, Z.I., *Positively Definite Berwald Spaces,* Tensor N.S., 35, 1981, 25–39.

[158] SZILASI, J., *Notable Finsler connections on a Finsler manifold*, Lectureas Matemáticas 19(1998), 7–34.

[159] TAMASSY, L., *Area and metrical connections in Finsler spaces*, In the vol.: Finslerian Geometries, Kluwer Acad. Publ. FTPH, vol. 109, 2000, 263–281.

[160] TAMASSY, L., *Point Finsler spaces with metrical linear connections*, Publicationes Mathematicae, Debrecen T. 56, fasc. 3-4, 2000, 643–655.

[161] TAMASSY, L., KOZMA, L., *Connections in K-vector Bundles*, Mathematica Panonica, 6/1 (1995), 105–114.

[162] TAMIM, A.A., YOUSSEF, N.L., *On Generalized Randers Manifolds*, Algebras, Groups and Geometries (Hadronic Press), 16,(1999), 115–126.

[163] ŢARINĂ, M., *Invariant Finsler connections on Vector Bundles*, An. St. Univ. A.I. Cuza, Iaşi, s. I-a Mat. 33(1987), 87–94.

[164] THOMPSON, A.C., *Minkowski Geometry*, Encycl. of Math. and its Appl. 63(1996).

[165] ŢIGĂERU, C., *v-Projective symmetries of fibered manifolds*, Archivum Mathematicum, T. 34, 347–352 (1988).

[166] VACARU, S.I., *Interactions, Strings and Isotopies in Higher Order Anisotropic Superspaces,* Hadronic Press, Palm Harbor, USA (1998).

[167] VAISMAN, I., *Lectures on the Geometry of Poisson Manifolds,* Birkhäuser, Verlag, Basel, (1994).

[168] VARGAS, J.G., *On the geometrization of Electrodynamics Foundations of Physics*, 21(4), 379–401, 1991.

[169] VARGAS, J.G., TORR, D.G., *Canonical Connections of Finsler Metrics and Finslerian Connections on Riemannian Metrics*, General Relativity and Gravitation 28(4), 451–469, 1996.

[170] VERSTAELEN, L., *Curves and Surfaces of finite Chen type*, Geometry and Topology of Submanifolds, World Scientific III (1990), 304–311.

[171] WATANABE, S., IKEDA, S., IKEDA, F., *On a metrical Finsler connection of generalized Finsler metric* $g_{ij} = \exp(2\sigma 9x, y))\gamma_{ij}(x)$., Tensor N.S. 40(1983), 97–102.

[172] YANO, K., DAVIES, E.T., *Contact tensor calculus*, Ann. Math. Pur. Appl. 37(1954), 1–36.

[173] YANO, K., MUTO, Y., *Homogeneous contact structure*, Math. Annalen 167(1966), 195–213.

[174] YANO, K., MUTO, Y., *Homogeneous contact manifolds and almost Finsler manifolds*, Koday Math. Sem. Rep. 21(1969), 16–45.

[175] YANO, K., ISHIHARA, S., *Tangent and Cotangent Bundles. Differential Geometry,* M. Dekker, Inc., New York, (1973).

[176] YASUDA, H., SHIMADA, H., *On Randers spaces of scalar curvature*, Rep. on Math. Phys. 11(1977), 347-360.

[177] ZET, GH., *Lagrangian Geometrical Models in Physics*, Math. Comput. Modelling, vol. 20, no. 4/5, (1994), 83–91.

[178] ZET, GH., *Unitary Symmetries and Gauge Theories* (Romanian), Ed. Gh. Asachi, Univ. Tehnică, Iaşi, (1998), (342 p.).

Index

Fundamental Theories of Physics

Series Editor: Alwyn van der Merwe, University of Denver, USA

1. M. Sachs: *General Relativity and Matter*. A Spinor Field Theory from Fermis to Light-Years. With a Foreword by C. Kilmister. 1982 ISBN 90-277-1381-2
2. G.H. Duffey: *A Development of Quantum Mechanics*. Based on Symmetry Considerations. 1985 ISBN 90-277-1587-4
3. S. Diner, D. Fargue, G. Lochak and F. Selleri (eds.): *The Wave-Particle Dualism*. A Tribute to Louis de Broglie on his 90th Birthday. 1984 ISBN 90-277-1664-1
4. E. Prugovečki: *Stochastic Quantum Mechanics and Quantum Spacetime*. A Consistent Unification of Relativity and Quantum Theory based on Stochastic Spaces. 1984; 2nd printing 1986 ISBN 90-277-1617-X
5. D. Hestenes and G. Sobczyk: *Clifford Algebra to Geometric Calculus*. A Unified Language for Mathematics and Physics. 1984 ISBN 90-277-1673-0; Pb (1987) 90-277-2561-6
6. P. Exner: *Open Quantum Systems and Feynman Integrals*. 1985 ISBN 90-277-1678-1
7. L. Mayants: *The Enigma of Probability and Physics*. 1984 ISBN 90-277-1674-9
8. E. Tocaci: *Relativistic Mechanics, Time and Inertia*. Translated from Romanian. Edited and with a Foreword by C.W. Kilmister. 1985 ISBN 90-277-1769-9
9. B. Bertotti, F. de Felice and A. Pascolini (eds.): *General Relativity and Gravitation*. Proceedings of the 10th International Conference (Padova, Italy, 1983). 1984 ISBN 90-277-1819-9
10. G. Tarozzi and A. van der Merwe (eds.): *Open Questions in Quantum Physics*. 1985 ISBN 90-277-1853-9
11. J.V. Narlikar and T. Padmanabhan: *Gravity, Gauge Theories and Quantum Cosmology*. 1986 ISBN 90-277-1948-9
12. G.S. Asanov: *Finsler Geometry, Relativity and Gauge Theories*. 1985 ISBN 90-277-1960-8
13. K. Namsrai: *Nonlocal Quantum Field Theory and Stochastic Quantum Mechanics*. 1986 ISBN 90-277-2001-0
14. C. Ray Smith and W.T. Grandy, Jr. (eds.): *Maximum-Entropy and Bayesian Methods in Inverse Problems*. Proceedings of the 1st and 2nd International Workshop (Laramie, Wyoming, USA). 1985 ISBN 90-277-2074-6
15. D. Hestenes: *New Foundations for Classical Mechanics*. 1986 ISBN 90-277-2090-8; Pb (1987) 90-277-2526-8
16. S.J. Prokhovnik: *Light in Einstein's Universe*. The Role of Energy in Cosmology and Relativity. 1985 ISBN 90-277-2093-2
17. Y.S. Kim and M.E. Noz: *Theory and Applications of the Poincaré Group*. 1986 ISBN 90-277-2141-6
18. M. Sachs: *Quantum Mechanics from General Relativity*. An Approximation for a Theory of Inertia. 1986 ISBN 90-277-2247-1
19. W.T. Grandy, Jr.: *Foundations of Statistical Mechanics*. Vol. I: *Equilibrium Theory*. 1987 ISBN 90-277-2489-X
20. H.-H von Borzeszkowski and H.-J. Treder: *The Meaning of Quantum Gravity*. 1988 ISBN 90-277-2518-7
21. C. Ray Smith and G.J. Erickson (eds.): *Maximum-Entropy and Bayesian Spectral Analysis and Estimation Problems*. Proceedings of the 3rd International Workshop (Laramie, Wyoming, USA, 1983). 1987 ISBN 90-277-2579-9
22. A.O. Barut and A. van der Merwe (eds.): *Selected Scientific Papers of Alfred Landé*. [*1888-1975*]. 1988 ISBN 90-277-2594-2

23. W.T. Grandy, Jr.: *Foundations of Statistical Mechanics*. Vol. II: *Nonequilibrium Phenomena*. 1988 ISBN 90-277-2649-3
24. E.I. Bitsakis and C.A. Nicolaides (eds.): *The Concept of Probability*. Proceedings of the Delphi Conference (Delphi, Greece, 1987). 1989 ISBN 90-277-2679-5
25. A. van der Merwe, F. Selleri and G. Tarozzi (eds.): *Microphysical Reality and Quantum Formalism, Vol. 1*. Proceedings of the International Conference (Urbino, Italy, 1985). 1988 ISBN 90-277-2683-3
26. A. van der Merwe, F. Selleri and G. Tarozzi (eds.): *Microphysical Reality and Quantum Formalism, Vol. 2*. Proceedings of the International Conference (Urbino, Italy, 1985). 1988 ISBN 90-277-2684-1
27. I.D. Novikov and V.P. Frolov: *Physics of Black Holes*. 1989 ISBN 90-277-2685-X
28. G. Tarozzi and A. van der Merwe (eds.): *The Nature of Quantum Paradoxes*. Italian Studies in the Foundations and Philosophy of Modern Physics. 1988 ISBN 90-277-2703-1
29. B.R. Iyer, N. Mukunda and C.V. Vishveshwara (eds.): *Gravitation, Gauge Theories and the Early Universe*. 1989 ISBN 90-277-2710-4
30. H. Mark and L. Wood (eds.): *Energy in Physics, War and Peace*. A Festschrift celebrating Edward Teller's 80th Birthday. 1988 ISBN 90-277-2775-9
31. G.J. Erickson and C.R. Smith (eds.): *Maximum-Entropy and Bayesian Methods in Science and Engineering*. Vol. I: *Foundations*. 1988 ISBN 90-277-2793-7
32. G.J. Erickson and C.R. Smith (eds.): *Maximum-Entropy and Bayesian Methods in Science and Engineering*. Vol. II: *Applications*. 1988 ISBN 90-277-2794-5
33. M.E. Noz and Y.S. Kim (eds.): *Special Relativity and Quantum Theory*. A Collection of Papers on the Poincaré Group. 1988 ISBN 90-277-2799-6
34. I.Yu. Kobzarev and Yu.I. Manin: *Elementary Particles. Mathematics, Physics and Philosophy*. 1989 ISBN 0-7923-0098-X
35. F. Selleri: *Quantum Paradoxes and Physical Reality*. 1990 ISBN 0-7923-0253-2
36. J. Skilling (ed.): *Maximum-Entropy and Bayesian Methods*. Proceedings of the 8th International Workshop (Cambridge, UK, 1988). 1989 ISBN 0-7923-0224-9
37. M. Kafatos (ed.): *Bell's Theorem, Quantum Theory and Conceptions of the Universe*. 1989 ISBN 0-7923-0496-9
38. Yu.A. Izyumov and V.N. Syromyatnikov: *Phase Transitions and Crystal Symmetry*. 1990 ISBN 0-7923-0542-6
39. P.F. Fougère (ed.): *Maximum-Entropy and Bayesian Methods*. Proceedings of the 9th International Workshop (Dartmouth, Massachusetts, USA, 1989). 1990 ISBN 0-7923-0928-6
40. L. de Broglie: *Heisenberg's Uncertainties and the Probabilistic Interpretation of Wave Mechanics*. With Critical Notes of the Author. 1990 ISBN 0-7923-0929-4
41. W.T. Grandy, Jr.: *Relativistic Quantum Mechanics of Leptons and Fields*. 1991 ISBN 0-7923-1049-7
42. Yu.L. Klimontovich: *Turbulent Motion and the Structure of Chaos*. A New Approach to the Statistical Theory of Open Systems. 1991 ISBN 0-7923-1114-0
43. W.T. Grandy, Jr. and L.H. Schick (eds.): *Maximum-Entropy and Bayesian Methods*. Proceedings of the 10th International Workshop (Laramie, Wyoming, USA, 1990). 1991 ISBN 0-7923-1140-X
44. P. Pták and S. Pulmannová: *Orthomodular Structures as Quantum Logics*. Intrinsic Properties, State Space and Probabilistic Topics. 1991 ISBN 0-7923-1207-4
45. D. Hestenes and A. Weingartshofer (eds.): *The Electron*. New Theory and Experiment. 1991 ISBN 0-7923-1356-9

Fundamental Theories of Physics

46. P.P.J.M. Schram: *Kinetic Theory of Gases and Plasmas.* 1991 ISBN 0-7923-1392-5
47. A. Micali, R. Boudet and J. Helmstetter (eds.): *Clifford Algebras and their Applications in Mathematical Physics.* 1992 ISBN 0-7923-1623-1
48. E. Prugovečki: *Quantum Geometry.* A Framework for Quantum General Relativity. 1992 ISBN 0-7923-1640-1
49. M.H. Mac Gregor: *The Enigmatic Electron.* 1992 ISBN 0-7923-1982-6
50. C.R. Smith, G.J. Erickson and P.O. Neudorfer (eds.): *Maximum Entropy and Bayesian Methods.* Proceedings of the 11th International Workshop (Seattle, 1991). 1993 ISBN 0-7923-2031-X
51. D.J. Hoekzema: *The Quantum Labyrinth.* 1993 ISBN 0-7923-2066-2
52. Z. Oziewicz, B. Jancewicz and A. Borowiec (eds.): *Spinors, Twistors, Clifford Algebras and Quantum Deformations.* Proceedings of the Second Max Born Symposium (Wrocław, Poland, 1992). 1993 ISBN 0-7923-2251-7
53. A. Mohammad-Djafari and G. Demoment (eds.): *Maximum Entropy and Bayesian Methods.* Proceedings of the 12th International Workshop (Paris, France, 1992). 1993 ISBN 0-7923-2280-0
54. M. Riesz: *Clifford Numbers and Spinors* with Riesz' Private Lectures to E. Folke Bolinder and a Historical Review by Pertti Lounesto. E.F. Bolinder and P. Lounesto (eds.). 1993 ISBN 0-7923-2299-1
55. F. Brackx, R. Delanghe and H. Serras (eds.): *Clifford Algebras and their Applications in Mathematical Physics.* Proceedings of the Third Conference (Deinze, 1993) 1993 ISBN 0-7923-2347-5
56. J.R. Fanchi: *Parametrized Relativistic Quantum Theory.* 1993 ISBN 0-7923-2376-9
57. A. Peres: *Quantum Theory: Concepts and Methods.* 1993 ISBN 0-7923-2549-4
58. P.L. Antonelli, R.S. Ingarden and M. Matsumoto: *The Theory of Sprays and Finsler Spaces with Applications in Physics and Biology.* 1993 ISBN 0-7923-2577-X
59. R. Miron and M. Anastasiei: *The Geometry of Lagrange Spaces: Theory and Applications.* 1994 ISBN 0-7923-2591-5
60. G. Adomian: *Solving Frontier Problems of Physics: The Decomposition Method.* 1994 ISBN 0-7923-2644-X
61. B.S. Kerner and V.V. Osipov: *Autosolitons.* A New Approach to Problems of Self-Organization and Turbulence. 1994 ISBN 0-7923-2816-7
62. G.R. Heidbreder (ed.): *Maximum Entropy and Bayesian Methods.* Proceedings of the 13th International Workshop (Santa Barbara, USA, 1993) 1996 ISBN 0-7923-2851-5
63. J. Peřina, Z. Hradil and B. Jurčo: *Quantum Optics and Fundamentals of Physics.* 1994 ISBN 0-7923-3000-5
64. M. Evans and J.-P. Vigier: *The Enigmatic Photon.* Volume 1: The Field $\boldsymbol{B}^{(3)}$. 1994 ISBN 0-7923-3049-8
65. C.K. Raju: *Time: Towards a Constistent Theory.* 1994 ISBN 0-7923-3103-6
66. A.K.T. Assis: *Weber's Electrodynamics.* 1994 ISBN 0-7923-3137-0
67. Yu. L. Klimontovich: *Statistical Theory of Open Systems.* Volume 1: A Unified Approach to Kinetic Description of Processes in Active Systems. 1995 ISBN 0-7923-3199-0; Pb: ISBN 0-7923-3242-3
68. M. Evans and J.-P. Vigier: *The Enigmatic Photon.* Volume 2: Non-Abelian Electrodynamics. 1995 ISBN 0-7923-3288-1
69. G. Esposito: *Complex General Relativity.* 1995 ISBN 0-7923-3340-3

Fundamental Theories of Physics

70. J. Skilling and S. Sibisi (eds.): *Maximum Entropy and Bayesian Methods*. Proceedings of the Fourteenth International Workshop on Maximum Entropy and Bayesian Methods. 1996 ISBN 0-7923-3452-3
71. C. Garola and A. Rossi (eds.): *The Foundations of Quantum Mechanics Historical Analysis and Open Questions*. 1995 ISBN 0-7923-3480-9
72. A. Peres: *Quantum Theory: Concepts and Methods*. 1995 (see for hardback edition, Vol. 57) ISBN Pb 0-7923-3632-1
73. M. Ferrero and A. van der Merwe (eds.): *Fundamental Problems in Quantum Physics*. 1995 ISBN 0-7923-3670-4
74. F.E. Schroeck, Jr.: *Quantum Mechanics on Phase Space*. 1996 ISBN 0-7923-3794-8
75. L. de la Peña and A.M. Cetto: *The Quantum Dice*. An Introduction to Stochastic Electrodynamics. 1996 ISBN 0-7923-3818-9
76. P.L. Antonelli and R. Miron (eds.): *Lagrange and Finsler Geometry*. Applications to Physics and Biology. 1996 ISBN 0-7923-3873-1
77. M.W. Evans, J.-P. Vigier, S. Roy and S. Jeffers: *The Enigmatic Photon*. Volume 3: Theory and Practice of the $\boldsymbol{B}^{(3)}$ Field. 1996 ISBN 0-7923-4044-2
78. W.G.V. Rosser: *Interpretation of Classical Electromagnetism*. 1996 ISBN 0-7923-4187-2
79. K.M. Hanson and R.N. Silver (eds.): *Maximum Entropy and Bayesian Methods*. 1996 ISBN 0-7923-4311-5
80. S. Jeffers, S. Roy, J.-P. Vigier and G. Hunter (eds.): *The Present Status of the Quantum Theory of Light*. Proceedings of a Symposium in Honour of Jean-Pierre Vigier. 1997 ISBN 0-7923-4337-9
81. M. Ferrero and A. van der Merwe (eds.): *New Developments on Fundamental Problems in Quantum Physics*. 1997 ISBN 0-7923-4374-3
82. R. Miron: *The Geometry of Higher-Order Lagrange Spaces*. Applications to Mechanics and Physics. 1997 ISBN 0-7923-4393-X
83. T. Hakioğlu and A.S. Shumovsky (eds.): *Quantum Optics and the Spectroscopy of Solids*. Concepts and Advances. 1997 ISBN 0-7923-4414-6
84. A. Sitenko and V. Tartakovskii: *Theory of Nucleus*. Nuclear Structure and Nuclear Interaction. 1997 ISBN 0-7923-4423-5
85. G. Esposito, A.Yu. Kamenshchik and G. Pollifrone: *Euclidean Quantum Gravity on Manifolds with Boundary*. 1997 ISBN 0-7923-4472-3
86. R.S. Ingarden, A. Kossakowski and M. Ohya: *Information Dynamics and Open Systems*. Classical and Quantum Approach. 1997 ISBN 0-7923-4473-1
87. K. Nakamura: *Quantum versus Chaos*. Questions Emerging from Mesoscopic Cosmos. 1997 ISBN 0-7923-4557-6
88. B.R. Iyer and C.V. Vishveshwara (eds.): *Geometry, Fields and Cosmology*. Techniques and Applications. 1997 ISBN 0-7923-4725-0
89. G.A. Martynov: *Classical Statistical Mechanics*. 1997 ISBN 0-7923-4774-9
90. M.W. Evans, J.-P. Vigier, S. Roy and G. Hunter (eds.): *The Enigmatic Photon*. Volume 4: New Directions. 1998 ISBN 0-7923-4826-5
91. M. Rédei: *Quantum Logic in Algebraic Approach*. 1998 ISBN 0-7923-4903-2
92. S. Roy: *Statistical Geometry and Applications to Microphysics and Cosmology*. 1998 ISBN 0-7923-4907-5
93. B.C. Eu: *Nonequilibrium Statistical Mechanics*. Ensembled Method. 1998 ISBN 0-7923-4980-6

Fundamental Theories of Physics

94. V. Dietrich, K. Habetha and G. Jank (eds.): *Clifford Algebras and Their Application in Mathematical Physics.* Aachen 1996. 1998 ISBN 0-7923-5037-5
95. J.P. Blaizot, X. Campi and M. Ploszajczak (eds.): *Nuclear Matter in Different Phases and Transitions.* 1999 ISBN 0-7923-5660-8
96. V.P. Frolov and I.D. Novikov: *Black Hole Physics.* Basic Concepts and New Developments. 1998 ISBN 0-7923-5145-2; Pb 0-7923-5146
97. G. Hunter, S. Jeffers and J-P. Vigier (eds.): *Causality and Locality in Modern Physics.* 1998 ISBN 0-7923-5227-0
98. G.J. Erickson, J.T. Rychert and C.R. Smith (eds.): *Maximum Entropy and Bayesian Methods.* 1998 ISBN 0-7923-5047-2
99. D. Hestenes: *New Foundations for Classical Mechanics (Second Edition).* 1999 ISBN 0-7923-5302-1; Pb ISBN 0-7923-5514-8
100. B.R. Iyer and B. Bhawal (eds.): *Black Holes, Gravitational Radiation and the Universe.* Essays in Honor of C. V. Vishveshwara. 1999 ISBN 0-7923-5308-0
101. P.L. Antonelli and T.J. Zastawniak: *Fundamentals of Finslerian Diffusion with Applications.* 1998 ISBN 0-7923-5511-3
102. H. Atmanspacher, A. Amann and U. Müller-Herold: *On Quanta, Mind and Matter Hans Primas in Context.* 1999 ISBN 0-7923-5696-9
103. M.A. Trump and W.C. Schieve: *Classical Relativistic Many-Body Dynamics.* 1999 ISBN 0-7923-5737-X
104. A.I. Maimistov and A.M. Basharov: *Nonlinear Optical Waves.* 1999 ISBN 0-7923-5752-3
105. W. von der Linden, V. Dose, R. Fischer and R. Preuss (eds.): *Maximum Entropy and Bayesian Methods Garching, Germany 1998.* 1999 ISBN 0-7923-5766-3
106. M.W. Evans: *The Enigmatic Photon Volume 5: O(3) Electrodynamics.* 1999 ISBN 0-7923-5792-2
107. G.N. Afanasiev: *Topological Effects in Quantum Mecvhanics.* 1999 ISBN 0-7923-5800-7
108. V. Devanathan: *Angular Momentum Techniques in Quantum Mechanics.* 1999 ISBN 0-7923-5866-X
109. P.L. Antonelli (ed.): *Finslerian Geometries A Meeting of Minds.* 1999 ISBN 0-7923-6115-6
110. M.B. Mensky: *Quantum Measurements and Decoherence Models and Phenomenology.* 2000 ISBN 0-7923-6227-6
111. B. Coecke, D. Moore and A. Wilce (eds.): *Current Research in Operation Quantum Logic.* Algebras, Categories, Languages. 2000 ISBN 0-7923-6258-6
112. G. Jumarie: *Maximum Entropy, Information Without Probability and Complex Fractals.* Classical and Quantum Approach. 2000 ISBN 0-7923-6330-2
113. B. Fain: *Irreversibilities in Quantum Mechanics.* 2000 ISBN 0-7923-6581-X
114. T. Borne, G. Lochak and H. Stumpf: *Nonperturbative Quantum Field Theory and the Structure of Matter.* 2001 ISBN 0-7923-6803-7
115. J. Keller: *Theory of the Electron.* A Theory of Matter from START. 2001 ISBN 0-7923-6819-3
116. M. Rivas: *Kinematical Theory of Spinning Particles.* Classical and Quantum Mechanical Formalism of Elementary Particles. 2001 ISBN 0-7923-6824-X
117. A.A. Ungar: *Beyond the Einstein Addition Law and its Gyroscopic Thomas Precession.* The Theory of Gyrogroups and Gyrovector Spaces. 2001 ISBN 0-7923-6909-2
118. R. Miron, D. Hrimiuc, H. Shimada and S.V. Sabau: *The Geometry of Hamilton and Lagrange Spaces.* 2001 ISBN 0-7923-6926-2

Fundamental Theories of Physics

119. M. Pavšič: *The Landscape of Theoretical Physics: A Global View*. From Point Particles to the Brane World and Beyond in Search of a Unifying Principle. 2001 ISBN 0-7923-7006-6

KLUWER ACADEMIC PUBLISHERS – DORDRECHT / BOSTON / LONDON